CLIMATE CHANGE AND THE GLOBAL HARVEST

CLIMATE CHANGE
AND THE
GLOBAL HARVEST

Potential Impacts of the
Greenhouse Effect on Agriculture

CYNTHIA ROSENZWEIG

DANIEL HILLEL

New York   Oxford • Oxford University Press   1998

Oxford University Press

Oxford   New York

Athens   Auckland   Bangkok   Bogota   Bombay   Buenos Aires
Calcutta   Cape Town   Dar es Salaam   Delhi   Florence   Hong Kong
Istanbul   Karachi   Kuala Lumpur   Madras   Madrid   Melbourne
Mexico City   Nairobi   Paris   Singapore   Taipei   Tokyo   Toronto   Warsaw

and associated companies in
Berlin   Ibadan

Copyright © 1998 by Oxford University Press, Inc.

Published by Oxford University Press, Inc.
198 Madison Avenue, New York, New York   10016

Oxford is a registered trademark of Oxford University Press

Library of Congress Cataloging-in-Publication data
Rosenzweig, Cynthia.
Climate change and the global harvest : potential impacts of the
greenhouse effect on agriculture / Cynthia Rosenzweig,
Daniel Hillel.
p.   cm.
Includes bibliographical references and index.
ISBN 0-19-508889-1
1. Climatic changes.   2. Greenhouse effect, Atmospheric.
3. Meteorology, Agricultural.   I. Hillel, Daniel.   II. Title.
S600.7.C54R67   1998
338.1'4—dc21   97-27058

3 5 7 9 8 6 4 2
Printed in the United States of America
on acid-free paper

ACKNOWLEDGMENTS

We acknowledge our many colleagues whose insightful work on the potential impacts of climate change informs this book. We are grateful specifically to the following for advice on their areas of expertise: Francesco Tubiello for carbon cycle and crop modeling, Vivien Gornitz for sea-level rise, Mark Hendel for the history of climate change science, and David Patterson for information on weeds. We further thank Christopher Shashkin for long-time logistical support, Katie Gallagher for dedicated search for bibliographic material, Jessy Thomas for helping with the index, and Sandra Cavalieri for assiduously acquiring the permissions. Finally, we cite the outstanding work of Andrei Gritsevskii and Tanya Ermolieva, who computerized the figures.

CONTENTS

CLIMATE CHANGE AND THE GLOBAL HARVEST

Climate Change: Portent
and Challenge

Mark Twain—or was it his friend Charles Dudley Warner?—said that everybody talks about the weather but nobody does anything about it. Now it appears that whoever said so was wrong. We have all been doing something about the weather, albeit—for too long—unwittingly. Indeed, we have been changing it, and not necessarily for the better. We have been doing this globally, especially during the last two centuries since the Industrial Revolution, by generating increasing amounts of radiatively active gases due to the burning of fossil fuels, clearing of forests, and expansion of cultivation. Thus, we have been modifying the composition of the atmosphere that envelops our earth, the domain of weather phenomena in which we live and grow our crops. The consequences of this modification may force far-reaching changes in agriculture in the coming decades.

So much has been said and written recently about the impending change in climate that many are confused by it all. The voluminous and rapidly proliferating scientific literature on this subject is highly technical, complex, fragmented, and still beset by disagreements. Some atmospheric scientists have warned that the earth's mean temperature is likely to increase by several degrees in the coming century and that such a significant warming will result in far-reaching changes. The changes envisaged include shifts in the zonation of vegetation and in the quantity and distribution of rainfall (and, hence, of river flow and groundwater recharge), melting of glaciers, expansion of ocean water, rise of sea levels, and inundation of coastal areas. Scientists generally hedge their warnings with admissions of uncertainty. Less restrained are advocates and journalists eager to dramatize the global warming issue, some representing it as potentially disastrous and some as insignificant. The exaggerations of alarmists should not drive us to frantic ill-considered action or inure us to a fatalistic acceptance of a presumed calamity. Nor should the complacency of nay-sayers cause us to ignore the issue entirely. Instead, we should strive to learn as much as we can about the real processes involved and to consider what can be done about them in a timely fashion.

Through burning of fossil fuels and eradication of forests, human activity has caused the carbon dioxide (CO_2) concentration of the atmosphere to increase by some 25% since the industrial revolution, and that increase continues. Despite its seemingly minute concentration (only 0.035%, i.e., one-third of one-tenth of 1%), CO_2 plays an important role in inhibiting the escape of the heat radiated by the earth. Other so-called greenhouse gases, which are present in even smaller concentrations but which similarly tend to trap heat, include methane, nitrous oxide, and a family of particularly insidious synthetic gases—the chlorofluorocarbons. (The latter have been insinuated into the atmosphere in just the last 50 years.)

So this much seems clear: if the buildup of greenhouse gases in the atmosphere continues without limit, it is bound, sooner or later, to warm the earth's surface. Such a warming trend cannot but affect the regional patterns of precipitation and evaporation, the biophysical processes of photosynthesis and respiration, and indeed the entire thermal and hydrological regimes governing both natural and agricultural ecosystems.

Beyond what is clear, however, lie great uncertainties: How much warming will occur, when and at what rate, and according to what geographical and seasonal pattern? What will be the specific consequences to natural ecosystems and to the agricultural productivity of different countries and regions? Will some areas benefit while other areas suffer, and who might the winners and losers be? And there are the practical questions: What can be done in advance by individual countries and by the international community as a whole to prevent potential damages to life-support systems? To the extent that such damages may be unavoidable, what can be done to adapt our practices so as to minimize or even overcome them? Upon our ability to answer these and other questions related to the environment may rest the welfare of nations in the coming century.

As yet, the evidence regarding the link between observed temperature trends and the enhanced greenhouse effect is not absolutely proven; the "signal" (i.e., the real long-term change) is difficult to discern from the "noise" (the apparent short-term fluctuations). The inherent spatial and temporal variability of weather tends to obscure the possible underlying trend. Volcanic events, such as the eruption of Mt. Pinatubo in the Philippines in 1991, send large quantities of sulfuric aerosols into the stratosphere, and these may reflect incoming solar radiation so as to lower global mean surface temperature for up to several years. Air pollution caused by industrial activity and biomass burning may be contributing a simultaneous cooling effect as well. Another interfering factor is the possible increase in the earth's cloud cover. Nonetheless, there are indications that the predicted warming may have already begun.

Because the earth's climate system is too large to allow controlled experiments, scientists have been employing mathematical models (encompassing the atmosphere, the oceans, and the continents) to assess the processes known to occur and their possible interactions. The results of such models are used to forecast the trend of climate over the coming decades. The strength of these models is that they integrate our best knowledge; their weakness is that they also, ipso facto, embody our ignorance. So their results are still tentative and should not be accepted uncritically. However, we should examine the implications of their predictions, while continuing to look for the emerging empirical evidence of changing climate.

Since such changes are likely to have an impact on agriculture, energy use, and the economies of both developed and developing countries, we need to take stock of the current state of knowledge regarding potential climate change, and consider actions and policies for dealing with it. The Intergovernmental Panel on Climate Change (IPCC) was established by the World Meteorological Organization and the United Nations Environment Programme in 1988 to assess the available scientific information on climate change and its environmental and socioeconomic impacts, and to evaluate response strategies. The Framework Convention on Climate Change, a global treaty initiated at the United Nations Conference on Environment and Development in Rio de Janeiro in 1992 and subsequently signed by over 100 nations, was the tangible start of the international response process.

PROVIDING SUFFICIENT FOOD for the world's people is becoming more difficult as our numbers are increasing and as land, water, and vegetative resources are progressively degraded through prolonged overuse. There is now concern that the difficulty will be exacerbated in the future by the process of global warming with its potential for affecting the climatic regimes of entire regions. Several complementary disciplines must be combined in order to implement the complex task of assessing potential climate change impacts on global agriculture. Among the relevant disciplines are atmospheric science, hydrology, soil science, crop physiology, and resource economics.

Climate variables can have significant impacts on agricultural production because heat, light, and water are major biophysical drivers of crop growth. Nonoptimal levels of these factors can reduce yields. The drought of the 1930s in the Southern Great Plains of the United States (the so-called Dust Bowl) caused some 200,000 farm bankruptcies in that region. The recent drought of 1988 in America's Midwest led to a 30% reduction in U.S. corn production. Above-normal temperatures accompanied both of these droughts. On the other hand, warmer and longer growing seasons in regions where crops are currently limited by cold but not by paucity of moisture (e.g., Canada and Russia) may enjoy increased productivity in a greenhouse world.

If atmospheric carbon dioxide accumulation were occurring without concomitant changes in temperature and water regimes, it might, indeed, be a blessing for crop production. CO_2 is an essential ingredient of the basic process of photosynthesis by which plants, powered by sunlight, combine that gas with soil-derived water to create carbohydrates and, ultimately, food for all animals, including humans. Plants commonly respond to higher levels of CO_2 with increased rates of photosynthesis because the CO_2 absorption is facilitated by the stronger concentration gradient between the external atmosphere and the air spaces inside the leaves.

As CO_2 concentrations increase in the ambient atmosphere, plants also exhibit partial closure of their stomates, thereby reducing transpiration per unit of leaf area. However, total crop transpiration may not change significantly because greater foliar growth (due to the higher rates of photosynthesis) partially compensates for the reduced transpiration per unit of leaf area. A net improvement may therefore occur in crop water-use efficiency, defined as the ratio between biomass accumulation and the amount of water transpired by the cropped field. Additional biophysical interactions involve the effects of changing climate and CO_2 on soil fertility and pests. Analyzing these simultaneous processes in combination is a complex task that requires a quantitative dynamic assessment of the relative magnitudes of both physiological and climatic changes.

Biophysical effects can lead to changes in the socioeconomic sphere within which farmers operate. The responses of individual farmers to changes of climatic regime might involve changes in the selection of crops and in practices of cultivation, irrigation, and pest control. Ultimately, the ability of farmers to adapt effectively can decide the success or failure of individual farms and, by extension, can affect regional and national economies highly dependent on agricultural production. Changes on the farm may, in turn, modify energy use, water demand, storage and transportation, and food trade.

National farm policy can be a critical determinant in the adaptation of the farming sector to changing conditions. In the United States, farm subsidies (currently costing on the order of $10 billion per year) may either help or hinder necessary adaptation to the eventuality of a changing climate. An important policy consideration is the assessment of risk due to weather anomalies. The drought of 1988, for example, cost U.S. taxpayers over $3 billion in direct relief payments to farmers. If drought frequency increases, the need for such emergency allocations will also increase. Anticipating the probability and potential magnitude of such anomalies can help make timely adjustments that may reduce such social costs.

Beyond national boundaries, changes in the global patterns of supply and demand may have far-reaching consequences. Because of the growing interdependence of the world food system, the impact of climate change on agriculture in each country depends more and more on what happens elsewhere. For example, the vulnerability of food-deficient regions to drought may work to the advantage of major grain producers such as the United States, but the intensified competition from still more favored regions (such as Canada and Russia) may limit that advantage. International trade policy issues, especially the movement to lower agricultural trade barriers, will be crucial in climate change response strategies.

AGRICULTURE IS NO LONGER perceived as a benign component of the environment. Not only does agricultural development, wherever it is carried out, replace natural ecosystems with artificially managed ones, but it also affects surrounding areas by its increased rates of erosion and runoff, and by its release of fertilizer and pesticide residues into surface water and groundwater. Moreover, large climate changes, should they occur, will force shifts in crop and livestock production zones. Farming may invade regions now primarily covered by forests and other types of less intensively managed ecosystems. Such interactions involving agriculture and the natural environment under a changing climate will reverberate throughout the world food system, thus altering rates of soil erosion, competition for water resources, the use of agricultural chemicals, and the fate of wildlife habitats.

THIS BOOK EXPLORES SOME of the complex interactions among biophysical and socioeconomic factors that might affect agriculture in a greenhouse world. We discuss the methodological issues involved in both simulation studies and field experiments, and we point to some of the limitations and the strengths of the methods employed by researchers of climate change impacts.

The overall goals of climate change impact research are to define more clearly the ranges of possible impacts, to determine which locations and farming systems may be most vulnerable, to identify their critical thresholds, and to explore adaptation strategies. Our specific goal is to inform our readers regarding the reciprocal relationship between agriculture and climate, as well as on how the practice of agriculture in the various regions of the globe may adapt to the eventuality of climate change and to the challenge posed by it.

The Greenhouse Effect
and Global Warming

This chapter defines the naturally occurring greenhouse effect that warms the surface of our planet, and the prospect of its enhancement due to anthropogenic modification of the atmosphere's composition. It specifies the different sources, sinks, and relative strengths of the various gases responsible for the projected global warming, and it describes the mathematical models used to predict their consequences on the earth's climate system. Such models predict not only global warming but also changing hydrological regimes and rising sea level. The chapter ends with a consideration of some unresolved questions regarding global warming, including the following: Has the anticipated process actually begun and can it already be discerned? What of the opposing (i.e., cooling) role of pollutant aerosols produced by industrial processes and biomass burning? What is likely to be the net effect of the earth's cloud cover? And will atmospheric CO_2 enrichment "green" the earth by spurring photosynthesis?

The Greenhouse Effect: Natural and Enhanced

Solar radiation received on the earth's surface provides the energy that drives practically all processes in the biosphere (Figure 1.1). Nearly all of the solar radiation is in the wavelength range of 0.15 to 4.0 micrometers (μm). Over 90% of solar radiation is in the form of visible light (0.3 to 0.7 μm). In passage through the atmosphere, solar radiation diminishes in intensity through reflection, scattering, and absorption caused by water vapor and other gases. It is also affected by suspended aerosols (fine particles with diameters in the range of 10^{-3} to 10 μm). On average, the atmosphere reflects about 30% of the solar radiation back to space—more where clouds are present and less where the sky is clear. In addition, the troposphere (the lower layer of the atmosphere) absorbs about 20% of incoming radiation (depending on its dust and aerosol content). (A fraction of the energy absorbed is transmitted to the earth's surface in the form of infrared radiation.) The remaining sunlight, the part that is neither reflected

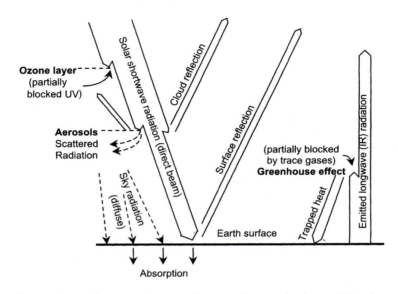

Figure 1.1 Radiation exchange in the atmosphere and at the earth's surface (Hillel and Rosenzweig, 1989).

nor absorbed in the atmosphere, reaches the earth's surface, and is either absorbed or reflected by it.

As a consequence of the net solar radiation it receives, the earth's surface warms up and emits its own radiation, but its radiation is in the infrared or thermal wavelength range—that is, 4 to 50 μm.[1] Since there is very little overlap between the two spectra, incoming solar radiation is commonly referred to as "shortwave" radiation, whereas the earth's emission is called "longwave" radiation. The atmosphere absorbs a fraction of the longwave radiation coming from the earth's surface and reradiates a portion of it back to the surface, the remainder being released to outer space. The relative magnitudes of the incoming and outgoing radiative energy streams of the earth vary both diurnally and seasonally. Over longer periods of time, however, incoming solar radiation absorbed by the earth is balanced by outgoing thermal radiation emitted from it, so the earth's surface is maintained at a more or less constant mean temperature.

If the atmosphere were transparent to the outgoing longwave radiation emitted by the earth, blocking none of it, the equilibrium mean temperature of the earth's surface would be a frigid −18°C. In reality, however, the absorption of the outgoing radiation by water vapor (which makes up about 1% of the atmosphere on average), by liquid water in stratiform cloud droplets, and by some of the trace gases (so called because they are present in small, or "trace" concentrations) raises the mean surface temperature to about 15°C, thus creating a much more hospitable environment for life on earth. The absorptive gases that occur naturally in the atmosphere are water vapor (H_2O), carbon dioxide (CO_2), ozone (O_3), methane (CH_4), and nitrous oxide (N_2O).

The warming caused by the partial trapping of emitted longwave radiation is known as the natural "greenhouse effect." It is called that by analogy to the major processes that keep a greenhouse warm (Figure 1.2). The glass shielding of a greenhouse is transparent to visible light, but partly opaque to infrared radiation. Sunlight entering the greenhouse is absorbed by the interior and converted into heat, then emitted as infrared radiation. That radiation, in turn, is partially blocked by the glass, so it is trapped, thus serving to retain heat that would otherwise escape. The interior of the greenhouse then warms up, until it reaches a temperature at which the intensity of the outgoing infrared radiation equals (in energy terms) the incoming radiation. With the incoming and outgoing radiative fluxes becoming equal, the greenhouse then attains an equilibrium state at a higher temperature than otherwise.[2]

Early Understanding of the Greenhouse Effect

The basic theory of the earth's greenhouse effect has been understood since the 19th century. Already in the first decades of the 1800s, Fourier (1824), a French physicist and mathematician, realized that the atmosphere allows solar radiation to reach the earth but inhibits some of the earth's heat from escaping. He compared the phenomenon to the action of a glass-covered container exposed to the sun that prevents the escape of "obscure radiative heat."

In the 1860s, a British professor of "Natural Philosophy," John Tyndall (1861, 1863) measured the radiative absorption of various gases and atmospheric constituents. He concluded that water vapor, rather than oxygen and nitrogen, is the primary selective absorber of thermal radiation in the air. He also recognized "carbonic acid" (CO_2) as a longwave absorber, and even speculated that changes in climate over geologic times may have been caused by variations in these atmospheric constituents. Emphasizing the warming effect of water vapor, Tyndall wrote, in his colorfully dramatic style:

> This aqueous vapour is a blanket more necessary to the vegetable life of England than clothing is to man. Remove for a single summer-night the aqueous vapour from the air which overspreads this country, and you would assuredly destroy every plant capable of being destroyed by a freezing temperature. The warmth of our fields and gar-

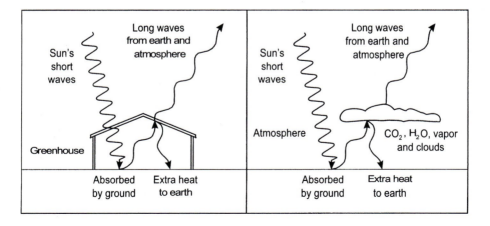

Figure 1.2 The greenhouse analogy (Gedzelman, 1980).

dens would pour itself unrequited into space, and the sun would rise upon an island held fast in the iron grip of frost. (1863, pp. 204–5)

Three decades later, Svante Arrhenius (1896), a Swedish chemist, focused on the role of "carbonic acid" in the greenhouse effect. He linked variations in quantity of atmospheric carbon dioxide to changes in temperature on the ground. Arrhenius advanced the theory that CO_2 could have contributed to the causation of alternating warm and cold periods in the course of the earth's geologic history. Furthermore, he recognized that industrial development (based at that time on the combustion of 500 million tons of coal per year) was supplying a significant increment of CO_2 to the atmosphere. He estimated, however, that the anthropogenic augmentation of atmospheric CO_2 was partially balanced by the formation of limestone and other mineral carbonates through weathering of silicates. Arrhenius made the original, albeit rough, calculation that a hypothetically doubled atmospheric CO_2 concentration could cause a mean warming of 5.5°C of the earth. Remarkably, this result is not far from the range of projections made by some of the sophisticated global climate models of today.

Evidence for the Natural Greenhouse Effect

Atmospheres of other planets besides the earth exhibit various degrees of transparency to incoming sunlight and to outgoing infrared radiation. "Greenhouse" atmospheres cause some planets' surfaces to be warmer than if those planets had no such atmospheres. The masses and compositions of the atmospheric gases, as well as the surface temperatures of the earth's closest planetary neighbors provide evidence for the greenhouse effect (Table 1.1). Venus is much warmer than the earth, with only part of the difference ascribable to that planet's closer proximity to the sun. The surface is calculated to be 523°C warmer than it would be if devoid of a thick CO_2-rich atmosphere. (Its surface pressure is 90 times greater than the earth's.) In contrast, Mars has a very thin atmosphere (its surface pressure only 0.007 of the earth's). Although its atmosphere is >80% CO_2, Mars's mean surface temperature is a frigid −47°C, due to its exceedingly thin atmospheric "blanket" of greenhouse gas. Still, that temperature is about 10°C higher than it would be in the absence of the CO_2 gas enveloping it.

Could the progressive augmentation of atmospheric CO_2 eventually trigger a

Table 1.1   Planetary evidence for the natural greenhouse effect (IPCC, 1990)

Planet	Surface pressure (relative to Earth)	Main greenhouse gases	Calculated surface temperature in absence of greenhouse effect (°C)	Observed surface temperature (°C)	Warming due to greenhouse effect (°C)
Venus	90	>90% CO_2	−46	477	523
Earth	1	~0.4% CO_2 ~1% H_2O	−18	15	33
Mars	0.007	>80% CO_2	−57	−47	10

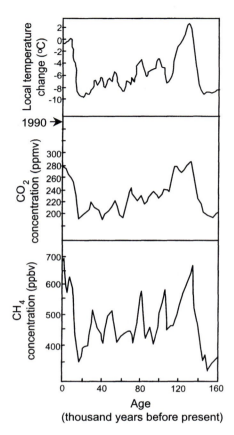

Figure 1.3 Methane and carbon dioxide concentrations and local air temperature over the last 160,000 years; 1990 carbon dioxide level is indicated (IPCC, 1990).

"runaway" greenhouse on the earth similar to that on Venus? The proposed mechanism for this nightmarish scenario is an escalation of the water vapor feedback resulting from the greater evaporation likely to take place in a warmer climate, until temperatures rise to such a degree that calcium carbonate rocks begin to decompose, thus releasing yet more CO_2. The oceans would then evaporate and our planet would become hot and dry (Gedzelman, 1980). Fortunately, such an occurrence appears to be highly unlikely, thanks to the greater distance of the earth from the sun and the concomitant diminution of the intensity of received solar radiation. When we speak of the enhanced greenhouse effect at present, we refer to a warming trend of only a few degrees, on average. However, even a few degrees of warming may produce significant and widespread impacts, to be described in subsequent chapters.

Paleoclimatic records of glacial-interglacial CO_2 concentration changes (evidenced in the air trapped in ancient ice) appear to corroborate the Arrhenius hypothesis concerning the earth's greenhouse effect. They show positive correlations between atmospheric CO_2 concentrations and the temperatures that prevailed in past geologic eras (Figure 1.3) (Delmas et al., 1980; Neftel et al., 1982; Barnola et al., 1987). Methane levels and temperatures are correlated as well (Chappellaz et al., 1990). Significantly high levels of CO_2 and methane occurred during warmer interglacial periods, whereas relatively low levels of CO_2 occurred during glacial periods over the last 160,000 years. However, a causative relation of CO_2 and CH_4 concentrations to temperatures in the past is difficult to prove, since changes in these factors may have ocurred in conjunction with simultaneous changes in ocean circulation and biogeo-chemical cycles.

The Enhanced Greenhouse Effect

Most of the greenhouse gases occur naturally in very small concentrations. However, human activity is causing the release into the atmosphere of several radiatively active gases, thus artificially enhancing the greenhouse effect. Prominent among these gases are carbon dioxide, methane, nitrous oxide, and the several chlorofluorocarbons (known as CFCs). (The latter are synthetic gases used as refrigerants, propellant sprays, and foaming agents.) By further blocking the escape of heat emitted from the earth's surface, these anthropogenic gases—in growing concentrations—have the potential to cause a shift in the planetary energy balance, leading to higher surface temperatures and to changed hydrological regimes. The projection of these climatic consequences, often called "global warming," poses certain dangers that may justify—or even re-quire—concerted societal action to control gaseous emissions in an effort to minimize the enhanced greenhouse effect and to moderate its negative impacts.

The Culprit Gases

In recent years, intensive scientific effort has been devoted to the task of defining the mechanisms by which the various trace gases contribute to the greenhouse effect. The IPCC periodically synthesizes relevant information about the causes and effects of the phenomenon (IPCC, 1990, 1992, 1995, 1996). Knowledge of the sources, sinks, and strengths of the greenhouse gases contributes to improved predictions of global warm-ing (Lashof and Tirpak, 1990). Tables 1.2 and 1.3 list the relevant attributes of the ma-jor greenhouse gases.

The most prevalent of the culprit gases is carbon dioxide. Though the other ma-jor radiatively active trace gases—namely, CH_4, N_2O, and the CFCs—are present in much smaller concentrations than CO_2, most of them are increasing at faster relative rates. Moreover, their specific absorptive capacities for longwave radiation are gener-ally greater than that of CO_2, so their combined influence has gradually increased and is now quantitatively comparable to that of carbon dioxide (Figure 1.4).

Radiative forcings are changes imposed on the planet's energy balance. The con-tribution of a particular greenhouse gas to radiative forcing is measured as the change in average net radiation in Watts (W) m^{-2} at the top of the troposphere caused by that gas. That contribution depends on the radiative wavelength at which the gas absorbs (within the range of longwave radiation), the strength of absorption per molecule, the concentration of the gas, and whether or not other gases present absorb in the same wavelength bands (Mitchell, 1989). Factors other than greenhouse gases may consti-tute radiative forcings, either positive (i.e., warming) or negative (i.e., cooling). These other factors include tropospheric aerosols caused by fossil fuel and biomass combus-tion, aerosols injected into the stratosphere by volcanic eruptions, changes in land-surface albedo caused by deforestation, and variations in the sun's irradiance.

Table 1.2 Sources of the principal greenhouse gases[1] (Lashof and Tirpak, 1990; IPCC, 1995)

Gas	Natural source	Anthropogenic source
Carbon dioxide (CO_2)	Terrestrial biosphere Oceans	Fossil fuel combustion (coal, petroleum); Cement production Land-use modification
Methane (CH_4)	Natural wetlands Termites Oceans and freshwater lakes	Fossil fuels (natural gas production, coal mines, petroleum industry, coal combustion) Enteric fermentation Rice paddies Biomass burning Landfills Animal waste Domestic sewage
Nitrous oxide (N_2O)	Oceans Tropical soils (wet forests, dry savannas) Temperate soils (forests, grasslands)	Nitrogenous fertilizers Industrial sources (adipic acid/nylon, nitric acid) Land-use modification (biomass burning, forest clearing) Cattle and feed lots
Chlorofluorocarbons[2] (CFCs)		Rigid and flexible foam Aerosol propellants Teflon polymers Industrial solvents

1. Sources listed in order of decreasing magnitudes of emission except where otherwise indicated
2. Sources of chlorofluorocarbons not in order of decreasing magnitudes of emission.

Table 1.3 Characteristics of the principal greenhouse gases (IPCC, 1995)

Characteristic	CO_2	CH_4	N_2O	CFC-12	HCFC-22[1]
Preindustrial concentration (1750–1800)	280 ppmv	700 ppbv	275 ppbv	zero	zero
Concentration in 1992	355 ppmv	1,714 ppbv	311 ppbv	503 pptv[2]	105 pptv
Recent rate of con- centration change per year (over 1980s)	1.5 ppmv yr^{-1}	13 ppbv yr^{-1}	0.75 ppbv yr^{-1}	18–20 pptv yr^{-1}	7–8 pptv yr^{-1}
Atmospheric lifetime (years)	0.4% yr^{-1} (50–200)[3]	0.8% yr^{-1} (12–17)[4]	0.25 yr^{-1} 120	4% yr^{-1} 102	7% yr^{-1} 13.3

1. A CFC substitute.

2. pptv = 1 part per trillion (million million) by volume.

3. No single lifetime for CO_2 can be defined because of the different rates of uptake by different sink processes.

4. This has been defined as an adjustment time which takes into account the indirect effect of methane on its own lifetime.

The net change in concentration of a gas depends both on the rate at which the gas is introduced to the atmosphere and on the concurrent rate at which it is removed. The atmospheric lifetime (the average time a molecule of the gas remains in the atmosphere) is thus an important characteristic. A *sink* is a chemical reaction (such as the reaction of methane with an OH^- radical or the photochemical decomposition of N_2O in the stratosphere) that results in the removal of a gaseous species from the atmosphere. When gases remain in the atmosphere for a long time, their radiative effect persists. This means that their atmospheric concentrations may not change rapidly in response to small changes in the rate at which new amounts are being generated. Rather, substantial reductions of emission rates are needed to lower the concentrations of persistent gases within a reasonable period of time. Nitrous oxide and the CFCs are relatively persistent gases, tending to remain in the atmosphere for over 100 years on average.

In contrast to the other gases, the case of carbon dioxide is rather complex, as carbon is cycled among several reservoirs in the carbon cycle, rather than being simply removed from the system altogether. Among these reservoirs are terrestrial and marine organisms and their residues, carbonic acid in aqueous solution, and carbonaceous rocks such as limestone. An "average" molecule of CO_2 remains in the atmosphere only about 4 years. Although carbon dioxide gas is highly soluble in water, it takes a long time for carbon to sink from the surface waters to the deep ocean. This slows the flux of CO_2 from the atmosphere into the ocean, delaying atmospheric stabilization of CO_2. Thus, it may take between 50 and 200 years for atmospheric CO_2 concentrations to recover after an abrupt change in emissions.

In order to hold the atmospheric concentrations of the persistent trace gases steady

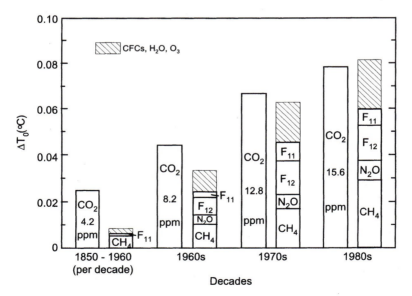

Figure 1.4 Estimated decadal additions to global mean greenhouse forcing of the climate system. ΔT_0 is the equilibrium temperature change computed with a one-dimensional radiative-convective model with no climate feedbacks. Forcings shown by dotted lines are highly speculative (Hansen et al., 1988).

at their present levels, current emissions would need to be reduced by over 60%. Even greater percentage reductions will be necessary if the emission rates of these gases continue to grow above current levels. For methane, which has a relatively short atmospheric lifetime of about 15 years, a reduction of about 10% of current emissions will be required, and atmospheric recovery will occur sooner than in the case of nitrous oxide and the CFCs (IPCC, 1996).

Carbon Dioxide and the Carbon Cycle

Carbon dioxide is the most abundant and, hence, the most important of the increasing trace gases, accounting for about two-thirds of the anthropogenic radiative forcing from 1880 to 1980 (Ramanathan et al., 1985), and about one-half during the decade of the 1980s (Hansen et al., 1988). The buildup of atmospheric CO_2 since the Industrial Revolution is attributed primarily to burning of fossil fuels (e.g., coal, oil, and natural gas) for industry, electricity, and transportation; and, to a lesser extent, to the oxidation of biomass and the decomposition of soil organic matter due to the conversion of forests to agricultural land. Tropical land-use changes now account for approximately 10 to 30% of the annual anthropogenic CO_2 emissions to the atmosphere (IPCC, 1995).

Altogether, human activities have contributed to an increase of some 25% in CO_2 concentration (from about 280 to more than 350 parts per million by volume [ppmv]) since the beginning of the Industrial Revolution (circa 1750). The early measurements of atmospheric composition were sporadic. Continuous direct measurements were begun in 1958 on Mauna Loa in Hawaii (Keeling et al., 1989). Since that time, average CO_2 concentration has increased from about 315 ppmv to about 358 ppmv in 1994 (IPCC, 1996). The continuing upward trend is shown in Figure 1.5, which also shows

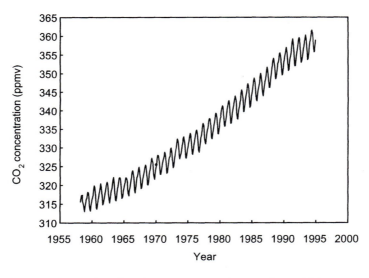

Figure 1.5 Carbon dioxide concentration (averaged monthly) in parts per million of dry air, observed continuously at Mauna Loa, Hawaii (Keeling, 1995, NDP-001,CDIAC).

the annual rise of CO_2 in Northern Hemisphere winter and its fall in summer, mainly due to the seasonal variation of photosynthesis and respiration by vegetation in the Northern Hemisphere. Overall carbon dioxide concentration is continuing to increase at a rate of about 0.4% per year, with the annual increment now estimated to be about 1.5 ppmv (IPCC, 1995). If this trend is allowed to continue, the CO_2 concentration will rise to nearly double its pre–Industrial Revolution level by the end of the 21st century.

Understanding the global carbon cycle is the key to understanding how atmospheric concentrations of CO_2 can change over time (Houghton and Skole, 1990). Chemical processes transmute carbon from one compound to another (including CO_2 gas in the atmosphere, carbohydrates in biomass, hydrocarbons and elemental carbon in fossil fuels, and carbonates in limestone and dolomite, as well as in shelled organisms). Physical processes transfer and redistribute the carbon among the reservoirs at varying rates (fluxes), as illustrated in Figure 1.6.

The processes involving carbon transformations can be described in the following sequence. Green plants capture some of the radiant energy of sunlight and change it into chemical energy contained in carbohydrates, in the process of photosynthesis. In turn, the energy stored in carbohydrates is released through respiration and combustion. Respiration of carbohydrates provides the short-term energy for most life on earth. Organic matter residues from plants buried in geologic deposits have been transformed over time into hydrocarbons commonly referred to as fossil fuels. Combustion of such materials (coal, oil, and natural gas) is used to power the energy-consuming activities of modern industrial society.

Since the Industrial Revolution, humans have dramatically expanded their uti-

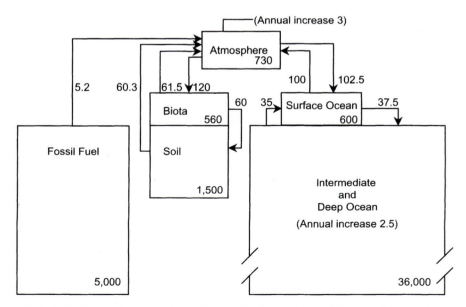

Figure 1.6   The major reservoirs and flows of carbon (GtC yr^{-1}) for the globe, during 1980–1989 (Houghton and Skole, 1990).

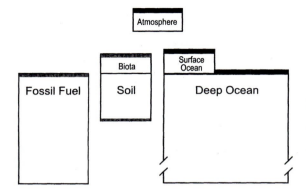

Figure 1.7 Relative changes in the major reservoirs of carbon between 1700 and 1985. Gray indicates a loss of carbon; black indicates a net accumulation of carbon (Houghton and Skole, 1990).

lization of the carbon stored in fossil fuels. At the same time, clearing land for agriculture needed to feed a growing population has caused the release of carbon formerly stored in vegetation and in the soil's organic matter. Thus, the carbon input to the atmospheric reservoir has increased, while the storage of carbon in the land reservoirs has diminished (Figure 1.7) (Houghton and Skole, 1990). However, not all of the added CO_2 remains in the atmosphere. Of the ~7 gigatons of carbon (1 Gt = 10^9 metric tons) currently released to the atmosphere each year from human activities, only about half remains in the atmosphere. Most of the remainder (3.2 +/− 0.2 GtC yr^{-1}) is absorbed by the oceans, while a small part (0.5 +/− 0.5 GtC yr^{-1}) is thought to be taken up by the regrowth of forests (especially in the northern mid and high latitudes) (Tans et al., 1990; IPCC, 1995).

However, exact quantification of the carbon stores and fluxes is extremely difficult to achieve. Sources of uncertainty include measurement errors of fluxes, incomplete knowledge of reservoir magnitudes, and inaccurate accounting of carbon utilization (both direct and indirect) by humans. Measurements of isotopic $^{13}C/^{12}C$ ratios in reservoirs and O_2/N_2 ratios offer some new approaches to assessing the global carbon cycle (IPCC, 1995).

Methane

Next to carbon dioxide in importance as a contributor to global warming, methane is a relatively short-lived greenhouse gas. Although its concentration (1.7 ppmv) is only 0.5% that of CO_2, its per-molecule absorption of infrared radiation is 20 times stronger. Methane is released into the atmosphere from both natural and anthropogenic sources, with the latter accounting for 60 to 80% of current emissions (IPCC, 1995). Major natural sources include anaerobic decomposition of organic matter in wetlands, metabolism of termites, release of gas dissolved in ocean and freshwater, and decomposition of methane hydrates[3]. The anthropogenic sources include coal mining, natural gas and petroleum extraction, rice paddies, enteric fermentation by domesticated ruminants, farm animal wastes, domestic sewage, landfills, and releases during biomass burning. Methane makes up 90% of natural gas, is present in the gas trapped in coal and petroleum, and is released in the processing of most fossil fuels.

Methane is removed from the atmosphere as it reacts with the hydroxyl radical

(OH^-) in the stratosphere, resulting in water vapor. In the troposphere, methane re-actions can lead to the formation of ozone. Both reaction products—stratospheric wa-ter vapor and tropospheric ozone—are greenhouse gases; thus, methane emissions contribute to the enhanced greenhouse effect both directly and indirectly. The aver-age atmospheric lifetime of methane is on the order of 12 to 17 years.

The concentration of methane in the atmosphere has more than doubled since the Industrial Revolution (having gone from 0.7 to 1.7 ppmv), most probably due to the spread of fossil-fuel processing and use, rice cultivation, animal husbandry, and solid-waste landfills. Methane increases were about 0.02 ppmv per year in the 1970s, but slowed to 0.01–0.013 ppmv in the 1980s (Blake and Rowland, 1988). Recent data suggest that growth rates of methane since 1993 may again be increasing (IPCC, 1996). The current growth rate is about 0.8% per year. Why methane emissions have fluctuated in recent years is not yet known.

Nitrous Oxide

Nitrous oxide is a persistent (mean residence time about 120 years) trace gas that is also a stronger infrared absorber than CO_2. The present atmospheric concentration of N_2O is about 0.31 ppmv, some 10% above its pre–Industrial Revolution concen-tration of 0.28 ppmv. That concentration is increasing currently at the rate of 0.0008 ppmv (0.25%) per year (IPCC, 1995). The sources of N_2O are not well quantified, but are known to include significant emissions from the oceans and soils. In the soil, ni-trous oxide evolves mainly from the nitrogen metabolism of soil microorganisms, through the processes of denitrification and nitrification (Sahrawat and Keeney, 1986). Development of pastures in tropical regions and applications of nitrogen fertilizers ev-idently enhance soil emissions of N_2O. Other relatively small anthropogenic sources of N_2O include biomass burning, the production of both adipic acid (used in making nylon) and nitric acid (primarily for fertilizer), and the combustion of fossil fuel. The major processes by which N_2O is removed from the atmosphere are photodissociation and photooxidation in the stratosphere.

Chlorofluorocarbons

Chlorofluorocarbons (CFCs) are synthetic industrial chemicals used as foaming agents, spray-can propellants, refrigerants, and solvents for the electronic industry. CFCs are both greenhouse gases and stratospheric ozone scavengers. Chemically, they are largely inert (noninteractive) in the troposphere; hence, then are wafted prac-tically unchanged to the stratosphere. Here the CFCs are subject to ultraviolet radia-tion that releases chlorine ions, which, in turn, react with ozone molecules. Since it is stratospheric ozone that selectively filters ultraviolet radiation, thereby preventing skin cancer and genetic mutation of plants and animals at the earth's surface, the de-struction of this ozone poses a serious danger to many organisms.

Chlorofluorocarbons have only a very small presence in the atmosphere, being measured in parts per trillion by volume (pptv). As greenhouse gases, however, the CFCs are extremely potent infrared absorbers, being on the order of 10,000-fold stronger per molecule than CO_2. Moreover, they can remain in the atmosphere for decades to centuries. The atmospheric concentrations of the major CFCs (so-called

CFC-11 and CFC-12) are about 300 and 500 pptv, respectively. The concentration of each of these gases is still increasing, notwithstanding the 40% reduction in global consumption of CFCs that took place from 1986 to 1991. That reduction was a consequence of the Montreal Protocol on Substances that Deplete the Ozone Layer (1987), an international agreement aimed at reducing and ultimately phasing out CFC production.

For refrigeration and air conditioning, the major CFCs, which are fully halogenated (having chlorine or fluorine at every attachment site) are being replaced with hydrochlorofluorocarbons (HCFCs) and hydrofluorocarbons (HFCs). The HCFCs and HFCs, while still greenhouse gases, are shorter lived (1 to 40 years) and thus pose a lesser threat of ozone depletion and future warming. Increases in these substitute gases range from 1.5 to 15% per year (IPCC, 1990).

CFCs are removed from the stratosphere by photolysis (disintegration under the action of sunlight) and HCFCs are removed by reaction with OH^- in the troposphere. Because of the long lifetimes of the CFCs, their influence will continue in the atmosphere for at least a century, even though their emissions have been curtailed.[4]

Ozone

Ozone (O_3) is more than a greenhouse gas, because it absorbs both solar ultraviolet and terrestrial longwave radiation. Ozone occurs naturally primarily in the stratosphere; only 10% of total ozone resides in the troposphere (Figure 1.8). As already mentioned, stratospheric ozone protects the earth's surface from harmful ultraviolet radiation that can cause skin cancer and genetic mutations. However, tropospheric ozone absorbs terrestrial longwave radiation; hence, it acts as a greenhouse gas, retaining heat and returning it to the earth's surface.

Observations show that the losses of stratospheric ozone have been as high as 90% over Antarctica (the so-called Ozone Hole), and up to 10% at middle to high latitudes (IPCC, 1992). No substantial depletion has been found thus far in the tropics. Ozone losses are apparently linked to processes involving CFCs and other halogenated compounds.[5]

The concentration of ozone in the lower atmosphere, a component of photochemical smog, seems to be growing in some regions (in Europe, it has been increasing at the rate of about 10% per decade). Ozone in the troposphere is a pollutant (as well as a greenhouse gas), contributing to smog and corrosion. Thus, while decreases in stratospheric ozone are worrisome because of loss of atmospheric protection from ultraviolet radiation, increases in tropospheric ozone are also a cause for concern.

Modeling the Climate System

While even an elementary understanding of the earth's energy balance is sufficient to predict in principle that progressive increases in the concentration of the greenhouse gases should lead to global warming, more sophisticated tools are needed to quantify the predictable effect in specific terms, considering the complexity of the numerous dynamic factors composing the earth's overall climate system. The Global Climate Models (GCMs) have been offered as such tools (Trenberth, 1992). GCMs are comprehensive mathematical models that attempt to calculate the dynamic temporal and

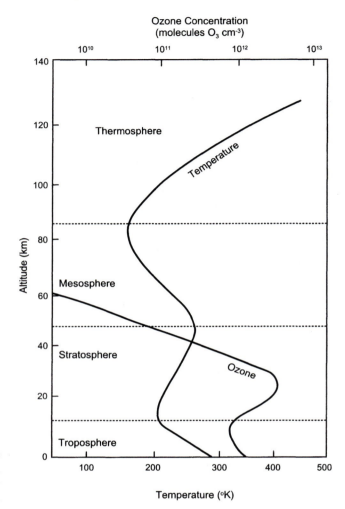

Figure 1.8 Ozone distribution and temperature profile in the atmosphere (Watson, 1986).

spatial transports and exchanges of heat, moisture, and momentum throughout the earth's atmosphere and its surface, including the continents and oceans.

To understand the usefulness and limitations of GCMs, we should consider the way they treat climate phenomena. Climate is defined as the prevalent pattern of the weather observed over a prolonged period of time. Climate variables (e.g., temperature, precipitation, wind speed) can be time-averaged on a daily, monthly, yearly, or longer basis. Associated with the average states of climate variables are indications of their oscillations or variations about their mean values.

Since the earth's climate is ever changing, it is important to define the time scale over which comparisons are made (Figure 1.9). Over long periods of time (say, geologic eras), the occurrences of glacial and interglacial periods appear to be related to changes in the earth's orbit, with associated long-term fluctuations in the patterns of

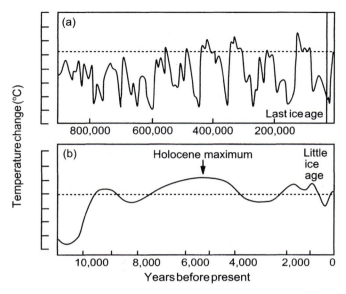

Figure 1.9 Global temperature variations for (a) the last 1 million years and (b) the last 10,000 years. The dotted line represents conditions near the beginning of the 20th century (IPCC, 1990).

ocean circulation and in the extents of major ice sheets. For shorter periods of time, climatic variables appear to fluctuate about a discernible mean. Accordingly, the term *climate change* refers to an overall alteration of mean climate conditions, whereas *climate variability* refers to fluctuations about the mean. However the distinction between these two terms is not absolute. The warming trend expected as a result of the enhanced greenhouse effect can be regarded as a "climate change" if viewed on the time scale of the coming decades or few centuries. Over longer time scales of millennia or geologic eras, however, the predicted trend could be regarded merely as a deviation from the much longer-term mean. Thus it is that, while we may indeed be nearing the end of an interglacial period and be headed on the long time scale toward another in the series of glacial periods, the appreciable effect of this long-term trend may not become widely apparent for several thousands of years (Berger, 1981). In the meantime, our concern is focused on the shorter-term trend that may be taking effect due to human activities.

The Climate System

The climate system consists of a series of fluxes and transformations of energy (radiation, heat, and momentum), as well as transports and changes in the state of matter (e.g., air, water, and aerosols). Received solar radiation is the major energy source that powers the entire system. The flows and transports occur between and within the main components of the system: the atmosphere, oceans, land, biota, and cryosphere (the domain of ice and snow) (Figure 1.10)—all interlinked via physical, biological, and chemical processes.

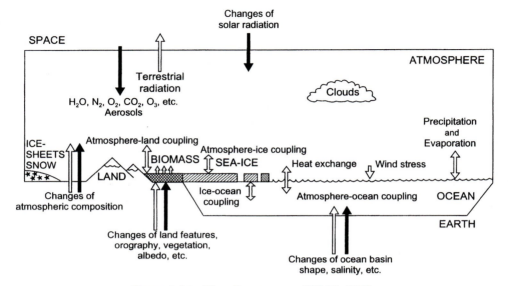

Figure 1.10   The climate system (WMO, 1975).

The atmosphere and the oceans are both fluids subject to internal movements as-sociated with random turbulence and systematic circulation. In the atmosphere, gases and aerosols absorb, scatter, or transmit shortwave and longwave radiation. Aerosols also promote the formation of clouds and precipitation. The oceans cover about 70% of the earth's surface, thereby providing the major medium of interaction with the at-mosphere. The spatially uneven heating of ocean water drives the ocean currents, which are in turn deflected by the earth's rotation. The salinity of the oceans is im-portant because it modifies the density of water and, therefore, the patterns of the ocean currents, which distribute heat around the globe.

Land covers only about 30% of the earth's surface, but it has a major effect on the atmosphere. On long time scales, the shifting positions of the continents modify the absorption of solar radiation and the differential heating of land masses and oceans, which in turn influence water and air circulation patterns. On shorter time scales, the land and its different vegetation zones affect the radiation balance through their re-flective and absorptive characteristics, the hydrological cycle (including evaporation and recycling of precipitated water), and momentum transfer (influenced by the areal variability of surface roughness). All these processes operate on different time scales (Figure 1.11).

Global Climate Models

Global climate models, being mathematical formulations of the processes that com-prise the climate system (Figure 1.12), simulate climate by solving, sequentially or si-multaneously, the fundamental equations for conservation of mass, momentum, en-ergy, and water (Table 1.4). For the boundary conditions relevant to the earth's geographic features and with the relevant parameters, these equations constitute numerical representations of radiation, turbulent transfers at the ground–atmosphere boundary, cloud formations, condensation and precipitation of moisture, transport of heat by ocean currents, and other salient physical processes (Figure 1.12).

These equations are solved for a number of vertical layers in the atmosphere and for gridpoints in finite difference models (or for waves in spectral models) at the surface of the earth. The atmosphere is usually divided into 2 to about 20 vertical layers. Horizontal resolution varies from about 300 to about 1000 kilometers between gridpoints. As more powerful computers are employed, the grids can be made denser, so the resolution can conform more closely to the finer features of the earth's surface.

Global climate models are used to analyze the effects of increasing greenhouse gas concentrations. Some of the interactions involved in the modeled processes may reinforce a greenhouse warming and are therefore termed positive feedbacks, whereas other interactions may counter the warming and thus constitute nega-

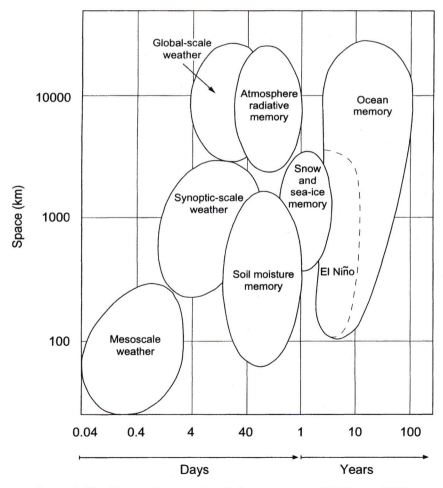

Figure 1.11   Time and space scales of climate processes (Dickinson, 1986).

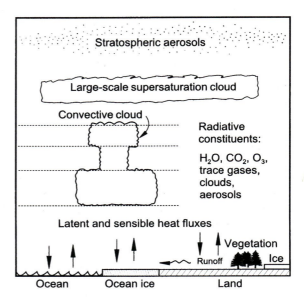

Figure 1.12 Schematic illustration of GCM structure at a single gridbox (Hansen et al., 1983).

tive feedbacks. (A feedback is the return of a portion of the output of a process or system to its input.)

An example of the intricate sequence of interactions is the following: the atmosphere, warmed by enhanced absorption of thermal radiation due to the greenhouse gases, is thereby able to hold more water vapor. Water vapor is itself a greenhouse gas; so the effect is for the atmosphere to become still warmer. A positive feedback thus comes into play. Another instance where a positive feedback occurs is in the case of ice and snow: as higher temperatures melt more ice and snow, the surface loses its reflective whiteness (i.e., its albedo is lowered), less of the incoming solar radiation is reflected to space, and the surface absorbs more radiation. The ground then converts the received radiation into heat, thus enhancing the warming of the atmosphere. For a doubling of CO_2, the predicted radiative warming is on the order of 1°C; however, the positive effects of water vapor and other feedbacks raise the estimated warming to between 1.5 to 4.5°C. This calculation ignores the negative feedback resulting from an increased presence of aerosols in the atmosphere.

The term "GCM" originally connoted "general circulation model," implying that the calculations were limited to the earth's atmospheric processes. Now the term is applied more generally to describe a *global climate model*, which includes a more comprehensive treatment of ocean and land processes. To obtain an inclusive global prediction of the rates, magnitudes, and regional distributions of climate changes under enhanced greenhouse conditions, models of oceanic and atmospheric processes must be coupled. The more sophisticated GCMs of that type are now able to simulate, at least partially, El Niño–Southern Oscillation (ENSO) events, which are driven by variations in sea–surface temperatures in the Pacific Ocean and are the cause of interannual climate variations over large areas (e.g., Meehl, 1990). Such coupled models have been used to project future transient greenhouse warming, as will be described subsequently.

Table 1.4 Fundamental equations represented in GCMs (Hansen et al., 1983)

Conservation quantity	Equation
Conservation of momentum (Newton's second law of motion)	$\dfrac{dV}{dt} = -2\,\Omega \times V - \rho^{-1}\nabla p$ $+ g + F$
Conservation of mass (continuity equation)	$\dfrac{d\rho}{dt} = -\rho\,\nabla \cdot V + C - D$
Conservation of energy (first law of thermodynamics)	$\dfrac{dI}{dt} = -p\,\dfrac{d\rho^{-1}}{dt} + Q$
Ideal gas law (approximate equation of state)	$p = \rho R T$

V = velocity relative to rotating earth

t = time

$\dfrac{d}{dt}$ = total time derivative $\left[= \dfrac{\partial}{\partial t} + V \cdot \nabla \right]$

Ω = planet's angular rotation vector

ρ = atmospheric density

g = apparent gravity [= true gravity $- \Omega \times (\Omega \times r)$]

r = position relative to planet's center

F = force per unit mass

C = rate of creation of (gaseous) atmosphere

D = rate of destruction of atmosphere

I = internal energy per unit mass [$= c_v T$]

Q = heating rate per unit mass

R = gas constant

c_v = specific heat at constant volume

Strengths and Weaknesses of GCMs

The advantages of GCMs as predictors of climate change are their consistent internal logic, inclusion of simultaneous and interacting processes, and global integration. They are better adapted to simulating temperature, which is a spatially continuous variable, than precipitation, which is spatially and temporally discontinuous.

GCM deficiencies arise from incomplete understanding of ocean circulation patterns, lack of knowledge concerning the formation and feedback effects of clouds (whether positive or negative),[6] simplistically formulated hydrological processes (ignoring land-surface and vegetation features), and coarse spatial resolution. Lack of understanding of cloud processes hinders projections of the magnitude of climate change, while the underdeveloped state of ocean models limits the ability of present-day models to predict the time rate of that change and its regional patterns. At currently used grid spacings, GCMs do not resolve atmospheric events such as fronts and severe storms that take place over small distances.

GCM simulations of temperature and precipitation over continental areas often agree well with observations at annual and seasonal time scales. Over much of the earth's land surface, the differences between observed temperature and precipitation, and predictions made by four GCMs were less than 2°C and 1 mm day^{-1}, respectively (Kalkstein, 1991). Regions with monsoonal climates are especially difficult to simulate accurately (Figure 1.13) (Robinson, 1991).

The margin of uncertainty regarding GCM simulations of current *regional* climates (e.g., Grotch, 1988) raises doubts over their predictions of future regional climates. Furthermore, most of the greenhouse warming predictions have been made

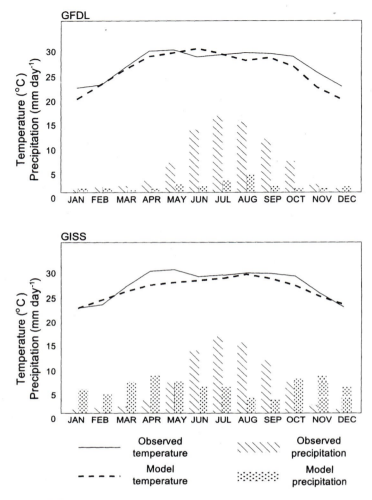

Figure 1.13   Observed versus GCM simulated monsoonal climate for
the region of Bangladesh (Robinson, 1991).

with GCMs still lacking fully interactive ocean-land-atmosphere models. Mitchell et
al. (1987) showed that differences in current climate representations affect the regional response of climate models to perturbations such as doubled CO_2 and sea-
surface temperature increases. This result again emphasizes the need for improving
the capability of GCMs to simulate present-day regional climate in order to gain confidence in their predictions of future climate.

Equilibrium and Transient Projections of Climate Change

At least 10 different GCMs have been developed by atmospheric scientists in various
research groups, and have been used to simulate the effects of greenhouse gas increases. Most calculations have been made in an equilibrium mode, meaning that the

Table 1.5 Results of selected GCM doubled-CO_2 equilibrium simulations with seasonal cycle and mixed-layer ocean (IPCC, 1990, 1992)

Group	ΔT (°C)	ΔP (%)
GISS	4.2	11
UKMO	5.2	15
	3.2	8
	1.9	3
CCC	3.5	4
GFDL	4.0	8
CSIRO	4.8	10

ΔT = equilibrium surface temperature change under doubled CO_2

ΔP = percentage change in annual precipitation

CCC = Canadian Climate Centre, Canada
CSIRO = Commonwealth Scientific and Industrial Research Organization, Australia

GFDL = Geophysical Fluid Dynamics Laboratory, USA

GISS = Goddard Institute for Space Studies, USA

UKMO = Meteorological Office, United Kingdom. Note: The three sets of numbers represent different parameterizations of cloud physics.

models are subjected to an abrupt (rather than gradual) doubling of CO_2 concentration and are then run to simulate an equilibrium climate under those hypothetical conditions. Those conditions are usually taken to represent the combined effects of all the greenhouse gases (including CO_2, CH_4, N_2O, CFCs, etc.) that would be equivalent to the radiative forcing of a doubled concentration of CO_2. Results from these numerical simulations show a mean global warming in the range of 1.5 to 5.0°C (Table 1.5). The IPCC has endorsed a range of 1.5 to 4.5°C for doubled CO_2, with a "best estimate" of 2.5°C based on modeling results, observations, and sensitivity analyses. When incorporating the possible effects of future changes of anthropogenic aerosol concentrations, the best estimate for 2100 is a temperature increase in the range of 1.0 to 3.5°C, somewhat lower than earlier projections (IPCC, 1996).

The GCM doubled-CO_2 simulations also predict an increase in global precipitation ranging from about 5 to 15%. In a warmer climate, precipitation increases because evaporation increases. As the saturation vapor pressure of the atmosphere rises with temperature, warmer air can hold more water vapor. Downward temperature fluctuations (the proximate cause of precipitation) then result in more condensation per unit of adiabatic cooling.

The GCMs further predict that:

1. The high latitudes in the Northern Hemisphere are likely to experience greater warming than the global mean warming, especially in winter. High latitudes are subject to greater-than-average warming in the GCM simulations because the

northward retreat of the limits of snow and sea ice tends to reduce the albedo. Reduction of the albedo implies greater absorption of solar radiation and, hence, a warming of the surface. Another factor contributing to greater warming at high latitudes is the relative stability of polar air. Because the air is colder in the higher latitudes, there is less tendency for vertical mixing; a given degree of warming of the air in contact with the ground surface is less likely to be dissipated by upward convection.

2. Mid-continental land areas in the middle latitudes may undergo a drying trend in summer, because of earlier snowmelt, increases in potential evaporation, and reduced cloud cover (Manabe and Wetherald, 1987; Kellogg and Zhao, 1988). Since soil moisture is a key variable for crops, production may suffer in such areas, which include some of the most important agricultural regions in the world (e.g., the Great Plains of North America and the wheat-growing areas of the former Soviet Union) (Hillel and Rosenzweig, 1989).

3. Land areas will warm faster than the oceans because of their lower heat capacity. Hence, changes in the climates of various continental regions are likely to differ considerably from the global mean. However, the exact regional distribution of such changes is very difficult to predict given the limited accuracy of the current GCMs.

Transient Climate Change

GCM scenarios specifying a sudden doubling of CO_2 concentration represent a synthetic climate. In reality, the changes in atmospheric composition are gradual. Some simulations have been made with combined atmospheric and ocean GCMs in a transient mode—that is, with progressively rising atmospheric greenhouse gas concentrations on the order of 1% per year (Table 1.6). These simulations account for heat transport by ocean currents. They generate large-scale patterns of change that gradually tend toward the equilibrium doubled-CO_2 simulations, but the warming in the high

Table 1.6 Results of selected CO_2 transient simulations with coupled ocean-atmosphere GCMs (IPCC, 1992)

Group	Control CO_2 (ppm)	CO_2 rate of increase	CO_2 doubling time (yr)	Simulated period	Warming at CO_2 doubling (°C)	$2\times CO_2$ sensitivity (with mixed-layer ocean) (°C)
GFDL	300	1% yr^{-1} compounded	70	100	2.3	4.0
MPI	390[1]	~1% yr^{-1} compounded	60	100	1.3	2.6
NCAR	330	1% yr^{-1} linear	100	60	2.3 (projected)	4.5
UKMO	323	1% yr^{-1} compounded	70	75	1.7	2.7[2]

1. Equivalent CO_2 value for trace gases.

2. Estimate from low-resolution experiment.

MPI = Max-Planck Institute, Germany

latitudes is delayed—especially over the oceans. The North Atlantic and the Southern Ocean near Antarctica show the least warming and in some cases even a cooling. This is attributed to the vertical mixing resulting from the upwelling of cold water from the deep ocean. The Southern Hemisphere tends to warm more slowly than does the Northern Hemisphere because of the greater areal coverage by oceans.

These simulations of gradual greenhouse forcing predict that, after an initial period, warming will take place at the rate of about 0.3°C per decade. The warming predicted to occur eventually, when the CO_2 concentration doubles, ranges from 1.3 to 2.3°C; this occurs after 60 to 100 simulated years. This amount of warming is about 60% of the equilibrium projections reported for the abrupt doubling of CO_2, evidently because the more realistic simulation of the deep ocean with its thermal inertia retards the warming process.

However, it is still difficult to associate the warming projected by the GCM transient simulations of climate change with actual years. One shortcoming is that the coupled atmospheric–oceanic models suffer from systematic errors tending to produce "climate drift"—the gradual straying of the current climate simulation away from presently observed climatology. In the simulations, the climate drift is compensated by inserting currently observed ocean–atmosphere fluxes of heat and water in order to prevent deviations from known conditions.

Variability

Under an enhanced greenhouse effect, changes will occur in both the mean values of climate parameters and the frequency and severity of extreme meteorological events. Such events may include spells of extra high temperature, torrential storms, or droughts. The relationship between changes in mean temperature and the corresponding changes in the probabilities of extreme heat spells tends to be nonlinear: relatively small changes in mean temperature can trigger relatively large increases in the frequency of extreme events (Mearns et al., 1984), such as days with very high temperatures that are deleterious to crop growth (Rosenzweig and Hillel, 1993a). The frequencies of such occurrences are evidently more sensitive to changes in the variability of temperature than to changes in its mean value (Katz and Brown, 1992). Global warming could thus result in disproportionate impacts, depending on the relative magnitudes and effects of the mean and variance changes.

Similarly, the increase in the probability of drought may be greater than the relative reduction in overall rainfall amount. This is especially so because a reduction of rainfall is generally accompanied by a rise of potential evapotranspiration, thus raising the demand for water by plants even while reducing the supply. The rise of potential evapotranspiration may induce drought conditions even where precipitation per se increases (Rind et al., 1990). A study of climate change scenarios pertaining to the U.S. Great Plains region suggests that under an enhanced greenhouse effect droughts may be even more severe than the Dust Bowl that occurred in the 1930s (Figure 1.14) (Rosenzweig and Hillel, 1993b).

Mearns et al. (1990) have analyzed GCM simulations of current climate and found some important shortcomings in their representation of observed variability. Such simulations commonly overestimate both temperature and precipitation vari-

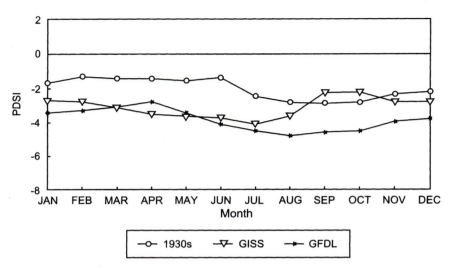

Figure 1.14   Palmer Drought Severity Index (PDSI) relative to 1951–1980 in Dodge City, Kansas, for the 1930s and the GISS and GFDL climate change scenarios (Rosenzweig and Hillel, 1993b). The PDSI is a widely used index for evaluating drought conditions in the United States based on deviations from normal precipitation and temperature conditions (Palmer, 1965).

ability. For the doubled CO_2 case, several of the GCMs are consistent in predicting that precipitation variability is likely to increase; temperature variability on both daily and interannual time scales may decrease. Rind et al. (1990) suggest that hydrological extremes — droughts and floods — may intensify with climate change, and Emanuel (1987) hypothesizes that hurricanes may become more intense.

Sea-level Rise

Another predicted consequence of the enhanced greenhouse effect is a rise of sea level, primarily due to the thermal expansion of ocean water, as well as the melting of continental ice. In the ocean, warming will reduce the density of the top layers of the water, thus increasing its volume. On land, melting of mountain glaciers is projected to occur, along with an uncertain input from polar ice sheets (as additional snowfall may in part offset the melting of polar ice). The IPCC (1996) has predicted that continuation of current trace gas and aerosol emission rates will cause mean sea level to rise at the rate of about 2 to about 9 cm per decade in the coming century, with a "best guess" of about 5 cm per decade. Such a rate of sea-level rise could result in a total rise of 20 cm by the year 2050 and of about 50 cm by 2100 (Figure 1.15). The frequency of damaging storm surges is thereby likely to rise. The effects of sea-level rise can be expected to vary from place to place, depending on local conditions such as elevation of the land and currently occurring geological subsidence. The effects of sea-level rise are likely to be most severe in low-lying areas such as deltas, estuaries, and certain island nations. Agricultural activities concentrated in such regions will be at risk.

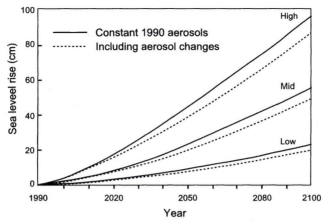

Figure 1.15   Sea-level rise predicted to result from IPCC Business-as-Usual emissions, with and without changing aerosols (IPCC, 1996).

Has Climate Change Begun?

This question has sparked a lively scientific debate. In the summer of 1988, following a succession of unusually warm years and in the midst of an extreme drought in the U.S. Midwest, Dr. James Hansen (1988) made the following statement to the U.S. Senate Committee on Energy and Natural Resources: "The global warming is now sufficiently large that we can ascribe (it) with a high degree of confidence . . . to the (enhanced) greenhouse effect." This statement was based on a comprehensive statistical analysis of observed land-based temperatures of the last 100 years and a comparison of the recorded warming with climate model simulations. The decade of the 1990s is the warmest on record (Figure 1.16). In the early 1990s, the global mean surface temperature reached about 15.4°C. The volcanic eruption of Mt. Pinatubo in the

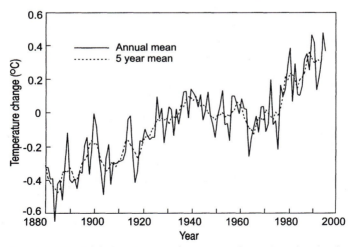

Figure 1.16   Global surface air temperature change based on land-surface measurements (Hansen, personal communication).

Philippines in 1991 is believed to have led to a temporary cooling in 1992 (Hansen et al., 1992), most likely caused by aerosols injected into the stratosphere, but global surface air temperatures began to rebound in 1993.

The comparison between observed global mean surface temperature and climate model calculations for the last hundred years shows a rough consistency between calculated and actual warming, but it is still difficult to ascertain how much of the warming has been caused by the enhanced greenhouse effect and how much may be due to natural, unforced climate variability. Enhanced greenhouse warming effects may be delayed by ocean-heat storage or may have been masked by opposing cooling effects of tropospheric aerosols caused by industrial pollution (as well as by the injection of aerosols into the stratosphere by volcanic eruptions).

Proving climate change is difficult not only because of the considerable natural variability of the climate system but also because of the human phenomenon of urbanization. Since cities are built of materials that absorb solar radiation (concrete and asphalt) and dissipate less heat by evaporation and wind than do rural areas, the former tend to be warmer (on average) than the latter. This is known as the *urban heat island* effect. Overall, the instrumental record shows that in the last century there has been an irregular increase in global surface temperature of 0.45 ± 0.15°C, while atmospheric carbon dioxide has risen from about 280 ppmv to about 356 ppmv. A small part of this warming, probably less than 0.1°C, is attributable to the fact that many of the measurements were made in or near cities.

While reasonably consistent with the amount of trace gas accumulation and the predictions of climate models, the global warming observed thus far in the surface measurements does not yet prove beyond all doubt that the enhanced greenhouse effect has actually begun to change the climate. The apparent change in global temperature from the surface measurements may lie within normal climate variability. Satellite microwave soundings used to measure temperature trends beginning in 1978 have not identified a similar warming trend (Spencer and Christy, 1992).

Another possibility is that other climate forcings such as tropospheric aerosols from industrial pollution and biomass burning may be negating or masking the greenhouse-induced warming trend (Figure 1.17). Anthropogenic smoke and sulfate aerosol emissions have occurred over roughly the same time period as the enhanced greenhouse gas emissions, since both are products of industrialization and, to some extent, of agricultural development. The observed warming is in better accord with model estimates when the increasing concentrations of tropospheric aerosols are taken into account (Figure 1.18).

In the 1970s, there was much speculation that increasing atmospheric turbidity caused by human action might actually cool the earth, because aerosols in the troposphere tend to reflect and scatter radiation, just as do stratospheric aerosols of volcanic origin. A further cooling effect of increased tropospheric aerosols may be to provide condensation nuclei leading to the formation of clouds, which in turn increase reflectivity.

An estimated 25 to 50% of tropospheric aerosols is caused by humans through the processes of smoke- and sulfur-emitting industrialization, urban air pollution, mechanized agriculture, and increased population, as well as land denudation in dust-prone semiarid lands (Hansen and Lacis, 1990). Aerosol injections with their cooling effect and greenhouse gas emissions with their warming effect may thus be counteracting

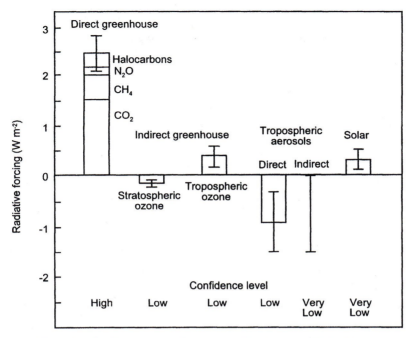

Figure 1.17 Comparison of major radiative forcings (IPCC, 1996).

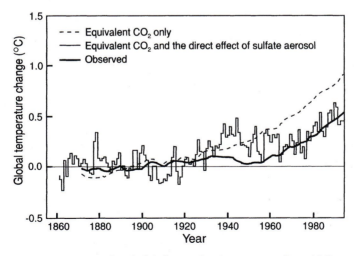

Figure 1.18 Simulated global annual mean warming from 1860 to 1990, allowing for increases in equivalent CO_2 alone, in comparison with combined increases in CO_2 and sulfates (Mitchell et al., 1995). The observed changes are from Parker et al. (1994). The anomalies are calculated relative to 1880–1920 (IPCC, 1996).

each other,[7] but differences in geographical and seasonal distribution and lifetimes (about 1 week for tropospheric aerosols and up to hundreds of years for long-lived greenhouse gases) of the forcings create a complex situation (Charlson et al., 1992).

In contrast to the growing emissions of the trace gases, however, the release of tropospheric SO_2 aerosols emanating from industrial regions has lessened somewhat, at least in the United States, since the Clean Air Act went into effect in 1970 to reduce acid rain. Currently, about 21 million tons of sulfur are released over the United States every year; the sulfate aerosols generated from these emissions may be keeping the mean U.S. temperature about 1°C cooler than it would be otherwise. If control of air pollution continues, the negative radiative forcing of tropospheric aerosols may diminish in the future, while greenhouse gas forcing continues.

Volcanic eruptions spew large quantities of sulfur dioxide and other compounds into the air. Once aloft, volcanic SO_2 is transformed into tiny droplets of sulfuric acid that are spread around the globe by stratospheric winds. High in the atmosphere, the SO_2 aerosols both reflect a fraction of incoming solar radiation back to space and absorb a fraction that is reradiated as thermal infrared radiation. The net effect is to cool the lower parts of the atmosphere for up to a few years following a large volcanic eruption. Large eruptions in the last two centuries have been followed by periods of 1 to 3 years during which the earth's average surface temperature declined by 0.2 to 0.5°C. Volcanoes located near the equator tend to have a greater and more extensive effect on climate than those at higher latitudes.

On June 12, 1991, the eruption of Mount Pinatubo in the Philippines ejected about 20 million tons of SO_2 to a height of up to 25 km into the stratosphere. This eruption, which may have been the largest volcanic eruption of the 20th century, provided a real-time test of the global climate models' ability to predict climate sensitivity to the presence of aerosols in the stratosphere. Accurate depiction by GCMs of the Pinatubo eruption's cooling effect could lend credence to the predictions of global warming by the same models.

Soon after the eruption, climate modelers entered an estimate of the amount of aerosols produced by Pinatubo into the model stratosphere and calculated the effects on temperature (Figure 1.19) (Hansen et al., 1992). In 1992, the year following the eruption, global surface temperatures did indeed cool by about 0.3°C on average, while Northern Hemisphere temperatures cooled by up to 0.5°C. The GCM-calculated cooling was of the same order as that observed. In 1993 and 1994, both model calculated and mean global surface temperatures began to rise, as the volcanic aerosols gradually settled from the stratosphere.

Monitoring the rise of mean surface air temperature relative to the random variability (known as the "signal" to "noise" ratio) is only one approach to detection of the enhanced greenhouse effect. Climatologists search climate records for "fingerprints"—patterns of regional or altitudinal change or relationships among climate variables that are uniquely relevant to climate change predictions. High-latitude temperatures are especially scrutinized since GCMs predict that warming should be higher in those regions of the Northern Hemisphere. This effect is predicted to be particularly strong in winter, due to feedback effects of the expectable melting of sea-ice and snow.

Large-scale analyses of observed precipitation indicate that precipitation has been increasing over land areas in the Northern Hemisphere mid latitudes as well as in the

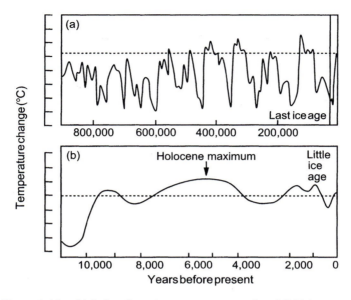

Figure 1.19 Global surface air temperature record and GCM simulations before and after the Mt. Pinatubo eruption (Hansen et al., 1993, and Hanen, personal communication).

Southern Hemisphere (Bradley et al., 1987; Diaz et al., 1989; and Vinnikov et al., 1990). However, it is even harder to discern the greenhouse effect from precipitation records than from temperature records, because precipitation is inherently more variable in both the temporal and spatial domains. Similarly, sea level has risen by 10–20 cm over the last century, but one cannot attribute its cause to the greenhouse effect unequivocally in view of the other interfering factors such as geological subsidence and the consequences of large-scale river basin management projects. An attempt to quantify the impacts of various human activities (including the damming of rivers and the utilization of their waters for irrigation) has recently been made by Gornitz et al. (1994). They showed that such activities may have tended to reduce the extent of sea-level rise due to the enhanced greenhouse effect.

What about climate variability? As of 1990, the IPCC reported no clear evidence that the climate had become more variable in recent decades on a global basis. Since that time, Karl et al. (1995) have shown that climate extremes in the United States have occurred consistently since 1976 and that climate trends accord with those anticipated from a greenhouse-enhanced atmosphere. But these trends are still not strong enough to prove a changing climate beyond doubt.

Water Vapor and Clouds: Positive or Negative Feedbacks

As atmospheric temperature rises in consequence of the increased absorption of thermal radiation, the warmed air can hold more water vapor. This creates a positive feedback, reinforcing the greenhouse effect. As simulated in global climate models, this feedback is responsible for raising the temperature for the projected doubled CO_2 by a factor of three, with the atmosphere becoming moister at all levels. Some scientists contend that this strong positive water–vapor feedback is exaggerated in GCMs, suggesting that tropical convection from a warmer earth surface will lift water vapor to higher atmospheric levels where the air is colder, thus able to hold less water vapor (Lindzen, 1990). While this may be true, it can be counteracted by increased upward moisture transport by the general circulation and by increased accumulation of water vapor and ice at cumulus cloud tops (Del Genio et al., 1991). Furthermore, satellite data suggest that the greenhouse effect is indeed reinforced by additional water vapor (Raval and Ramanathan, 1989; Rind et al., 1991).

When water vapor condenses to form clouds, dual effects come into play. Clouds have a cooling effect on the earth's climate because they reflect incoming solar radiation; at the same time, they have a warming effect because they absorb longwave radiation emitted from the earth's surface. Satellite studies have shown that, in the current climate, these two effects are roughly in balance, but with a slight net cooling effect (Ramanathan et al., 1989). Because atmospheric vapor content is predicted to increase with warmer temperatures and to result in denser and thicker cloud cover, it may exercise more of a negative feedback on greenhouse warming. However, the sign of the net feedback depends on how clouds change with global warming: if clouds at low altitudes become thicker and reflect more light, cooling will ensue. However, if more convective thunderstorms send water vapor higher into the atmosphere where absorptive ice crystals form, a positive feedback may prevail. The IPCC (1992) has concluded that the water–vapor feedback will be largely positive, augmenting the greenhouse warming beyond its expected levels based on trace gas emissions alone.

Will More CO_2 Enrichment Green the Earth?

Because carbon dioxide is a major component of photosynthesis, some scientists have hypothesized that higher atmospheric concentration of the gas will boost net carbon assimilation and storage in ecosystems. This promotion of plant growth may provide a natural self-correction to the potentially negative impacts of climate change, thus, in effect serving to "green the earth." Other processes associated with higher atmospheric CO_2 that may also be beneficial to vegetation include inhibition of transpiration and, hence, reduction of water stress during dry spells. While such positive responses to atmospheric CO_2 enrichment are well documented in enclosed greenhouses and in small-scale experimental settings in the field, they have yet to be demonstrated in long-term large-scale open-field studies for either natural or managed ecosystems.

Furthermore, the so-called fertilization effects of CO_2 enrichment are projected to occur simultaneously with global warming, bringing concomitant shifts in the geographical distribution of ecosystems with different levels and patterns of carbon assimilation and storage. Thus, the potential consequences of CO_2 rise and of climate change must be considered conjunctively in order to allow better predictions regarding the possible future patterns of natural vegetation and crop growth.

Studies performed with ecosystem models have examined the effects of equilibrium and of transient climate change scenarios, the subsequent redistribution of vegetation types, and the potential for CO_2 fertilization on carbon cycle dynamics. The results suggest that initially there may be a large net release of carbon (about 200 GtC) to the atmosphere, as current biomes respond to warming with die-back (Smith and Shugart, 1993; Dixon et al., 1994). Subsequently, ecosystems more suited to the changed climate will develop. Most studies predict an eventual enhancement of carbon uptake and storage of carbon (on the order of several hundred gigatons) due to the ecosystem response to both warmer temperatures and CO_2 fertilization (e.g., Alcamo et al., 1994).

Summary

Uncertainties about the exact magnitudes, rates, and regional patterns of climate change abound. The natural variability of climate makes it difficult to distinguish between real long-term trends and short-term fluctuations. The rate at which oceans will absorb more heat is unknown, and projections of sea-level changes are similarly uncertain.

Although the magnitude, timing, and regional distribution of the potential climate change are not yet known, preliminary studies suggest that much of the world is likely to be impacted significantly. Not only is the mean temperature likely to rise, but so is the incidence of extreme events such as heat spells, droughts, and floods. A warmer atmosphere will hold more water vapor, a positive feedback that may contribute to yet more warming and an intensification of the hydrological cycle. Overall, global precipitation is predicted to increase, but not uniformly. In some regions, higher temperatures and stronger evaporative demand may be associated with aridification.

To obtain more reliable projections of future climate, we need to improve our understanding of the greenhouse effect in relation to the complex processes of the climate system, especially those involving the roles of clouds and oceans. Uncertainties in the workings of the carbon cycle must also be resolved. We must continue and improve the monitoring of temperature, precipitation, sulfate aerosols, and other key climate-related variables throughout the world. Continuing studies of past changes in the earth's climate and atmospheric constituents may further illuminate the interactions among those variables. Since climate change is a global phenomenon, it must be considered in an international context, with free exchange of expertise and data across al' political and regional boundaries.

2

Agricultural Emissions
of Greenhouse Gases

The vital dependence of agriculture on the nature and stability of climate has been recognized since the sowing of crops and the domestication of livestock began some 10 millennia ago. It is only in recent decades, however, that agriculture's reverse role as a potential contributor to destabilizing climate has come to light. Clearing forests for fields and pastures, transforming virgin soil into cultivated land, flooding areas for rice and sugarcane production, burning crop residues, raising ruminant animals, and applying nitrogen fertilizers — all are implicated in the release of greenhouse gases to the atmosphere. The gases emitted are primarily carbon dioxide, methane, and nitrous oxide, three of the principal greenhouse gases. The agricultural sources are now estimated to constitute about 20% of total anthropogenic emissions of greenhouse gases. Land-use modification, often for agricultural development, accounts for an additional 14% (Figure 2.1). This chapter describes the processes by which agricultural activities contribute to the release of carbon dioxide, methane, and nitrous oxide and considers some options for reducing those agricultural emissions.

Carbon Dioxide

Carbon dioxide, the most prominent greenhouse gas, is released as a consequence of land-use change, biomass burning, and energy production. As areas originally covered with natural vegetation are transformed into cultivated fields, much of the vegetative biomass originally present is converted to carbon dioxide. The material aboveground is either burned or decomposes rapidly, while the organic matter in the soil does so more slowly, but over a period of time releases significant quantities of CO_2. Energy-intensive modern agriculture depends largely on the combustion of fossil fuels to provide the energy needed to drive farm machinery and to produce agricultural chemicals.

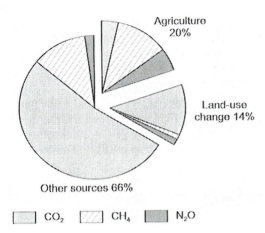

Agriculture
20%

Land-use
change 14%

Other sources 66%

CO₂   CH₄   N₂O

Figure 2.1 Contributions of agricultural activities and land-use change to global radiative forcing in the 1990s (IPCC, 1996).

Land-Use Change and Biomass Burning

Land-use change generally causes the release of CO_2 that had previously been sequestered in plant biomass and soil organic matter, both of which, in turn, resulted from cumulative prior photosynthesis. In photosynthesis, plants extract CO_2 from the atmosphere, later respiring part of it back and storing the remainder as fixed CO_2 in their tissues or as organic residues in the soil. On a global basis, soil is second only to the ocean in the amount of carbon it contains, holding 1.5 to 3 times as much carbon as living terrestrial vegetation. When land supporting a natural ecosystem is converted to agricultural use, a large percentage of the carbon present in the ecosystem is burned or rapidly decomposed, and the organic matter in the soil is gradually oxidized as the soil is cultivated and cropped (Table 2.1). Natural forests store 20 to 40 times more carbon in their biomass per unit area than do most crops (Houghton et al., 1983), and much of that storage is released when forests are cleared for cultivation.

The amount of carbon sequestered in crop stands may range from 1 to 5 t ha^{-1} (Houghton et al., 1983). If this range is multiplied by the 4.7×10^9 ha of global crop and pasture land, the resulting total carbon sequestration by crops ranges from 4.7 to 23.5 10^9 t C. This estimation is considerably higher than estimates of cropland carbon based on carbon content of total crop production, probably due to the inclusion of belowground biomass in the larger estimate. A recent global estimate for carbon in

Table 2.1 Average carbon loss from conversion to agricultural land (Schlesinger, 1986)

Ecosystem type	No. of studies	Mean loss (%)	Range (%)
Temperate forest	5	34.0	3.0 to 56.5
Temperate grassland	29	28.6	−2.5 to 47.5
Tropical forest	19	21.0	1.7 to 69.2
Tropical savanna	1	46.0	n.a.

harvested crops is on the order of 165–400 million (1.65–4.0×10^8) t C per year (Jackson, 1993). Estimates of total soil carbon (including litter and soil organic matter) are about 1.4–1.6×10^{12} t (Schlesinger, 1991; IPCC, 1995). Agricultural manipulation affects only a part of the large pool of carbon contained in the soils' organic matter fraction.[1]

When the original land cover is cleared, the organic matter contained in the plant material may be left on the surface, incorporated into the soil, burned in situ, or removed and burned elsewhere. The end product is CO_2, either from combustion or from decay through microbial processes. Biomass burning also results in emissions of methane, nitrous oxide, and carbon monoxide.

Over time, clearing land in order to facilitate food production has probably made the largest contribution to the greenhouse gas buildup of any agricultural activity. Since 1850, conversion to cropland has taken some 15% of the world's forested area (Figure 2.2) (Burke and Lashof, 1990). Approximately one-third of the earth's land surface (4.7×10^9 ha) is now devoted to agriculture. About 30% of agricultural land is cropland and about 70% is pasture (FAO, 1993). During the 18th and 19th centuries, the release of carbon from land conversion to agriculture in temperate regions is estimated to have averaged on the order of 0.5×10^9 t C released per year as CO_2 (Figure 2.2) (Houghton et al., 1983; IPCC, 1990). Agricultural expansion was therefore a major driver of the rise in atmospheric CO_2 concentration up until about 1960, after which accelerated fossil fuel burning came to dominate emissions (Wilson, 1978; Houghton et al., 1983).

Conversion of natural ecosystems into agricultural land continues apace today, with cropland and pasture still expanding globally (Figure 2.3) (FAO, 1993). Land clearing, being the second largest source of CO_2 emissions after fossil fuel combustion, currently accounts for approximately 10 to 30% of total net CO_2 emissions (IPCC, 1995). Much land clearing occurs in areas of tropical forests. Approximately 11 million hectares of tropical forest are now being converted each year, while only about 1 million hectares are reforested (Burke and Lashof, 1990). Conversion of trop-

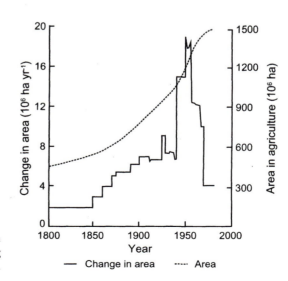

Figure 2.2 Worldwide area in cropland and rate of clearing (Houghton et al., 1983).

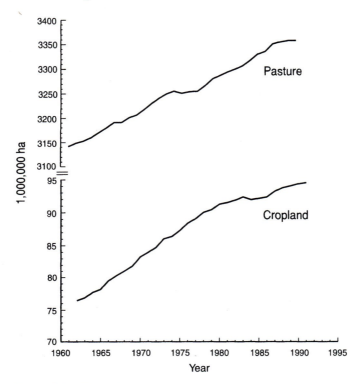

Figure 2.3 Global land area in crop production and pasture (FAO, 1993).

ical forests to agricultural uses currently releases an amount of carbon estimated to range between 0.6 and 2.6×10^9 t yr^{-1} (IPCC, 1995). Lack of precise knowledge regarding the rate of forest clearing and the amount of standing biomass removed account for the imprecision of these estimates. In 1980, half of the world's emissions from deforestation in the tropics were generated in five countries: Brazil, Colombia, Indonesia, the Ivory Coast, and Thailand (Houghton et al., 1987).

Land is also being retired from agriculture and returned to grassland and forest, or it is used for residential and industrial development. Such changes must also be taken into account in calculating the net effect of land-use change on carbon and nitrogen emissions. More accurate estimates of the gains and losses of carbon and nitrogen per hectare associated with changes in land use are needed to improve global calculation of CO_2 and N_2O fluxes.

As crop production continues over time, soil organic matter declines still further (albeit at a slowing rate), resulting in more CO_2 releases, until a steady state is reached or until the field is abandoned (Figure 2.4) (Houghton et al., 1983). However, the release of CO_2 from different agricultural practices varies. Curtailed tillage practices such as minimum or no-till, efficient crop rotation, strip cropping, and fallowing tend to reduce CO_2 fluxes.

Agriculture often involves biomass combustion. Common practices are the burn-

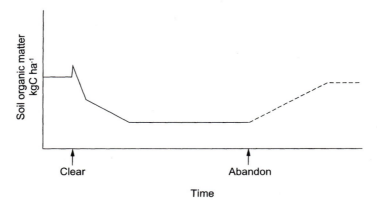

Figure 2.4   Change per unit land area in carbon content of soil following clearing of forest, cultivation, and abandonment (Houghton et al., 1983).

ing of crop residues to facilitate "clean" cultivation or to eradicate weeds and pests, and burning tropical savannas to promote grass regeneration, as well as clearing forests for crops or pasture. Andreae et al. (1988) have measured elevated concentrations of CO, CO_2, O_3, and NO in the biomass-burning plumes and haze layers over the central Amazon Basin. In addition to CO_2, biomass burning may release CH_4 and N_2O. The latter gas is produced because of the incomplete combustion and the nitrogen content of the plant material. Biomass burning of all types apparently accounts for about 5 to 15% of current global methane emissions and roughly the same percentage of global nitrous oxide emissions (IPCC, 1992). Some biomass burning occurs spontaneously. However, agricultural practices are estimated to be responsible for more than half of all biomass burning on an annual basis (Crutzen et al., 1979). There are great uncertainties in the estimates cited here, however, because data about the types and amounts of biomass that are burned are limited and because the emissions per unit of biomass burned are highly variable.

Energy

Energy is required for all agricultural operations, but the energy requirements of different agricultural systems vary. In traditional farming, animal labor, human labor, and biomass are the primary energy sources. Modern intensive agriculture requires much greater energy inputs than traditional farming (Table 2.2), as it relies on burning of fossil fuels for machinery and electricity. Agriculture's use of fossil fuels thus contributes to CO_2 emissions, but these emissions are hard to quantify. One estimate is that fossil fuel usage, primarily of liquid fuels and electricity, by modern agriculture constitute only 3 to 4% of total consumption in the developed countries (CAST, 1992).

Opportunities for reduction of energy use by agriculture and concomitant declines in greenhouse gas emissions include the use of no-till or reduced tillage and improved insulation and ventilation of farm buildings. Beyond primary farm operations, there is also potential for increased energy use efficiency in the food sector as

Table 2.2 Energy use in crop and animal production (Jackson, 1993)

Agricultural system	Energy demands (GJ ha^{-1})
Traditional	0–5
Transitional	6–14
Intensive	15–40

a whole (processing, packaging, refrigeration and other modes of preservation, and distribution).

A great deal of energy is required to manufacture agricultural chemicals. Fertilizer production is especially energy intensive: nitrogen, phosphorous, and potassium fertilizers require from 10 to 72 MJ of energy per ton (Table 2.3), primarily supplied by fossil fuels. Thus, fertilizer production, which has increased steadily since World War II (Figure 2.5) (FAO, 1993), releases CO_2 to the atmosphere. Of the three main types of fertilizers produced, nitrogen fertilizers are applied the most and are the greatest consumers of energy; hence, they are responsible for by far the greatest amount of CO_2 released (Table 2.4). Optimizing efficiency of fertilizer use, particularly of nitrogen fertilizers, is an indirect measure that will serve to mitigate CO_2 emissions. Incidently, it will also help reduce environmental pollution resulting from the residues of excessive or injudicious fertilizer application.

Alternatives to the reliance on fossil fuels in agriculture need to be examined. Crop plants and plant wastes are renewable sources of energy, and their use can replace that of nonrenewable fossil fuels. Use of biomass energy involves recycling of carbon and nitrogen as they are first fixed in crops and then released during energy utilization. Ideally, this decreases CO_2 emissions from fossil fuels, and creates neither a net gain nor a net loss of carbon. However, the exact extent of carbon recycling in biomass fuel production and use has not yet been quantified. Solar, wind, and nuclear energy devices do not create greenhouse gas emissions during use, but do require some fossil fuel energy for their infrastructure and servicing. Hydroelectric power similarly does not emit greenhouse gases, but dams may affect the carbon and nitrogen fluxes of the land affected by flooding on the one hand and by reduced water flow on the other (Jackson, 1993).

Table 2.3 Energy requirements for fertilizer production (FAO, 1993; Jackson, 1993)

Fertilizer type	MJ t^{-1}
Nitrogen (N)	72
Phosphorus (P$_2$O$_5$)	13
Potash (K$_2$O)	10

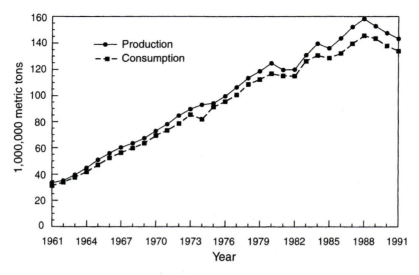

Figure 2.5   Global fertilizer production and consumption, including N, P_2O_5, and K_2O (FAO, 1993).

Slowing Deforestation

Forested lands, especially in tropical areas, are not very fertile and tend to degrade rapidly after removal of their natural vegetation. More often than not, they are utilized inappropriately and unsustainably. Pressure for expansion of agriculture at the expense of forested lands is prevalent today around the world and is likely to continue and may even intensify due to population growth and economic development. Activities such as large-scale conversion of forest to pasture, or to shifting cultivation without allowing sufficient time for forest regeneration, contribute to deforestation. This results in rapid leaching of nutrients, decomposition of unreplaced organic matter, and accelerated soil erosion. Such lands are eventually abandoned with consequent increasing pressure on the remaining exploitable lands. Formulating effective policies to protect the world's remaining forests is a difficult task since deforestation is driven by a complex set of economic and social causes that vary from region and region.

Table 2.4   Maximum potential CO_2 emissions (expressed as tons of C) due to fertilizer production (FAO, 1993; Jackson, 1993)

	Global fertilizer consumption (10^9 ton)			Carbon emissions[1] (10^{12} ton)		
Year	N	P_2O_5	K_2O	N	P_2O_5	K_2O
1985	70	33	26	120	10	6
1986	71	35	26	123	11	6
1987	76	37	27	130	11	6
1988	80	38	28	137	12	7

1. Calculated using CO_2 emission factor for coal (23.8 g C MJ^{-1}).

In principle, stopping large-scale deforestation requires reducing the consumption of wood for fuel, in construction, and as raw materials for various products (e.g., paper). Action is simultaneously needed to eliminate economic or political incentives that encourage clearing, especially for agriculture and urbanization. Where shifting cultivation does not allow forest regeneration, measures are needed to relieve the population pressure. This depends on alternative employment opportunities or the availability of land that can be farmed perennially and sustainably. Agroforestry, a sustainable system in which trees are grown in intensively managed combinations with food or feed crops, can provide food products, including fruit and nuts, animal fodder, fuelwood, resins, and other products such as medicinals and spices. Other activities worthy of being encouraged are improving the efficiency of cookstoves to reduce the fuelwood required and recycling wood and paper products.

Carbon Sequestration and Conservation

Both forests and agricultural soils can be managed to sequester carbon fixed by photosynthesis and thereby help to offset carbon emissions. Many plans have been proposed for reforestation aimed at balancing CO_2 emissions, since rapidly growing trees absorb and store considerable amounts of CO_2 and can be used to produce biofuels to substitute for fossil fuel combustion (Sedjo, 1989; Dixon et al., 1993).

Large areas of newly forested land would be needed to extract significant amounts of CO_2 from the atmosphere. Sedjo (1989) estimates that 465 million ha (an area greater than half the contiguous United States) of new fast-growing forest would be required to offset the annual increment of 2.9×10^9 t of carbon now emitted (Figure 2.6). According to Sedjo, if all carbon emissions were to be sequestered, the global costs of such a control strategy would be on the order of 5 to 10% of the U.S. annual gross national product (GNP), depending on whether the tree planting is done in tropical or temperate regions. While this seems costly, other mitigation strategies—such

Figure 2.6   Area required (465×10^6 ha) to sequester all annual anthropogenic emissions of carbon (2.9×10^9 tons) (Sedjo, 1989).

as installations to absorb CO_2 from smokestacks and exhausts or conversion to nuclear power—might be even more costly.

Another strategy for promoting the sequestration of carbon in forests is to increase the amount of biomass in existing forest areas, either through increasing growth (e.g., through greater use of fertilizers) or to reduce the harvesting of forest products. Designation of areas as forest reserves aids in the latter goal. While some forests may naturally expand with climate warming and may store more carbon as atmospheric CO_2 rises, this may do little to mitigate the greenhouse effect, as other forests may simultaneously retreat owing to aridification.

A promising strategy among others for controlling CO_2 emissions is the growing of short-rotation, high-intensity wood energy crops to provide renewable fuel (Table 2.5) (Sampson and Hamilton, 1992). The feasibility of this practice depends on the availability of suitable land, proper selection of the trees to be grown, and appropriate management in each location. Nations with large areas of land suitable for establishment of forest plantations include Brazil, China, India, Mexico, Russia, the United States, and Zaire (Dixon and Andrasko, 1992), so, globally speaking, there would appear to be extensive possibilities for carbon sequestration by means of afforestation. However, the processing and utilization of wood products from such efforts are also important considerations. While wood is a primary energy source in many developing countries, it is not now competitive with fossil fuels in industrialized countries.

Electric utilities in both the United States and Europe have begun to sponsor forest-sector carbon-offset projects around the world to mitigate greenhouse gas emissions (Dixon et al., 1993). Such projects are known as "joint implementation" activities because they team developed and developing countries in CO_2 reduction efforts. The first forestry project specifically intended to offset the CO_2 emissions of a particular industrial source began in 1988 (Figure 2.7) (Trexler et al., 1992). An independent U.S. power producer, Applied Energy Services, began construction of a new coal-fired

Table 2.5 Opportunities to mitigate U.S. carbon emissions through land conversion and fossil energy conservation using trees and forests (Sampson and Hamilton, 1992)

Mitigation activity	Carbon storage[1] (million t C yr^{-1}) (low/high estimate)	Fossil energy conservation[2] (million t C yr^{-1}) (low/high estimate)	Total impact (million t C yr^{-1}) (low/high estimate)
Converting marginal crop and pasture land to forests	33/143	11/37	44/180
Growing short-rotation wood energy crops	29/58	71/141	100/199
Establishing windbreaks and shelterbelts	2/3	1/4	3/7
Urban and community tree planting	2/6	8/32	10/38
Total opportunity	66/210	91/214	157/424

1. Includes storage in forests (vegetation and soils) and in products.

2. The carbon conserved by using woody crops for the production of electricity with current technology.

CO$_2$ emissions

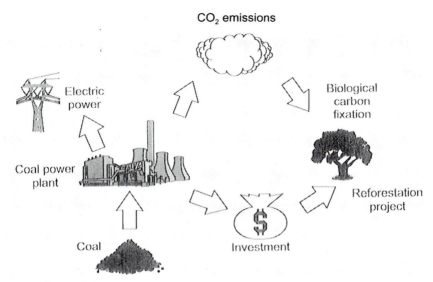

Figure 2.7 A carbon offset project (World Resources Institute, 1990).

power plant in Connecticut, while simultaneously funding a reforestation project in Guatemala through the international relief organization, CARE. The power plant is expected to emit about 17 million tons of carbon as CO$_2$ over its 40-year lifetime, while the total long-term carbon sequestration is projected to be about 20 million tons C, thus fully offsetting the CO$_2$ emissions from the power plant and providing a net benefit of about 3 million tons of carbon extraction from the atmosphere (Table 2.6) (Trexler et al., 1992).

The efficacy of joint-implementation forestry projects, such as the Guatemala CARE project, to offset greenhouse gas emissions should be examined in light of their reliability and cost-effectiveness (Dixon et al., 1993). The projects' success depends, among many factors, on whether existing trees are protected or new trees are planted, how the wood from the trees is ultimately used, and whether the utilization of the trees affects fossil fuel use positively or negatively (Trexler et al., 1992). The location, species, maintenance, and management of the trees also affect the amount of carbon sequestered. Integrated development projects may combine land in forest reserves, fuel wood plantations, agroforestry, and intensive agriculture, with the goals of sustained delivery of food, fuel, and fiber, as well as economic returns. Successful programs need to take into consideration local needs for goods and services, as well as national and international goals of greenhouse gas reduction.

In the agricultural sector, crops can also be grown as renewable fuel sources, and improved management can increase soil carbon. Crops available for production of fuel include grain for fuel alcohol and oilseed for ethyl or methyl esters (known as "biodiesel"). Among the options for boosting carbon in currently cropped soils and slowing its decline on newly converted land are conservation tillage practices, establishing shelterbelts and windbreaks, intercropping (the growing of a mixture of tree, agronomic, and horticultural crops in the same field), and retiring of erodible lands

Table 2.6 Carbon sequestration and emissions offset of the Guatemala project (from Trexler et al., 1992).

Project element	Carbon (million tons)
Carbon Sequestration	
Net addition to standing inventory of biomass carbon	2.9
Standing forest carbon retained as a result of demand displacement	15.8
Carbon added to project soils	0.4
Standing forest carbon protected through fire brigades	0.8
	19.9
Total Long-Term Carbon Sequestration	
40-year emissions of powerplant	17.1
Projected net benefit	2.8

Note: These values are based on highly simplified assumptions of tree planting densities, growth rates, harvest percentages, carbon flows, and project lifetime.

from cultivation. Controlling soil erosion also helps conserve carbon content. Conservation tillage affects soil aeration, heating, and drying, often retarding bacterial decomposition of soil organic matter. In some cases, organic carbon can be augmented up to 50% through no-till or reduced tillage practices (IPCC, 1996).

Enhancing soil fertility and maintaining neutral pH encourage the restoration and growth of carbon pools in agricultural soils, as do high residue-production rates and use of perennial forage crops. Sustainable management of grasslands (i.e., light rather than heavy grazing) is also to be encouraged, because grasslands can accumulate a large pool of soil carbon, which may remain stable for hundreds of years. Restoration of agricultural lands degraded from overuse and salinization can help to achieve carbon storage. In the United States, the Conservation Reserve Program encourages farmers to stop cultivating highly erodible land and to establish herbaceous cover crops. Protection of wetlands from conversion to cropland also helps achieve these goals.

Methane

Methane is the second most important greenhouse gas after carbon dioxide. Although it is very much less prevalent than CO_2, methane is about 20 times more powerful per molecule in terms of radiative forcing (IPCC, 1990). Hence, its rising concentration is a matter of concern. Agriculture is the largest anthropogenic source of methane, being responsible for about 40% of emissions (Figure 2.8). The major agricultural sources of methane are fields of wetland rice known as rice paddies, herds of ruminant animals, and biomass burning. The conversion of natural wetlands to shallow impoundments for irrigation water management tends to increase methane emissions as well (Harriss et al., 1988). In both rice paddies and ruminant animals, methane is produced in the process of carbohydrate (including cellulose) decomposition carried out by anaerobic bacteria (Cicerone and Oremland, 1988; Bouwman, 1990). Methane releases from combustion are associated with tropical forest clearing for agriculture and with burning of crop residues. Methane releases from rice paddies, enteric fermentation, and biomass burning are poorly quantified, but are estimated to total 20–150, 65–100, and 20–80 million tons per year, respectively (IPCC, 1992).

An interesting recent discovery is that some soils can serve as a sink for methane, via methanotrophs—bacteria that are able to grow using methane. Since methanotrophs require oxygen, methane uptake tends to occur in moist rather than saturated or flooded soils, or at the boundary between aerobic and anaerobic conditions (Knowles, 1993). Methane uptake by soils has been found in savanna and forest ecosystems (Schutz and Seiler, 1990). Globally, soils are estimated to remove from 15 to 45 million tons of atmospheric CH_4 per year (IPCC, 1992). However, agricultural development and nitrogen fertilization may diminish the sink capabilities of grassland soils and thereby lead to higher atmospheric concentrations of both methane and nitrous oxide (Mosier et al., 1991).

Paddy Rice Cultivation

Anaerobic decomposition of biomass in flooded rice paddies releases considerable quantities of CH_4. The annual global flux of methane from rice is estimated to be between 20 and 150 million tons CH_4 per year, with a "best" estimate of 60 million tons (IPCC, 1992). The wide range in the estimates is caused by large uncertainties in emissions among different regions, soil types, rice varieties, management practices, and seasons.

World rice cultivation in 1992 covered an area about 147 million ha and produced

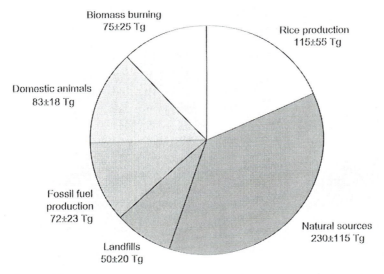

Figure 2.8 Global annual emissions of CH_4 by source estimated for the 1980s (Burke and Lashof, 1990).

some 525 million tons of grain (FAO, 1993). About 90% of that production takes place in Asia; China and India are the two largest producers, accounting for over 50% of the world rice area and production (FAO, 1993). Small amounts of rice are grown in the other continents. Methane emissions from rice fields in China are found to be considerably higher than emission rates from rice fields in the United States and Europe (Khalil et al., 1991). Total rice production nearly tripled in the period between 1950 and 1984, while the area of cultivation expanded by 40%. Over the same period, average yields doubled from 1.6 to 3.2 t ha^{-1} (Burke and Lashof, 1990). Current average yield is 3.6 t ha^{-1} according to FAO statistics (FAO, 1993). Both the area of rice cultivation and the productivity per unit area are still growing.

Rice is grown under many different management systems: irrigated wetlands (~50% of the total harvested area), rainfed wetlands (~27%), deep flooding (~8%), and upland areas that are not flooded (~15%) (Table 2.7) (Neue et al., 1990). Wetland rice production is more prevalent because it ensures an abundant water supply (rice is very susceptible to drought), relatively easy land preparation in saturated soft soils, effective weed control, and increased availability of nutrients (Bouwman, 1991). Nitrogen is provided by aquatic nitrogen fixers (algae). Rice production permits the

Table 2.7   Distribution of rice management systems (Neue et al., 1990; Bouwman, 1991)

Location	Total harvested rice area (1000 ha)	Type (%)			Yield (t ha^{-1})			1990 Production (1000 t)
		UP	IR	RF	UP	IR	RF	
South and Central Asia	89,380	13	33	54	1.0	3.4	1.5	185,464
China	32,400	2	93	6	1.0	5.9	1.5	179,496
Rest of centrally planned Asia[1]	8,550	13	38	50	1.0	6.1	1.4	26,613
South America	6,857	15	50	35	1.0	3.7	1.5	17,211
Japan	2,097	0	100	0	1.0	6.2	—	12,938
Africa	5,532	50	15	35	1.0	5.8	1.5	10,511
North America	1,087	0	100	0	1.0	6.5	—	7,011
Former Soviet Union	654	0	100	0	1.0	3.9	—	2,524
Europe[2]	482	0	100	0	1.0	4.5	—	2,475
Central America	728	15	50	35	1.0	4.8	1.5	2,250
Middle East[3]	669	0	38	62	1.0	4.4	1.5	1,746
Oceania	114	0	100	0	1.0	7.7	—	876
Total	148,550							468,575

UP = Upland rice; IR = Irrigated wetland rice; RF = rainfed wetland rice

1. Including Turkey.

2. Afghanistan, Iran, Iraq, Israel, Jordan, Kuwait, Lebanon, Oman, Saudi Arabia, Syria, United Arab Emirates, North Yemen, South Yemen.

3. North Korea, Vietnam, Cambodia, Mongolia.

utilization of naturally flooded lowlands (e.g., riverine floodplains), and it aids in soil conservation where it is produced on hillside terraces.

At the two extreme types of rice-growing conditions, very little methane is emitted from unflooded upland rice fields or from deep-water rice fields (in which the water depth exceeds 60 cm where so-called floating rice is grown). In the latter, the rise of methane to the surface is inhibited by deep ponding. In contrast, most methane is emitted from rice paddies flooded to a moderate depth of, say, 10 to 50 cm. Management systems encompass one to three rice crops per year, depending primarily on climate. Triple rice crops account for 60% of the harvested area, double for 15%, and single for 25% (Matthews et al., 1991). Methane emissions vary with the total duration of rice-growing periods per year, as well as with many other factors including temperature.

Methane Production in Rice Paddies

Methane production depends on the ratio of oxidizing to reducing capacity of the flooded soil (Watanabe, 1984). The oxidizing capacity depends on the presence of O_2, NO_3^-, Mn^{4+}, and Fe^{3+}; the reducing capacity depends on the presence of ammonium. Methanogenesis requires reduction-oxidation (redox) potentials of -150 to -190 mV (corrected to a pH of 7) (Neue et al., 1990).

After a soil is flooded and oxygen supply is cut off, bacteria decompose organic matter anaerobically in a series of steps. The processes follow a thermodynamic sequence in which the soil's redox potential falls as the oxidized compounds are chemically reduced (Table 2.8; Figure 2.9).

The sequence of processes is:

- aerobic respiration
- nitrate reduction
- general fermentations
- sulfate reduction
- methane fermentation

Table 2.8 Sequential reduction of oxidized soil compounds in a soil after inundation (Bouwman, 1991)

Redox reaction		pE_0^{7} [1] $(V)^2$
$O_2 + 4H^+ + 4e$	$\Rightarrow 2H_2O$	13.8
$NO_3^- + H_2O + 2e$	$\Rightarrow NO_2^- + 2OH^-$	12.66
$MnO_2 + 4H^+ + 2e$	$\Rightarrow Mn^{2+} + 2H_2O$	6.8
$Fe(OH)_3 + 3H^+ + e$	$\Rightarrow Fe^{2+} + 3H_2O$	-3.13
$SO_4^{2-} + 10H^+ 8e$	$\Rightarrow H_2S + 4H_2O$	-3.63
$CO_2 + 8H^+ + 8e$	$\Rightarrow CH_4 + 2H_2O$	-4.14

1. $pE_0^7 = -\log(E_0)$.
2. Corrected for pH = 7.

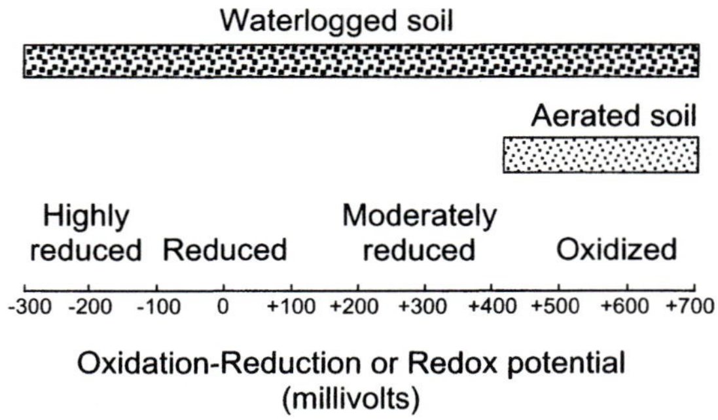

Figure 2.9　Redox
potential range of soils
(Patrick and DeLaune,
1977).

After the inorganic hydrogen acceptors have been exhausted, bacteria continue to reduce the organic matter to form a mixture of CO_2 and CH_4. The relative amounts of the two gases produced depend on the degree of oxidation of the original organic material — that is, on the ratio of oxidizing capacity to reducing capacity.

These processes may take place simultaneously in separate layers in paddy rice soil (Bouwman, 1991). The layers follow the values of redox potentials given in Table 2.8 and consist of:

1. A thin aerobic surface layer that is 0.5 to 10 mm thick. In this layer, consumption of methane produced in the reduced zone may occur.
2. The second layer is reduced with Fe^{3+}, Mn^{4+}, and NO_3^- still present.
3. The third layer is where sulfate reduction occurs.
4. The fourth layer is the main zone of methane generation.

Most methane is brought to the surface of rice paddies through diffusive transport in the rice plants themselves, rather than via rising bubbles and diffusion in the water (Figure 2.10) (Cicerone and Shetter, 1981). Rice plants have tissues with elongated cells and large intercellular spaces, known as aerenchyma, that facilitate gaseous exchange and maintain tissue buoyancy. The root system of the rice plant is critical in methane processes: roots apparently can absorb methane independently of water uptake (Bouwman, 1991).

Factors Affecting Rice Paddy Methane Emissions

Methane emissions from rice paddies vary with season, soil type, nutrients, redox potential, pH, and temperature. Further, they are affected by agronomic practices such as irrigation and drainage, depth of water layer, cropping system, fertilization, and additions of manure or rice straw.

Seasonality.　Methane emissions vary with the developmental stage of the rice plants, and thus emissions from rice paddies show strong seasonality (Figure 2.11) (Cicerone et al., 1983; Khalil et al., 1991; Yagi and Minami, 1991). High rates of methane emissions have been observed when rice plants are actively growing, with maximum emission rates between tillering and flowering (Holzapfel-Pschorn and Seiler, 1986). When rice plants are growing rapidly, root exudates provide a supply of easily decomposable organic substrates for methane-producing bacteria and there is efficient gas

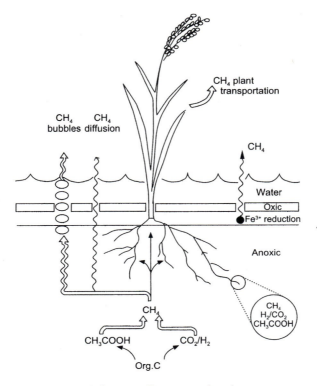

Figure 2.10 Schematic illustration of methane production and its release (Takai and Wada, 1990).

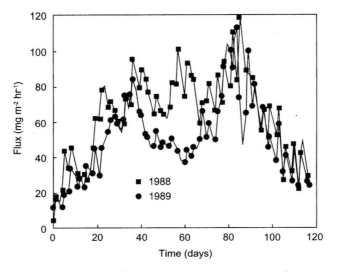

Figure 2.11 Seasonal variation of methane emissions from rice fields from planting to harvest (4/28/1988 to 8/23/1988 and 5/01/1989 to 8/25/1989) (Khalil et al., 1991).

transport through the root system. Other emission peaks have been observed at the end of the rice-growing season and are related to root decay. As paddy fields are allowed to dry out, the soil may release some additional methane that had been trapped in bubbles while the soil was saturated (Bouwman, 1991).

Soil Factors.   As a soil is flooded, changes in water content and aeration, availability of organic matter, prevalence or paucity of electron acceptors (nitrate, iron, manganese, sulfate) and of nutrients (N, P, S, K, and trace elements) all influence the generation of methane in the soil and its consequent release to the atmosphere. Over the long term, the migration of iron and manganese, along with the change in the soil's ion exchange complex and even in mineralogy, may influence methane reactions. Hydromorphic soils, in which restricted aeration causes chemical reduction to take place naturally, have the greatest potential for methane production. Entisols, Inceptisols, Alfisols, Vertisols, and Mollisols with a significant content of organic matter are especially likely to generate methane production under flooded conditions.

Methane production tends to be lower in soils that have (a) acidic pH; (b) ferritic, gibbsitic, ferruginous, or oxidic mineralogy, with high proportion of kaolinitic or halloysitic clays; and (c) under 18% clay if the water regime is epiaquic[2] (Neue et al. 1990). Oxisols, Ultisols, and some Aridisols, Entisols, and Inceptisols display these characteristics. Methane generation is also inhibited in soils with electrical conductivity (EC) exceeding 4 mS cm^{-1} when flooded.

Though soil redox potential is apparently the more critical factor, soil pH also affects methanogenesis. The range of pH at which methanogenesis occurs is between 6 and 8, with an optimum near neutrality at about pH 7 (Lindau et al., 1993). Methane production also tends to be positively correlated with soil temperature (Figure 2.12) (Watanabe, 1984; Holzapfel-Pschorn and Seiler, 1986; Khalil et al., 1991).

Agronomic Practices.   Farmers manipulate the soil–water system in rice paddies in several ways. Among these are puddling and compaction to minimize water percola-

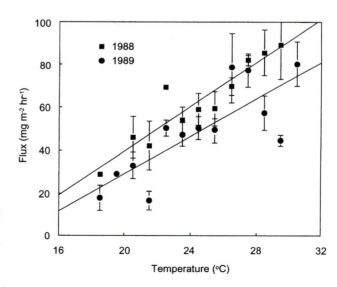

Figure 2.12
Relationship between methane flux and soil temperature (Khalil et al., 1991).

tion losses, temporary drainage during the rice-growing season, and reduction of the flooding period (Bouwman, 1991). Repeated tillage and trampling by humans and animals form a slowly permeable (but not totally impervious) subsurface layer, with percolation rates diminishing over time. Maintaining a minimal rate of percolation and leaching is necessary in rice paddies to prevent an accumulation of toxic substances and to restore nutrient imbalances. On the other hand, excessively high percolation rates tend to inhibit methane generation by bringing fresh supplies of oxygenated water and thus favoring chemical oxidation over reduction.

Emission rates are related to the depth of ponding in the rice paddies, increasing linearly up to depths of about 10 cm (Bouwman, 1991). Temporary drainage of rice fields and reduction of water levels during the growing season, as is done in several rice management systems, tends to restrict production and subsequent emissions of methane. However, the paddies cannot be drained out completely during the latter part of the vegetative growing period or during flowering, when rice plants are very susceptible to drought, without risking a decrease of yield (Doorenbos and Kassam, 1979). Rice yields also decline when the soil is maintained for too long in an excessively saturated condition. While temporary drainage may reduce methane emissions, however, it may well enhance nitrification-denitrification processes and thus boost emissions of nitrous oxide, another greenhouse gas. All this calls for careful management if the overall emissions of greenhouse gases from rice paddies is to be reduced significantly.

If rice plants are seeded directly in the field (rather than pregerminated in nursery plots and then transplanted), the period of flooding and the subsequent methane emissions are reduced (Bouwman, 1991). However, transplanting, while labor intensive, is advantageous to the extent that it produces greater yield, saves seed and water, controls weeds more effectively, and facilitates multiple cropping per year.

Nitrogen fertilizers are widely applied to rice paddies because nitrogen availability is often limited in flooded soils. The effect of nitrogen fertilizer application on methane production, however, is not well understood. Ammonium sulfate $[(NH_4)_2SO_4]$ and urea $[(NH_2)_2CO]$ are the most commonly used nitrogenous fertilizers in South and Southeast Asia (Bouwman, 1991). The application of ammonium sulfate to rice paddies has been observed to either boost or reduce emissions, depending on method of application (e.g., whether spread over the surface or incorporated into the soil; whether with or without rice straw). Urea applications have similarly produced variable results: in some cases enhancing and in other cases inhibiting the generation of CH_4, depending on mode and depth of application (Lindau et al., 1993).

Organic matter in rice paddies consists of rice straw, root biomass, root exudates, aquatic biomass (algae), and weeds. Addition of organic matter, such as rice straw, tends to increase methane emissions (Lindau et al., 1993). This occurs both with and without an accompanying application of mineral fertilizer. Because high-yielding varieties of rice have a higher grain-to-straw ratio and a shorter growth cycle (i.e., earlier maturation), they tend to produce less methane per crop (Burke and Lashof, 1990). Composted material with low C:N ratios appears to generate less CH_4 than uncomposted rice straw with high C:N ratios (Bouwman, 1991). However, the incorporation of organic residues into the soil provides an important and inexpensive source of nutrients for many farmers and helps restore soil structure after drying. Burning of rice straw, an alternative practice, also causes release of methane, but at substantially lower rates than field incorporation. Burning also contributes carbon monoxide (as well as carbon dioxide) and other pollutants to the atmosphere.

Limiting Methane Emissions from Rice Paddies

Limiting CH_4 emissions from rice paddies involves adjustment of cultural practices, including crop, water, and nutrient management. Agronomists are studying the effects of different rice varieties, tillage techniques, planting dates, fertilization, and modes of irrigation with the goal of reducing methane emissions while maintaining or increasing yields. The entire rice-growing system rather than any particular practice must be taken into account. For example, the use of concentrated fertilizers rather than cellulose-rich residues may lower emissions, but may incidentally increase emissions of nitrous oxide as well as exacerbate groundwater pollution. More efficient fertilization practices, such as precise placement during transplanting of the rice, may avoid such consequences. Where the incorporation of crop residues induces emissions, it may be advisable to find alternative uses for such residues (as animal fodder, building material, or fuel in biogasification systems).

If rice production is intensified by growing several crops per year, the longer periods of standing water may elevate emissions. The use of rice varieties with greater grain-to-straw ratio, and hence, lower residues, can lower emissions while increasing productivity. Greater production of upland rice rather than paddy rice would also reduce emissions. Temporary drainage and other water management techniques may likewise reduce emissions, but may also pose the risk of diminishing yields. Such measures may only be effective in regions with flat topography, high water availability, and control over water supplies. Specific technologies for reducing emissions (e.g., wind-driven aeration of paddies and chemical inhibitors incorporated into fertilizer) are now being developed.

Research needs include an improved understanding of how such factors as rice cultivars, cultivation practices, and management regimes contribute to methane emissions. Isotope studies of the organic matter in wetland rice soils are useful for measuring methane-producing capacity because anaerobic bacteria produce methane with a very low δ^{13} value, causing in turn ^{13}C-enriched residues (Neue et al., 1990). Such knowledge may help narrow the uncertainties in emissions estimates and develop effective approaches for their reduction.

Future Emissions of Methane from Rice Paddies

Methane emissions are projected to rise as rice cultivation expands in the future. Rice production is expected to grow from about 500 to 760 million t yr^{-1} by 2020, and this may result in methane emissions of 130 million tons per year, compared to 90 million tons estimated for emissions from wetland rice fields in the late 1980s (Lashof and Tirpak, 1990). Another estimate is that methane emissions from rice paddies will rise to as much as 145 million t yr^{-1} by 2025 (Anastasi et al., 1992). While most of the added production will come from increasing productivity per unit of land, there are still some 18 million ha of tidal wetlands and other lands available for expansion of rice cultivation in South and Southeast Asia (Bouwman, 1991).

Methane Emissions from Livestock

Ruminant animals having compound stomachs (cattle, sheep, goats, camels, and buffaloes) are primarily responsible for methane production via the process of enteric fermentation in the breakdown of cellulose (Crutzen et al., 1986). This process generates methane, which is released into the air. Emissions of methane from domestic livestock and wild animals are estimated to be between 65 to 100 million t yr^{-1} (about 15% of total methane emissions) (Crutzen et al., 1986; Lerner et al., 1988; IPCC, 1992). Improving the accuracy of these estimates requires quantifying emission rates per animal as affected by feeding habits and metabolism, as well as better data on animal populations around the world.

Microbial decomposition of manure in anaerobic conditions also produces methane. Estimates of global methane emissions from animal wastes range from 0.8 to 15 million tons CH_4 per year (Jackson, 1993). Manure collected in lagoons, dry pits, or feedlots releases both carbon dioxide and methane. A small amount of the latter is captured for use as fuel.

Livestock production in various parts of the world differ in type of animals, composition of feed, and primary end products, all of which affect methane emissions. Most of the available data regarding methane have been collected on commercial livestock herds in developed countries, where animals are intensively managed for either meat or dairy production. Under these conditions, animals are fed primarily grain, as opposed to forage, resulting in higher yields of meat or dairy product per animal. These systems result in higher methane emissions per head, though less methane per unit of product (Burke and Lashof, 1990).

In developing countries, ruminant animals are often part of mixed crop and livestock systems, as they provide fertilizer, fuel, wool, leather, and power in addition to meat and dairy products. The diet of livestock in such systems is primarily forage-based, and the animals tend to be smaller. Because this diet is richer in cellulose, methane emission rates appear to be higher per unit of feed consumed in this type of system.

As stated, ruminants are the main animal producers of CH_4 (figure 2.13). (Nonruminants such as horses, pigs, and humans emit only small amounts of

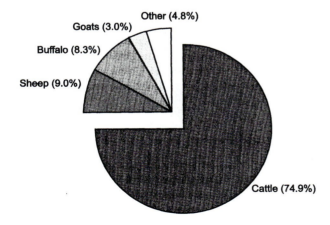

Figure 2.13   Annual global emissions of CH_4 from domestic animals estimated for the 1980s (Burke and Laschof, 1990).

Table 2.9 Methane production rates and livestock populations in 1983 (from Crutzen et al., 1986)

Population	Methane production rate (kg animal^{-1} yr^{-1})	Population ($\times 10^6$)	Total methane production (Tg yr^{-1})
Cattle			
developing countries	35	652.8	22.8
developed countries[1]	55	572.6	31.5
Sheep			
developing countries[2]	5	737.6	3.7
developed countries	8	399.7	
Buffalo	50	124.1	6.2
Goats	5	476.1	2.4
Camels	58	17.0	1.0
Horses	18	64.2	1.2
Pigs			
developing countries	1.0	444.8	0.4
developed countries	1.5	328.8	0.5
Mules and asses	10	53.9	0.5
Humans	0.05	4669.7	0.3
		Total	73.7

1 Tg = 10^{12}g = 1 million metric tons.

1. Includes Brazil and Argentina.

2. Includes Australia.

methane.) Of the total emissions from enteric fermentation, beef cattle account for approximately 57%, dairy cows 19%, and sheep 10% (Crutzen et al., 1986; Lerner et al., 1988). Methane emission rates and population data for methane producers are shown in Table 2.9. Dairy and beef cattle are most important, because of both their high emission rates per animal (ranging between 35 and 55 kg CH_4 per animal per year) and their high populations worldwide (over 1.25×10^9 in the late 1980s).

According to FAO (1993) developed countries had 390 million head of cattle and 512 million sheep in 1992 (FAO, 1993). In the developing countries, however, animal populations are much higher: some 895 million cattle and 600 million sheep in 1988. Livestock in Brazil, China, India, the former Soviet Union, and the United States are apparently responsible for about half the global methane emissions by domestic animals (Lerner et al., 1988).

Factors Affecting Methane Emissions from Livestock

Methane production varies with feed quality and amount, animal age, weight, genetic traits, activity, and enteric ecology. When animals convert plant carbon to methane rather than to animal tissue or dairy product, a loss of efficiency ensues. Energy loss in the form of released CH_4 varies from 3 to 8% of gross energy intake for cattle, depending mainly on the quality and quantity of feed (Burke and Lashof, 1990).

Overall, emissions tend to be higher for heavier animals, greater amounts of feed, and diets high in cellulose forage (Burke and Lashof, 1990). Working (draft) animals tend to emit more CH_4 because they generally require more feed to produce energy for work, and some of the additional feed is converted to methane. Higher quality feed contains a higher proportion of protein and starch relative to cellulose; thus, grain-fed cattle tend to emit less methane per unit of feed consumed, but more methane overall, than range-fed cattle whose diet contains more cellulose. Therefore, increasing the efficiency of feed conversion should both increase livestock productivity and decrease CH_4 emissions.

Methane emissions also vary directly with consumption level (i.e., the more feed consumed per animal the higher the emissions). At low consumption levels, a high-cellulose forage diet actually produces less methane than does a high-grain diet. On the other hand, at high consumption levels, a high-grain diet produces less methane than a high-cellulose diet (Blaxter and Clapperton, 1965). While this implies that the highest emissions occur with high-cellulose diets at high rates of consumption, it also shows that both subsistence and intensive livestock production are relatively efficient in terms of feed conversion.

Limiting Methane Emissions from Livestock

Relatively little research has been conducted on how to limit methane emissions from livestock. For intensively managed dairy cattle, genetic improvement, bovine growth hormone, and improved feed formulations are projected to reduce methane emissions on the order of 25% below present levels (CAST, 1992). Improved management could especially help to reduce emissions from the high ruminant populations in developing countries, but achieving such reductions may be difficult given the role of many of the animals in local subsistence farming. In India, for example (the country with the world's largest population of cattle, numbering about 200 million bovines), cattle are used primarily for traction, fertilizer, and fuel, rather than for meat.

Feeding supplements known as ionophores (carboxylic polyether ionophore antibiotics) modify rumen fermentation and tend to lower methane emissions (Bergen and Bates, 1984). Such additives can be fed to beef cattle in feedlots during "finishing" for market. Other feed additives, such as methane-inhibiting bacteria, charcoal, halogenated organic chemicals, and fatty acids, may also reduce methane emissions. Such dietary supplements are likely to be expensive, however, and may have side effects on animal metabolism.

The possibilities for breeding for decreased methane emissions should be evaluated, as well as the use of feed additives in nonintensive systems. Research aimed at better management and disposal of animal manures should take methane generation into account.

In the developed countries, cattle populations have hardly been growing in recent years. This trend reflects a diminishing demand for meat and dairy products in the United States and Europe, as human dietary preferences shift to lower amounts of meat and to food with lower fat content for health reasons. Other causes of the cattle population decline include enhanced productivity through biotechnology, selective breeding, and improved feed management. However, in developing countries such as Brazil where rising living standards tend to bring increased consumption of meat, cattle populations are growing at a fast pace. Therefore, global methane emissions from ruminant animals are still projected to increase in the future.

Nitrous Oxide

Nitrous oxide is present in the atmosphere in even smaller quantities than methane. The current concentration is 0.31 ppmv compared to 355 ppmv for carbon dioxide and 1.72 ppmv for methane (IPCC, 1995). The radiative forcing of the N_2O molecule is about 200 times greater than that of CO_2, and its lifetime in the atmosphere is about 120 years. The annual rate of rise of nitrous oxide is 0.2 to 0.3%. Besides being a greenhouse gas, nitrous oxide also plays another important role in the atmosphere: it is further oxidized into NO_x in the stratosphere, where it acts to deplete ozone.

Nitrous oxide emissions are part of the complex global nitrogen cycle (Figure 2.14), by which various forms of nitrogen are both transported and transmuted in soils, plants, animals, and the atmosphere (Sprent, 1987; Brady, 1990; Schlesinger, 1991). There are large uncertainties in the estimations of nitrous oxide emissions from both

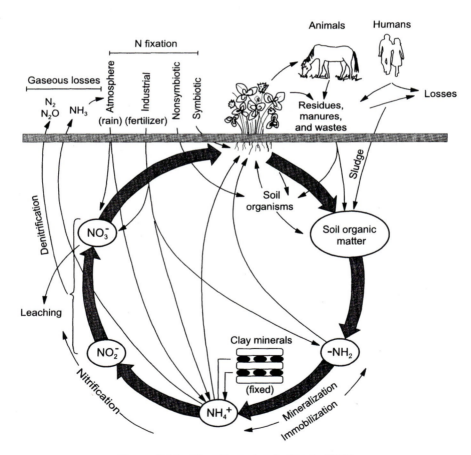

Figure 2.14   The nitrogen cycle (Brady, 1990).

natural sources and human activities because of the high degree of spatial and temporal variations exhibited by many small sources. Soils appear to be the major medium that generates N_2O through microbial processes (IPCC, 1992, 1995). Human activities that contribute to nitrous oxide emissions include the use of nitrogenous fertilizers, the burning of biomass, the conversion of forests to pastures in tropical regions, and the combustion of fossil fuel. Of these, agriculture plays a dominant role in fertilizer use, biomass burning, and land-use change.

Surface burning of vegetation is estimated to release 0.2 to 1.0 million tons of nitrogen per year as N_2O (IPCC, 1992), and to lead to enhanced biogenic soil emissions of N_2O for an extended period of time (up to 6 months) subsequently (Anderson et al., 1988). Tropical land-use changes, and to a lesser extent mid-latitude fertilizer and fossil-fuel use, are apparently contributors to atmospheric N_2O increases. To the extent that agriculture uses petroleum and coal-generated electricity for energy, it also adds to the emissions from fossil-fuel sources. Estimates of N_2O emissions (expressed as million tons N) from fertilizer applications are highly uncertain, varying from 0.03 to 3.0 million tons N per year out of an estimated 5 to 16 million tons N total annual emission (IPCC, 1992). Uncertainties are due to lack of measurements of N_2O fluxes, the complexity of the biogeochemical interactions, and the heterogeneity of the land surface.

Nitrous Oxide from Soils

Nitrous oxide is released under natural conditions from water bodies and unfertilized soils via the sequential microbial processes of nitrification and denitrification, nitrate assimilation, and dissimilation. These processes are important stages in the global nitrogen cycle (Sprent, 1987). Since plants cannot directly absorb the elemental nitrogen gas (N_2) that is abundant in the atmosphere, they must rely on certain bacteria in the soil and in aquatic systems to fix atmospheric nitrogen and convert it to ammonia or nitrate. It is in the latter forms that plants can absorb the nitrogen they need to synthesize proteins.

Nitrification is the biochemical oxidation of ammonium (NH_3) to nitrate ions (NO_3^-), predominantly by aerobic autotrophic bacteria (Figure 2.15). Denitrification (which is the reversal of nitrification) is the biochemical reduction of nitrate ions, first to nitrite ions (NO_2^-), and then to the gaseous compounds, nitric oxide (NO), nitrous oxide (N_2O), and nitrogen (N_2), by anaerobic microorganisms. Nitrous oxide is an intermediate stage of both nitrification and denitrification, but it is the latter that constitutes the main process through which nitrous oxide is released from soils. Under field conditions, denitrification produces both N_2O and N_2. N_2O emission is favored when there is ample nitrite ion (NO_2^-) present in the soil.

Nitrous oxide emissions vary widely with soil conditions. Decomposable soil organic matter is generally needed for nitrification and denitrification processes to occur. Other factors that affect emissions, besides soil organic matter, include soil moisture, available oxygen, temperature, and pH. Emissions of N_2O are also affected by the type of fertilizer; the amount, technique, and timing of application; the presence of other chemicals; the amount of residual nitrogen present in the soil; tillage methods; irrigation practices; and crop rotation.

The terrestrial nitrogen cycle is linked to water availability (Sprent, 1987).

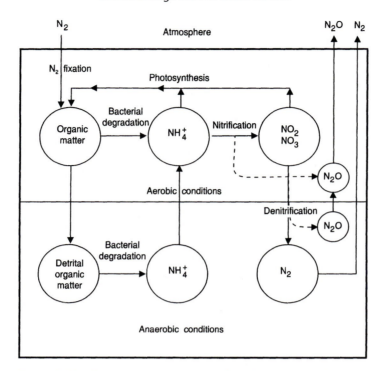

Figure 2.15   Schematic representation of nitrification-denitrification reactions (Wollast, 1981).

Production of N_2O is generally stimulated by precipitation, although prolonged flooding inhibits it. Alternating wet and dry periods have been shown to increase emissions. Soil aeration diminishes emissions, because aerobic conditions promote complete nitrification and the production of nitrates as an end product rather than N_2O.

Warmer temperatures also promote N_2O emissions. Denitrification has a broad temperature range of 25 to 65°C. N_2O production shows strong diurnal and seasonal cycles. Low pH values tend to arrest the denitrification process at the N_2O production phase before complete reduction occurs and forms N_2 gas. Therefore soil acidity tends to promote N_2O emissions.

Nitrous Oxide and Land-Use Change

Nitrogen in the original vegetative cover is lost when virgin land is converted to agriculture. Global N_2O loss from soils due to the conversion of land to agricultural uses has been estimated to be 0.4 +/− 0.2 million tons N per year (Seiler and Conrad, 1987), but this is highly uncertain. While tropical forests and savannas already emit significant amounts of N_2O per unit area (due probably to warm temperatures favoring high turnover rates) (Keller et al., 1986; Matson and Vitousek, 1987; Hao et al., 1988), deforestation and conversion to agricultural uses emit even more. Keller et al. (1988) have shown that addition of fertilizer to tropical forest soils boosted emissions of N_2O, and Luizao et al. (1989) found that a pasture that had been converted from tropical forest had threefold greater annual N_2O flux than a paired forested site, partly due to increased frequency of wetting and drying. Analysis of globally distributed, high-frequency surface measurements suggests that tropical sources of N_2O emissions contribute to the increasing N_2O trend in the atmosphere, probably resulting from human disturbance of tropical ecosystems (Prinn et al., 1990).

Nitrous Oxide and Fertilizer

Nitrogenous fertilizers are taken up by crops, though often incompletely. The residues are either leached from the soil profile by downward percolation or volatilized (following denitrification) in the form of nitrous oxide.

Significant amounts of nitrous oxide are emitted from agricultural soils, particularly from poorly aerated soils fertilized with organic manures or having a high organic matter content (Sahrawat and Keeney, 1986; Coyne et al., 1994). Synthetic fertilizers also contribute to N_2O emissions. Nitrous oxide is released from soils during nitrification of ammonium and ammonium-producing fertilizers under aerobic conditions, as well as by denitrification of nitrates under anaerobic conditions (Bremner and Blackmer, 1978).

Because much N_2O is produced naturally by soils, it is difficult to calculate anthropogenic contributions to emissions. The use of synthetic fertilizers for agricultural crops has grown widely since their advent in the late 1800s. Since the mid-1980s, more than 70 million metric tons of nitrogen have been applied to crops each year (FAO, 1993). Nitrogenous fertilizer has been estimated to contribute between 0.14 to 2.4 million tons of the total annual N_2O emissions of 8 to 22 million tons (Lashof and Tirpak, 1990). Another set of estimates is shown in Table 2.10.

The industrial fixation of nitrogen gas for fertilizer is now almost equal to the natural bacterial fixation (Erlich, 1990). This represents a large change in a natural cycle. High rates of nitrogen fertilizer often exceed plant uptake and result in leaching of nitrates to groundwater or in runoff to surface waters. High nitrate levels in soils and water bodies provide a substrate for denitrification, the major process that releases N_2O to the atmosphere.

Both natural fertilizers (animal manures and crop residues) and manufactured fer-

Table 2.10 Global conversion of fertilizer nitrogen to nitrous oxide in soils (Jackson, 1993)

Year	Global consumption of N fertilizer[1]. ($\times 10^6$ t)	Estimated emissions of N_2O due to N fertilizer (10^6 t N)
1985–1986	69.803	0.007–1.43
1986–1987	71.555	0.007–1.47
1987–1988	75.511	0.008–1.55
1988–1989	79.580	0.008–1.63

1. FAO. 1989. *Fertilizer Yearbook*. Vol. 39. Food and Agriculture Organization of the United Nations. Rome.

Table 2.11   Daily average N_2O emissions per kilogram nitrogen applied per hectare (from Eichner, 1990)

Fertilizer type	Daily average emissions[1] $(g N_2O-N ha^{-1} d^{-1})$	Daily average emissions per kg N applied per hectare $(mg N_2O-N ha^{-1} d^{-1} kg N^{-1})$
Anhydrous ammonium	44.0	200.9
Ammonium nitrate	4.5	40.1
Ammonium type	4.6	44.4
Urea	1.6	10.6
Ca, K, Na nitrate	1.5	15.2

1. Regardless of quantity of fertilizer applied.

tilizers tend to promote N_2O production and release, but the differential impacts of the various types of N fertilization are yet undefined. Synthetic fertilizers are assumed to be the major sources of emissions, due to the often large quantities applied and the non-optimal timing and placement of the applications (Erlich, 1990), but fluxes from manured soils, especially after rain, can also be high (Coyne et al., 1994).

Attempts have been made to estimate the relative amounts of nitrous oxide emissions from various fertilizers (Table 2.11). It appears that daily average emissions of N_2O per kg N applied per ha are highest for anhydrous ammonium, followed by ammonium chloride and sulfate; ammonium nitrate; calcium, potassium, and sodium nitrates; and urea (Eichner, 1990). Emissions tend to be higher when these fertilizers are applied in solution rather than in solid form. The average N_2O release from fertilized soils due to N fertilizer measured in one study varied from about 0.1 to 1.5% of applied nitrogen, with variations in release depending on soil type (Byrnes et al., 1990).

Limiting Agricultural Emissions of Nitrous Oxide

Finding ways to curtail nitrous oxide emissions is difficult since the exact details of the processes involved are not well understood. Nitrogenous fertilization is known to promote N_2O generation, but nitrogen applications to crops are necessary to maintain and expand productivity. Even so, the following practices could be modified in an effort to limit emissions: agricultural burnings of crop residues can be reduced, and fertilizer use can be made more efficient by avoiding unnecessary applications and choosing forms of fertilizer that emit less N_2O (see Table 2.11).

Overuse of nitrogen fertilizer is common in North America and Western Europe. One estimate states that a 20% decrease in fertilizer use is practical in those regions (CAST, 1992). Fertilizers that release nitrogen slowly in the soil should result in lower emissions. Proper timing and deep placement of fertilizer should also be effective. Nitrification inhibitors, slowly dissolving fertilizers, and fertilizer pellet coatings may also inhibit emissions (see, e.g., Singh and Prasad, 1985). Controlling irrigation frequency may also help lower N_2O emission from irrigated upland soils. Improving nitrogen fixation in leguminous crops and developing nitrogen fixation in nonlegumes may reduce the need for the currently massive application of nitrogenous fertilizers.

Fertilizer applications are growing at rates of over 1% per year in developed countries, and over 4% per year in developing countries (Burke and Lashof, 1990). If these trends continue unabated, N_2O emissions are likely to increase significantly. Projections of nitrogen fertilizer use suggest a doubling of the agricultural contribution to global N_2O flux over the next 30 years (Duxbury et al., 1993). Since biomass burning and land conversion to agriculture are also rising, emissions of N_2O from all major agricultural sources are likely to accelerate if mitigation actions are not taken.

Agricultural Emissions and Mitigation: A Look Ahead

Raising crops and livestock alters the natural fluxes of both the carbon and nitrogen cycles through production inputs (e.g., energy and fertilizers) and outputs (e.g., grain yield, meat, and waste products including crop residues and manure). Agriculture is both a source and a sink for carbon (Figure 2.16). On the source side, the fossil fuel inputs that power agricultural operations release CO_2, fertilizer production uses energy and releases CO_2, and the manufacture and transport of pesticides and herbicides also use fossil-fuel energy. On the sink side, as modern technology (including increased fertilization) raises yield, more atmospheric carbon and nitrogen are sequestered in agricultural products. When the carbon and nitrogen dynamics of agriculture are considered overall, the net effect of agriculture on greenhouse forcing may be low due to the sequestration of carbon and nitrogen in agricultural products (Jackson, 1992, 1993).

Quantitative data on the balances of these emissions and sequestrations, including the rates at which the carbon and nitrogen remain in agricultural products and soils, are needed to analyze the exact contribution of agricultural production to global climate change. The net effect, if not negligible, has been much smaller than the effects of the energy and CFC industries. Research needs include better measurement methods, identification of controlling factors of trace gas emissions, and continued development of simulation models of carbon dioxide, methane, and nitrous oxide formation and emission from agricultural systems.

Agricultural greenhouse gas emissions in the future will depend not only on agricultural practices but also on population growth and economic development. Since agricultural activities must continue to expand to feed the 10 billion people expected

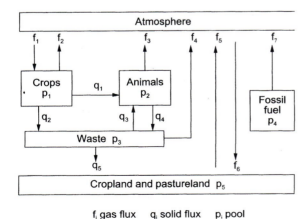

Figure 2.16 Carbon pathways in agroecosystems (Jackson, 1992).

MINIMUM DATA TABLES 4 AGRICULTURE

4 A & B   Enteric Fermentation & Animal Wastes

SOURCE AND SINK CATEGORIES	ACTIVITY DATA	EMISSION ESTIMATES		AGGREGATE EMISSION FACTOR	
	A	B		C	
Sector Specific Data (units)	Number of Animals	Enteric Fermentation	Animal Wastes	Enteric Fermentation	Animal Wastes
	(1000)	(Gg CH$_4$)		(kg CH4 per animal)	
4 AGRICULTURE				C=(B/A) x 1000	
A & B Enteric Fermentation & Wastes					
1 Cattle					
i Beef					
ii Dairy					
2. Goats					
3 Sheep					
4 Pigs					
5 Horses/Mules/Asses					
6 Buffalo					
7 Camels And Llamas					
8 Other					

4 C   Rice Cultivation

Source And Sink Categories	Activity Data		Emission Estimates	Aggregate Emissions Factor
	A	B	C	D
	Area cultivated in hectares	Hectare-Days of Cultivation	Emissions of CH_4	CH_4/N_2O Average Emission Factor
Sector Specific Data (units)	(Mha)	(Mha-days)	(Gg CH_4)	(kg CH_4 per ha-day)
				D=C/B
C Rice Cultivation				
1 Flooded Regime				
2 Intermittent Regime				
3 Dry Regime				

Figure 2.17   Inventory worksheets for greenhouse gas emissions from agricultural sources (IPCC, 1994).

to be living in the middle of the 21st century, greenhouse gas emissions from agriculture are likely to increase significantly unless determined and concerted efforts are made to limit them. This situation presents a major challenge to agricultural scientists—first to understand emissions from agricultural practices and to calculate emissions on a country-by-country basis, and then to devise approaches for reducing greenhouse emissions while expanding production to feed the world's growing population. If climate is changing at the same time, complex interactions are likely to ensue.

As part of the IPCC process, guidelines for calculating national inventories of

Table 2.12 Options for direct and indirect mitigation of greenhouse gas emissions from agriculture (IPCC, 1996)

Option	CO_2	CH_4	N_2O
Land conversion and management			
reduced deforestation rate	H	M	M
pasture immediately after deforestation	M		L
conversion of marginal agricultural land to grassland, forest, or wetland	M		L
Agricultural land utilization and management			
restoring productivity of degraded soils	H	L	
more intensive use of existing farmland	M	L	L
restricted use of organic soils	H		L
conservation tillage	M		L
reduction of dryland fallowing	M		
diversified rotations with forage crops	M		L
Biofuels			
energy crops for fossil fuel substitution	H		
agroforestry	L		
windbreaks and shelterbelts	L		
agroindustrial wastes for fossil fuel substitution	L		
Recycling of livestock and other wastes			
recycling of municipal organic wastes	L	L	M
biogas use from liquid manures		M	
Animal husbandry			
supplementing low quality feed		M	
increasing feed digestibility		L	
production enhancing agents		L	
Rice cropping systems			
irrigation management		M	L
nutrient management		M	L
new cultivars & other		M	L
Plant nutrient management			
improved fertilizer use efficiency	L		H
nitrification inhibitors			M
legume cropping to bolster system productivity			M
integrating crop and animal farming			L
Minimizing overall N inputs			
reduced protein inputs in animal feed			M
reduced protein consumption by society		M	M

L = low, M = medium, H = high.

greenhouse gas emissions have been developed (IPCC, 1994). These provide, in simple workbook form, formulas for calculating CH_4 emissions from animals and animal wastes, CH_4 from rice cultivation, and CH_4, CO, N_2O, and NO_x from agricultural burning (Figure 2.17). Scientists in all the countries (over 100 nations) that have ratified the Framework Convention on Climate Change are conducting inventories of trace gas emissions. This is a challenging task, given the multiple sources and the uncertainties involved in our knowledge of many of the processes causing the emissions, and a very important one.

Complete elimination of greenhouse gas emissions from agricultural sources is obviously impossible, given the nature of the agroecosystem and the importance of intensifying crop production. Forgoing dependence on rice production for sustenance in Asia and on nitrogen fertilizer applications to maintain high crop yields in many regions is simply not feasible. Reductions in methane emissions are needed, while productivity levels in rice production must be maintained. The efficiency of livestock production around the world is likely to improve, but perhaps too slowly to offset methane emissions by domesticated animals.

Furthermore, for mitigation strategies to reduce agricultural contributions to greenhouse gas emissions to be at all reasonable, at the same time that they reduce emissions, they must maintain or enhance agricultural production, allow additional benefits to farmers (via reduced labor, reduced inputs, or more efficiency), and ensure quality and desirability of agricultural products. For example, higher efficiency of nitrogen fertilizer, improved rice cultivation methods, and increased meat and milk production in beef and dairy cattle should simultaneously reduce methane and nitrous oxide emissions while raising crop and animal productivity. Such mitigative strategies are the most promising (Table 2.12).

If reductions in some gases prove to be more easily achievable, efforts may be concentrated preferentially on reducing emissions of some gases rather than others. For CO_2 mitigation, the greatest gains are probably to be made through the reduction of deforestation and land conversion to agriculture in the tropics and through increased utilization of biofuels. At the same time, reforestation programs can provide a sink for atmospheric CO_2. Decreasing the relatively low fossil-fuel carbon consumption from the agricultural sector offers only low potential for emission reductions.

3

Carbon Dioxide, Climate Change, and Crop Yields

A changed climate will alter the biophysical environment in which crops grow. The main factors subject to change are atmospheric CO_2, temperature, precipitation, and evapotranspiration. Crop response to changes in these factors is the first stage in the cascade of consequences leading to potentially profound changes in the agricultural economy of entire regions. This chapter reviews the physiological effects of CO_2 enrichment, as well as the primary effects of temperature and precipitation on crop growth and yield. To anticipate and evaluate crop responses to changes in these variables, we must examine how crop plants are adapted to current CO_2 and climate conditions. Are crops now limited by CO_2 availability, and may elevated CO_2 levels boost productivity? Are crops in different regions now growing at, below, or above the optimum levels of climate variables? How does the variability of climate affect crop growth and how might a change in variability alter crop performance? These are a few of the questions relevant to the topic of this chapter.

Physiological Effects of CO_2 Enrichment

Plant responses to higher concentrations of atmospheric CO_2 may be considered on different scales, ranging from the microscopic cellular level to the macroscopic agro-ecosystem level. The scaling up of plant responses in time and space from one level to another is complicated. Photosynthesis, respiration, and transpiration are the plant processes most directly affected by changing levels of CO_2. A host of interactive changes in crop growth flow from these primary effects, some resulting in positive feedbacks and others in negative ones. Generalized responses to elevated CO_2 are depicted in Figure 3.1.

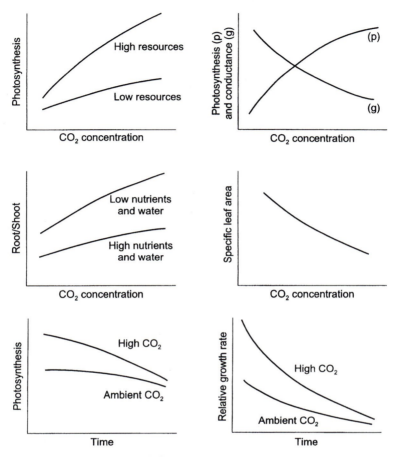

Figure 3.1   Responses of plants to CO_2 concentration (Bazzaz, 1990).

Photosynthesis and Respiration

Photosynthesis and respiration influence the direction of net carbon flow between the atmosphere and terrestrial ecosystems, both natural and managed.

Photosynthesis.   If atmospheric accumulation of CO_2 were occurring without concomitant changes in temperature and water regimes, it might indeed be a blessing for agriculture. CO_2 is an essential component of the process of photosynthesis, upon which life on earth ultimately depends. An enhanced CO_2 concentration in the atmosphere tends to increase the gradient between the external air and the air spaces inside the leaves, thus promoting diffusive transfer and absorption of CO_2 into the chloroplasts and its conversion to carbohydrates. Higher rates of photosynthesis are found in measurements of single leaves, as well as in entire canopies placed in a CO_2-enriched atmosphere (Lemon, 1983; Acock and Allen, 1985; Drake and Leadley, 1991).

Plant species vary in their response to CO_2, in part because of differing photosynthetic mechanisms (Figure 3.2). In some species, photosynthesis occurs via a "C3 pathway," so called because the first product in the sequence of biochemical reactions has three carbon atoms. Other species, in contrast, follow a "C4 pathway," as the first product contains four carbon atoms.[1] A comprehensive discussion of these photosynthetic pathways is given by Tolbert and Zelitch (1983).

C3 plants use up some of the solar energy they absorb in photorespiration.[2] A fraction of the carbon initially reduced from CO_2 and fixed into carbohydrates is reoxidized to CO_2, thus releasing the chemical energy that the plant had originally taken in as solar radiation. This process causes C3 crops (such as wheat, rice, and soybeans) to exhibit lower rates of net photosynthesis than do C4 crops such as maize at current CO_2 levels (~350 ppmv) because in C4 plants loss to photorespiration is minimal. However, at elevated levels, photosynthesis rates of C3 crops may exceed those of C4 plants due to suppression of photorespiration (Figure 3.3). In general, C3 crops are more responsive to CO_2 enrichment than are C4 crops and may therefore become more competitive than at present in relation to C4 crop species.

In C4 plants, CO_2 is first captured in the mesophyll cells as malic and aspartic acids, which are then transmitted to the Rubisco enzyme in the bundle-sheath cells. There the acids release CO_2, naturally raising the CO_2 concentration and promoting the carboxylase over the oxygenase enzymatic reaction. In this manner, photosynthesis is favored over photorespiration. Although C4 plants are more efficient photosynthetically than C3 plants under present CO_2 levels, they are likely to be less affected by CO_2 increase. C4 crops (developed primarily from tropical grasses) that are of

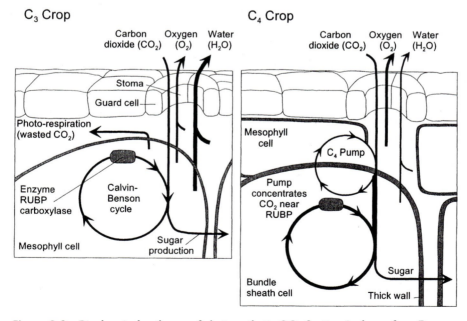

Figure 3.2 Biochemical pathways of photosynthetic CO_2 fixation (redrawn from Bazzaz and Fajer, 1992).

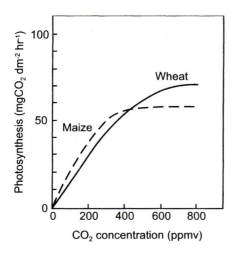

Figure 3.3 Effect of atmospheric CO_2 concentration on rate of photosynthesis of maize and wheat in controlled-environment experiments (adapted from Akita and Moss, 1973).

economic importance include maize (corn), sorghum, sugarcane, and millet. In a CO_2-enriched atmosphere, such crops may be more vulnerable to increased competition from C3 weeds.

Respiration. The response of photosynthetic carbon uptake to atmospheric CO_2 is reasonably well understood and can be mechanistically modeled across several scales. Not so the response of respiration to CO_2 concentration, which is less well understood on both short (seconds to minutes) and long (seasonal) time scales. An increase in photosynthesis, growth rate, and substrate levels should increase respiration rate per unit of land area, because higher biomass requires higher energy supply for maintenance and growth. On the other hand, increased CO_2 in the air (which promotes the inward diffusion of CO_2) inhibits the diffusive release of CO_2 by the plant and reduces the proportionate increase in respiration (Amthor, 1989).

The exact nature of the interactions of these two opposing effects is unknown. Some researchers have reported that doubling the CO_2 concentration causes an immediate but reversible decrease in the net rate of CO_2 efflux from plant tissue; others have found increases or zero effects. Increased carbon-use efficiency (the ratio of daily growth to daily photosynthesis) at higher CO_2 concentrations has also been reported. In wheat, this appears to be small (Gifford and Morison, 1993). There are also data that indicate a long-term decrease in respiration during growth at elevated CO_2 concentrations in both controlled environments and field studies. However, the extent to which the long-term decreases may be related to permanent changes in tissue composition rather than to a lingering short-term response has not been determined precisely.

Acclimation of Photosynthesis and Respiration. While the rate of photosynthesis is known to rise immediately in response to elevated CO_2 levels, this initial response is often found to diminish or disappear under long-term exposure (Stitt, 1991). The process of acclimation is defined as an organism's adjustment to new environmental conditions; such adjustments are not necessarily heritable (Hale and Orcutt, 1987). Plants acclimate to higher CO_2 by adjusting their rates of photosynthesis and perhaps respiration, but the degree to which plants acclimate is yet imperfectly defined. Photosynthetic acclimation is associated with reductions in the production of the Rubisco (ribulose-1,5-biphosphate) carboxylase-oxygenase enzyme, which, in turn, are linked to higher levels of carbohydrates in the leaves (Stitt, 1991).

One important determinant for sustained higher photosynthesis rates appears to be the physiological capacity of a plant to effectively store the products of photosynthesis. In other words, the persistence of high rates of photosynthesis depends on sink size (the size of the grains, kernels, and tubers), allowing for storage of the added fixed carbohydrate. Even with acclimation, leaf-level photosynthesis is generally increased in elevated CO_2. However, a better understanding of long-term acclimation of photosynthetic capacity and respiration is needed for a reliable prediction of potential changes at the agroecosystem level and in the carbon cycle in response to higher CO_2 concentrations.

Biomass and Yield

Enhanced photosynthesis in higher atmospheric CO_2 levels naturally promotes biomass accumulation (Kimball, 1983; Cure and Acock, 1986; Poorter, 1993). Reviews of experiments in controlled (enclosed) environments show a wide range in the magnitude of response to a doubling of CO_2, usually tested experimentally from a "current" level (300–350 ppmv CO_2) to a "doubled" level (600–700 ppmv CO_2). Most yield responses have been positive, and only a few have been slightly negative (Table 3.1). The few open-field studies that have been done generally confirm the positive results obtained in controlled environments (Lawlor and Mitchell, 1991; Hendrey, 1993).

Responses to elevated CO_2 vary among different crops and even among varieties of the same crop. The varying responses depend in part on environment (water and nutrient availability) and in part on genetics. Among varieties of the same crop, it is often hard to distinguish the relative contributions of genetic differences, experimental techniques, and operative feedbacks to the variation in response. Among crops, the dichotomy between the C3 and C4 photosynthetic pathways appears to contribute the most to the differences in overall response.

In Kimball's (1983) review of over 70 published experiments of crop responses to CO_2 enrichment, yields increased in 90% of the experiments, with the increases varying from 0 to as much as 100%. A few negative results have also been reported, though they are unexplained. Gifford and Morison (1993) suggested that negative results may be due to contamination of the experimental CO_2 with other gases, such as ethylene. On average, however, C3 crop yields in the experiments reviewed by Kimball (1983) and by Cure and Acock (1986) increased an average of about 33% under doubled CO_2 conditions, whereas C4 crops showed increases of about 10%. For C3 crops this is equivalent to a dry matter increase of 0.1% per ppmv CO_2. The increases in C4 plants may be due in large part to improvements of water-use efficiency as described in the next section, rather than on photosynthetic response per se. While few experiments have been done on yield quality, there appears to be little effect of high CO_2 on this important characteristic.

Increased yields of wheat are due primarily to more tillering and branching, as well as to greater numbers of seeds and tubers, rather than to increases in the sizes of

Table 3.1 Crop yield responses to elevated CO_2 (from Cure and Acock, 1986)

Response category	Wheat[1]	Barley[2]	Rice	Corn	Sorghum	Soybean	Alfalfa	Cotton[2]	Potato	Sweet potato	Weighted average[3]
Yield	+35 ± 14 (17, 8)	+70 ± (2, 1)	+15 ± 3 (6, 3)	+29 ± 64 (3, 1)	—	+29 ± 8 (28, 12)	—	+209 ± (2, 1)	+51 ± 111 (6, 3)	+83 ± 12 (3, 1)	+41

Data represent the percentage change at 680 ppm CO_2 compared with controls (300–350 ppm) ±95% confidence limits, as estimated by regression analysis. The values in parentheses are the number of relative response values used in each regression and the number of studies that supplied those values.

1. Based on a quadratic model.

2. In cases where results were based on only two data points, error degrees of freedom were 0 and confidence limits could not be calculated.

3. For the weighted average, each predicted response value was multiplied by the number of studies associated with it, then these were summed and divided by the total number of studies in the response category row.

these structures (Lawlor and Mitchell, 1991). Experimental CO_2 enrichment has also been observed to cause changes in plant organs. Such changes include increases in stem and root lengths, as well as greater leaf area and thickness. Greater leaf thickness associated with increased starch levels leads to smaller specific leaf surface area (the ratio of leaf surface to volume). These two effects tend to reduce potential assimilation per leaf and thus may act to diminish the photosynthetic enhancement of atmospheric CO_2 enrichment. Hastened leaf turnover rates (i.e., earlier abscission of the lower leaves) are possibly due to greater self-shading resulting from enhanced growth in a CO_2-enriched atmosphere.

Virtually all studies that have examined root dry weight found it to increase under elevated atmospheric CO_2; this effect occurred across many different species and study conditions (Rogers et al., 1994). In cotton grown in high CO_2 (550 ppmv) both tap and lateral roots exhibited increases in dry weight and volume (Prior et al., 1995). Root:shoot ratios, a key indicator of morphological changes in response to CO_2, both increased and decreased in various high CO_2 experiments, although positive responses tended to predominate. Increases in root:shoot ratios may be caused by reduction in the amount of nitrogen required to sustain maximum leaf photosynthetic rates at elevated CO_2. Hence, additional nitrogen can be allocated to the roots for anabolic processes. The variability in the variously reported root:shoot ratios at high CO_2 may result from the occurrence of nutrient, water, or pot volume limitations in some trials. Nutrient and water deficiencies would likely promote root growth at elevated CO_2 relative to that observed under current CO_2 levels. Phosphorus may also become a limiting factor at faster growth rates.

Stomatal Resistance and Water-Use Efficiency

Another important physiological effect of CO_2 enrichment is the partial closure of stomates, the small openings in leaf surfaces through which CO_2 is absorbed and water vapor is released. (The opposite effect, namely the widening of stomatal aperture, is seen in lower than ambient CO_2 concentrations.) Accordingly, a rise in atmospheric CO_2 may reduce transpiration (evaporation from foliage) even while promoting photosynthesis. A reduction in transpiration on the order of 30% has been observed for crop plants grown in doubled-CO_2 atmospheres (Table 3.2). The dual effect of promoting photosynthesis while suppressing transpiration is likely to improve water-use efficiency (WUE), defined as the ratio between biomass accumulation and the amount of water transpired by the crop. An improvement in WUE on the order of 70 to 100% was reported by Kimball and Idso (1983) and by Morison (1985). C4 crops, although less responsive to higher CO_2 photosynthetically, also tend to close their stomates in its presence, thereby improving WUE. The relative CO_2 growth response is greater for both C3 and C4 plants when water supply is limited than when it is abundant, due to the positive feedback of stomatal closure.

At the leaf level, a wide range of stomatal responses to CO_2 has been observed in over 50 types of plants—in plants with C3 and C4 photosynthetic pathways, in gymnosperms and angiosperms, and in dicots and monocots (Morison, 1985). However, the physiological mechanisms involved in the stomatal response to CO_2 are still not completely understood. Contrary to earlier notions that the stomates of C4 plants were more responsive than those of C3 plants, it now appears that stomatal conductances

Table 3.2 Crop transpiration responses to elevated CO_2 (from Cure and Acock, 1986)

Response category	Wheat	Barley	Rice	Corn	Sorghum	Soybean	Alfalfa	Cotton	Potato	Sweet potato	Weighted average[1]
Transpiration	-17 ± 17 (4, 2)	-19 ± 6 (7, 3)	-16 ± 9 (7, 3)	-26 ± 6 (15, 6)	-27 ± 16 (6, 2)	-23 ± 5 (19, 8)	—	-18 ± 17 (7, 3)	-51 ± 24 (3, 1)	—	-23

Data represent the percentage change at 680 ppm CO_2 compared with controls (300–350 ppm) $\pm 95\%$ confidence limits, as estimated by regression analysis. The values in parentheses are the number of relative response values used in each regression and the number of studies that supplied those values.

1. For the weighted average, each predicted response value was multiplied by the number of studies associated with it, then these were summed and divided by the total number of studies in the response category row.

of both plant types at elevated CO_2 concentration are linearly related to the initial stomatal conductance, with an average reduction of about 40% for doubled CO_2 conditions (Figure 3.4). This response has not been shown to acclimate (i.e., to diminish over time under normal conditions), but may be diminished at either high or low temperatures.

Stomatal response to CO_2 varies with many factors, including leaf age, light intensity, atmospheric humidity, and temperature (Morison, 1985). Complete characterization of stomatal response to CO_2 is difficult in view of the complex interactions among CO_2, temperature, and plant hormones such as abscisic acid (ABA) and indoleacetic acid (IAA). The precise interaction between stomatal regulation and gas exchange is similarly difficult to define. Morison (1985) cites evidence to suggest that while the effect of CO_2 on stomatal conductance per se is large, it may not be so large in regulating gas exchange.

Feedbacks occur in the response of stomatal aperture to CO_2, just as in the response of photosynthesis. For instance, one result of the closure of stomates and the concomitant reduction of transpiration (and, hence, of latent heat loss) is a rise in leaf temperature. Consequently, the intercellular vapor pressure rises and transpiration increases. In controlled-environment chambers, leaf and canopy temperatures were found to rise from 1 to 3°C when plants were exposed to elevated concentrations of CO_2 (Chaudhuri et al., 1986; Idso et al., 1987). The reduction of transpiration associated with observed stomatal closure (25–35%) is less than commensurate with the reduction of stomatal conductance (40%) (Allen et al., 1985; Morison, 1985), evidently because the raised vapor pressure in the leaf leads to a greater leaf-to-air vapor pressure gradient, which partially compensates for the reduced conductance. Higher

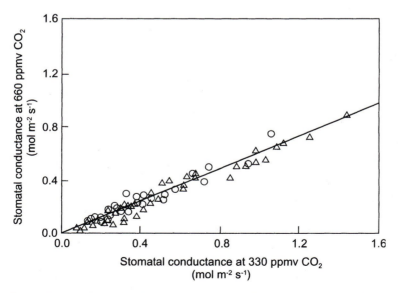

Figure 3.4 Relationship between stomatal conductance at present and doubled CO_2 concentrations. Triangles = C3 species; circles = C4 species (Morison, 1985).

leaf temperature may also lead to a higher rate of plant metabolism (including respiration and physiological development) and to accelerated aging of leaf tissue.

The combined photosynthetic and stomatal closure effects can be considered at three levels: single leaf, whole plant, and crop stand. At the leaf level, increases in the instantaneous "transpiration efficiency" (A/E, mol CO_2 uptake/mol water loss, where A is assimilation and E transpiration) resulting from CO_2 enrichment have been observed to vary between 60 and 160% in certain plants due to both an increase in assimilation and a decrease in transpiration (Morison, 1985).

At the whole plant level and over longer periods of growth, increases in water-use efficiency (mass of dry matter produced[3]/transpiration) are not as high as the increases in A/E seen in single leaves. This occurs because of the photosynthetic acclimation and the negative temperature feedback that affect growth, yield, and transpiration. Under doubled CO_2, Morison (1985) suggests that an increase of approximately 30% in plant growth is associated with a commensurate decrease in transpiration, which together raise the water-use efficiency of whole plants by about 70 to 100%.

At the crop canopy level—in the open field—water-use efficiency is defined as the ratio between crop biomass accumulation and the total amount of water consumed by the crop (i.e., evapotranspiration). The latter term generally includes both transpiration, which is affected by stomatal closure, and evaporation from soil, which is not directly affected by CO_2 enrichment but which may change due to changes in canopy cover, temperature, and moisture uptake by the crop.

Another alteration in hydrological regime under elevated CO_2 is greater rainfall interception by the enlarged leaf surface. This leads to more direct evaporation from the crop canopy, which can be a significant occurrence in dry environments. Thus, more complex interactions and feedbacks operate on water-use efficiency at the field canopy level. Morison (1985) cites an experiment in which the improvement in water-use efficiency of a canopy of wheat grown in dense and large stands in pots was on the order of 30%, down from ~80% for isolated plants. He estimated that increases in water-use efficiency for crops in the field or for natural vegetation may be 30 to 50% for doubled CO_2. Contrariwise, Allen et al. (1985) projected water use per unit land area to be scarcely reduced.

Positive and Negative Feedbacks

A positive feedback in plant growth with CO_2 enrichment occurs when the plant allocates CO_2-stimulated growth into expanding leaf area, which in turn absorbs more light and CO_2, thereby stimulating still greater growth. The extent to which this positive feedback operates depends on the plant's supply of or competition for other essential resources (besides CO_2), especially light, water, and soil nutrients. Thus, single plants growing in the absence of aerial or subterranean competition (as is often the case in experiments) can expand their roots and shoots readily, provided they are endowed with adequate water and nutrient supplies (Gifford and Morison, 1993). In the field, however, where crops grow in communities, the opportunities for such a high positive feedback are limited by competition for those supplies.

The situation is different where root growth is restricted by lack of water, nutrients, aeration, or space, as well as by salinity or mechanical impedance. With such restrictions, a plant may not fully respond to the stimulation of high CO_2, and a negative feedback may ensue. This is known as a sink feedback, since the growth of roots, being a sink for photoassimilates, is the cause of the restriction. Among other factors, adjustments to the root restrictions caused by pots or by interplant competition may induce a decline in leaf photosynthesis rate, a process termed "down-regulation." Many early high-CO_2 experimental results may have been affected by limited root growth of plants grown in small pots (Arp, 1991). Stitt (1991) similarly argues that the inhibition of photosynthesis under enhanced CO_2 observed during gradual accli-mation is related to such a sink effect. Therefore, experimental differences observed in the long-term response to CO_2 may be explained by differences in the sink-source status, which in turn depends on species, developmental stage, and growing condi-tions.

Positive and negative feedbacks on plant growth often occur simultaneously. While higher CO_2 concentrations increase crop growth rate, leading to increased leaf area, there is also a concurrent increase in respiratory loss of CO_2, because larger plants with faster growth rates respire more (Gifford and Morison, 1993). Another neg-ative feedback involves self-shading. While higher leaf area allows for greater light in-terception, an increased proportion of the leaves will shade each other, thus reducing the average rate of photosynthesis per unit leaf area. The combined effect of all these positive and negative feedbacks is the actual canopy growth rate, which, in most ex-periments done to date, is still positive.

CO_2 Interactions

There is abundant evidence to demonstrate that elevated atmospheric CO_2 concen-trations increase the total plant dry weight of C3 plants, and to a lesser extent of C4 plants, through enhancement of leaf CO_2 assimilation rates and improved plant wa-ter relations. Greater growth may occur even where stress from lack of water, low nu-trients, and high temperatures limit growth. As stated, the magnitude of crop responses to elevated CO_2 depends on the availability of essential resources other than CO_2, among them light, water, and nutrients. Studies have shown that *absolute* yield in-creases due to high CO_2 are lower when other resources are limiting, but that *relative* increases may be equal or even larger than responses under "optimal" conditions. Experiments that test the interactive responses of CO_2 with other environmental vari-ables are especially useful for projecting CO_2 effects in the "real world" where condi-tions for crop growth are often nonoptimal.

Light.   Higher CO_2 appears to improve a plant's utilization of radiation in several ways. At the biochemical level, CO_2 concentration steepens the initial slope of the photosynthetic light response curve, known as the quantum yield (Figure 3.5). In plants with the C3 photosynthetic pathway, quantum yield increases with increasing CO_2 due to the suppression of photorespiration. Under low light levels, where CO_2 uptake is sensitive to quantum yield, rates of photosynthesis per unit leaf area are like-wise stimulated by increased CO_2 concentration. Elevated CO_2 raises the efficiency of conversion of intercepted radiation into dry matter at high as well as low levels of radiation. This has been called by the greenhouse industry the "mutually compensat-ing effect of CO_2 and light" (Gifford and Morison, 1993, p. 328). Its evidence in prac-tice is the success achieved in Europe of growing lettuce in greenhouses enriched with

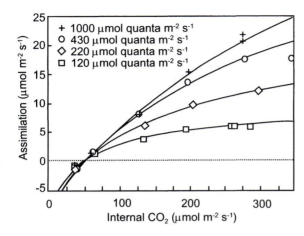

Figure 3.5 The dependence of soybean leaf assimilation rate on internal CO_2 concentration for a range of photosynthetically active radiation (PAR) flux densities (Norman and Arkebauer, 1991).

CO_2 during the short and dim days of winter. An interesting morphological response that may explain observed changes in photosynthesis is an additional layer of palisade cells reported by Acock (1991).

Temperature. In theory, plant growth should respond positively to a rise in both CO_2 and temperature, but experimental results have not always shown this to be the case. Since elevated atmospheric CO_2 reduces photorespiration in C3 plants, and since the rate of net photosynthesis relative to photorespiration increases with temperature, the two effects should work in tandem. Maximum rates of electron transport and carboxylation are temperature dependent and should also lead to improved growth. Finally, as temperature rises, the opportunity for photosynthate sinks (e.g., roots and tillers) to develop may also increase in crops with the ability to take advantage of longer growing seasons.

Despite this theoretical reasoning, however, there is no solid body of experimental evidence to support a strong positive interaction between high CO_2 and high temperature in regard to plant growth. Various experiments have shown both positive and negative responses to combined high CO_2 and high temperature, and little plant response to CO_2 enrichment at low temperatures (below about 15°C for many species), but the mechanisms of these interactive responses to CO_2 and temperature are not sufficiently understood (Allen et al., 1990; Bazzaz, 1990). Other plant processes or factors may be more important in overall plant response and may mask the response to CO_2 and temperature elevations. Among those factors may be tissue chemistry, phenology (especially senescence), photoperiodic control, acclimation of respiration, and interactions with pathogens.

For example, high CO_2 combined with high temperature may promote vegetative growth in soybeans, but may suppress reproductive growth (Allen, personal communication). Also, combined high CO_2 and high temperature may lead to positive effects up to crop temperature threshold levels, at which high temperature damage occurs at the cellular level. Baker et al. (1992) found that rice yields were affected much more strongly by temperature than by CO_2 level. Similarly, Reddy et al. (1995) found that at high temperatures (~35°C) the positive effects of elevated CO_2 (700 ppmv) on cotton were masked by apparent high-temperature injury that limited growth of all plant organs, particularly fruiting structures.

Water. As described here, high CO_2 can help reduce water stress due to partial closure of stomates. Under water-limited conditions, the conservation of water owing to the antitranspirant effect of elevated CO_2 enhances water-use efficiency more than it does under wet conditions. While absolute yield levels may still be low under water stress, the relative response to high CO_2 tends to be greater in water-stressed plants (Gifford, 1979). However, Schonfeld et al. (1989) conducted an experiment on winter wheat at the vegetative stage under increased atmospheric CO_2 and water limitation and found that, although CO_2 enrichment had positive effects on growth and development at tillering, they were insufficient to counterbalance the debilitating effects of water stress. Chaudhuri et al. (1989) found that high CO_2 did appear to compensate for water stress in wheat.

Nutrients. The interactions of atmospheric CO_2 and nutrient supply have only been investigated to a limited degree. It appears that positive responses to CO_2 enrichment can occur even in nutrient-limited conditions (Table 3.3) (Sionit et al., 1981; Sionit, 1983; Cure et al., 1988). However, at low nutrient levels, crop growth is restricted even under high CO_2 levels. Since the degree of CO_2 stimulation tends to rise with nutrient availability, it is clear that adequate fertilization will be required to take full advantage of elevated CO_2 levels.

According to Gifford and Morison (1993), gradual nitrogen deficiency does not totally eliminate the response to CO_2 but merely reduces it by limiting foliar growth. If CO_2 encourages root growth as has been seen in many experiments, the expanded root system should scavenge more nitrogen from the soil. Root growth may not be enhanced, however, in soils that are severely nitrogen deficient. Lower shoot (excluding grain) nitrogen concentrations have been found in wheat plants grown in higher CO_2 (Gifford and Morison, 1993).

Symbiotic nitrogen fixation in nodulated legumes has been found to increase with higher CO_2 under favorable conditions. This is apparently due to the greater

Table 3.3   Seed yield of soybean grown at 350 and 700 ppm CO_2 and five nitrogen concentrations (from Cure et al., 1988).

NO_3^- concentration (mM)	Seed yield (g plant^{-1})	
	350 ppmv CO_2	700 ppmv CO_2
0.05	0.2	0.2
1.00	14.2	20.5
2.50	40.3	56.5
5.00	49.9	76.0
10.00	64.1	94.3

availability of carbon providing an improved substrate for the nitrogen-fixing bacteria.

CO_2 and Other Environmental Factors

Crop responses to carbon dioxide will vary with a host of other environmental factors besides light, temperature, water, and nutrients. These factors, crucial to high-quality, high-yielding agricultural production, include soil physical conditions, salinity, air pollution, weed competition, and pests.

Soils.   Plants grown at elevated CO_2 generally contain higher carbon:nitrogen (C:N) ratios in their litter material, resulting in a reduction in quality. Such material is less readily decomposable, and the process of its decomposition may further immobilize soil nitrogen. This will negatively affect the efficacy of crop residues intended to improve soil fertility. On the other hand, inputs of readily decomposable carbonaceous substances in the rhizosphere are often increased where plants are grown under elevated CO_2. The altered supplies of these labile plant products (including root exudates) to the rhizosphere may partially counteract the effect of reduced litter quality on decomposition rate (see Chapter 5 for further discussion of soil processes).

Salinity.   Where evaporation exceeds downward percolation (especially where high water table conditions exist), salts may accumulate near the soil surface and cause damage to crop plants. High CO_2 has been found to reduce salinity damage (i.e., to increase salt tolerance) in crops in several experiments (Acock and Allen, 1985). However, the mechanism for the increase in salt tolerance is still unknown. It is possible that the higher level of photosynthates may in part offset the greater respiration demand to provide energy to exclude the salt through osmotic regulation. An alternative explanation is that the suppression of transpiration rate (due to partial stomatal closure) reduces the amount of salt taken up by the plant.

Air Pollution.   Because the narrowing of stomatal aperture reduces the amount of airborne pollutants entering the interior of the leaf, higher levels of atmospheric CO_2 should lessen air pollution damage. Elevated CO_2 has been demonstrated in several studies to reduce injury from such atmospheric pollutants as O_3, SO_2, NO, and NO_2. Allen (1990) calculated that the reduction of stomatal conductance by doubled CO_2 could reduce the harmful yield effects of ambient O_3 and SO_2 by 15%.

Weed Competition.   Weeds compete with crop plants for resources, thereby hindering productivity (see Chapter 4). As mentioned, C3 weeds may become more aggressive, especially in fields of C4 crops, due to their higher photosynthetic response to elevated CO_2. Conversely, C4 weeds are likely to become less competitive in fields of C3 crops. This has been demonstrated experimentally (Patterson and Flint, 1990). Changes in weed growth will entail modification of weed-control practices by chemical or nonchemical (e.g., biological or mechanical) means.

Pests. CO_2 enrichment may modify insect–crop relations. Changes in carbon and nitrogen partitioning in crops grown under elevated CO_2 conditions may affect the nutritional quality and attractiveness of foliage toward various insects. Nonstructural carbohydrates (starches and sugars) in leaves generally concentrate under high CO_2, while relative protein content diminishes. Thus, food quality and digestibility of forage by herbivorous consumers decline as C:N ratio rises. Studies have shown that insect feeding rates may rise and that insects may be weakened by the lower protein levels of leaves grown in such conditions (Lincoln et al., 1984).

CO_2 and Major World Crops

To date, most of the high CO_2 experiments have been conducted on a relatively small number of major food and fiber crops, primarily soybean, maize, wheat, and cotton, with very little information available on the majority of other crops. Field experiments are especially lacking. Data regarding crops that are important in tropical and subtropical developing countries are underrepresented in the experimental literature. These crops include rice, sorghum, millet, and tuber crops. Extended research programs concentrating on key crops that are economically and/or nutritionally important are vital to address these lacunae.

Experimental Techniques

High CO_2 experiments have been carried out under various conditions: in greenhouses, controlled environment chambers, branch exposure chambers, leaf cuvettes, and continuous stirred reactors (Drake et al., 1985). CO_2 enrichment has long been known to spur plant growth under near-optimal conditions in greenhouses, where it is often used as a technique for enhancing crop production (Enoch and Kimball, 1986). It is difficult to extrapolate results from these settings to what may happen in open fields, however, particularly because of strikingly different aerodynamics (Conway and Pretty, 1991). Experimental conditions inside greenhouses and controlled environment chambers, in particular, cannot reproduce evaporative conditions realistically.

More realistic experiments have been done in field chambers (often made of plastic), in which CO_2 is pumped into walled, open-top enclosures set in the ground, leaving the plants otherwise exposed to the ambient environment. Such chambers have also been used in the study of CO_2–temperature interactions. However, open-top structures also modify the crop environment by increasing temperature and relative humidity, while limiting radiation, precipitation, and windspeed.

The most realistic set of experiments designed to date is known as the free-air CO_2 enrichment system (FACE) (Figure 3.6) (Hendrey, 1993). In this system, CO_2 is vented from a series of pipes that form a circle of up to 20 meters in diameter. A computer program based on instrumental monitoring of wind speed and direction controls the release of CO_2 into the circle. This system is the best currently available approach for gathering data to assess the impacts of elevated CO_2 on crops (although not CO_2–temperature interactions). Though much improved over other methods, it also creates a modified environment in the form of an "island" of high CO_2 that alters the water balance and energy transfer. The high CO_2 island setting is analogous to a patch of well-watered vegetation located in the midst of a dry region (the so-called oasis effect), where advection of sensible heat from a surrounding region boosts evaporation to greater amounts than would be generated by local net radiation.

Early experiments raised CO_2 levels during a part of the growing period of the crop rather than throughout the season. The sudden introduction and later withdrawal

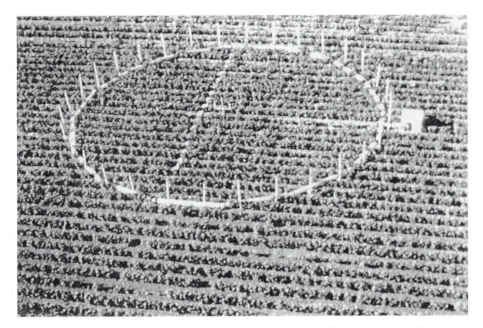

Figure 3.6   Free-air CO_2 enrichment (FACE) experiment, in Phoenix, Arizona (Rogers, personal communication).

of CO_2 enrichment constituted an abrupt "shock" to the plants. In more recent experiments, enrichment is applied over the entire lifetime of annual crops, a much more realistic approach. Even so, this is still not completely realistic, as in actual conditions the rising CO_2 levels will occur gradually, thus allowing adaptation to high CO_2 to take place over succeeding generations of plants.

Experimental Evidence and the Real World

Since the Industrial Revolution began in the late 1700s, CO_2 in the atmosphere has increased from about 280 to 360 ppmv. Based on the crop experiments just cited, these increases should have already augmented yields on the order of 7.5%. Has such a yield response actually occurred? While agronomists have conjectured that CO_2 has contributed to the overall increase in yields, such contribution is difficult to prove due to weather variability and to modifications of crop management (e.g., improved varieties, nutrient applications, better water management, and weed and disease control). Augmentation of nitrogen supply via acid rain may also have benefited crops. Therefore, the evidence of past increases in yield may not be a reliable indicator of future benefits from rising CO_2 levels. Instead, our projections rely primarily on enhanced CO_2 experiments in controlled environments and open-topped field chambers, CO_2 enrichment in greenhouses, and the limited data obtained to date from open-field experiments. Nonoptimal conditions (e.g., weeds and pests) encountered in farmers' fields might constrain the beneficial response seen in carefully tended experimental settings.

Jarvis and McNaughton (1986) have proposed that higher CO_2 levels may have more of an effect on transpiration in forests, which are more strongly "coupled" to atmospheric turbulence, than on field crops that are often only weakly coupled to the atmosphere. The conductance of the canopy boundary layer influences the degree of change in transpiration. In crop fields where canopies are not as strongly influenced by atmospheric turbulence, local vapor pressure deficit (VPD) may depend in part on canopy evaporation. A negative feedback then occurs as increased CO_2 reduces stomatal conductance, thereby reducing transpiration and leading to an increase in VPD and a consequent opening of the stomates in response. This hypothesis can only be proved by means of large-scale experiments on managed and natural vegetation types. The challenge here is to devise experiments so as to avoid interference from equipment on evaporative conditions.

Research Needs

There is still much uncertainty regarding the magnitude and significance of the improvements in yield and in water-use efficiency to be expected in the field under the combined influences of higher atmospheric CO_2 levels and a changing climate. The necessary testing of the interactive effects of atmospheric composition, temperature, water, nutrients, and light on various crops in field settings has only begun. Because it requires elaborate equipment and much time, such research is very expensive. Nonetheless, comprehensive experiments must be carried out to define the roles of short-term acclimation and long-term adaptation of crop processes to higher CO_2 concentrations. The responses of different genotypes of the same crop have not been adequately analyzed. Such analysis is needed to facilitate selection of crops and varieties for future environmental conditions. Aspects requiring further study include belowground processes (root systems and the soil in which they grow), pest interactions, carbon and nutrient allocation within the plant, and yield quality of the agricultural products grown under high CO_2. The few studies done to date suggest that yield quality (especially of grain crops) may be relatively unchanged by high CO_2, notwithstanding the modifications in C:N ratio noted earlier; yet the question remains open. Potential changes in water use on a regional scale are still unknown.

Future research may include the use of improved experimental techniques such as stable and short-lived isotopes, noninvasive exposure methods, and nuclear magnetic resonance (NMR). Interactions of CO_2 with other environmental factors such as tropospheric ozone and ultraviolet-B radiation (which are changing simultaneously with atmospheric CO_2) must also be studied in order to project future crop behavior realistically. Finally, research efforts are also needed to better utilize experimental data in dynamic-process crop models. Such models are often used to project high-CO_2 responses of crops at regional scales (see Chapter 7). Comprehensive meteorological, crop, and soil data are needed for validation of any hypothesis or model.

Temperature Responses

High-temperature stress is among the least understood of all plant processes. Soil and air temperatures often vary considerably and not necessarily in tandem, causing roots and shoots to grow in differing environments. The metabolism of crop plants is strongly affected by changes in these environmental temperatures. Specifically, temperature is known to affect morphology, partitioning of photosynthetic products, and the root:shoot ratio. Critical temperature parameters used empirically to describe effects on plant growth include mean, minimum, and maximum daily temperatures, as well as the cumulative heat load above a defined threshold during the growing period. Different physiological processes and stages of plant development have different temperature dependencies.

While considerable attention has been focused on the effects of higher atmospheric CO_2 levels on crops, less research has investigated crop responses to high temperature per se. Because the temperatures of crop canopies and roots fluctuate, field studies of high-temperature effects are difficult both to carry out and to interpret. Most data on crop and temperature relationships come from controlled-environment experiments, often without realistically varying day and night temperatures. In the field, climatic factors are difficult to separate, as higher temperatures are usually associated with higher radiation levels and enhanced water demand.

Global climate models can provide estimates of the number of days with temperatures that may exceed the optimal or the known threshold of tolerance for any particular crop. These may be used in the analysis of the potential impacts of high temperatures on crop yields.

Cardinal Temperatures and Plant Processes

Above a threshold minimum level, the response of a crop to a rise of temperature tends to be positive up to a characteristic optimum. The optimum value (actually a range of values) is usually considered to cover the temperature span at which plant growth or yield is within 10% of the maximum attainable when all other variables are also optimized (e.g., soil moisture, aeration, and nutrients, as well as aboveground light and ambient CO_2) (Fitter and Hay, 1987). When that optimal range is exceeded, however, crops tend to respond negatively, resulting in a steepening drop in net growth and yield (Figure 3.7). The optimal temperatures vary among different crops. For example, maize—which is of subtropical origin—has a higher temperature optimum for photosynthesis than wheat, which originated in a more temperate climate (Figure 3.8). Generally, crops can maintain the temperature of their foliage within their optimum range (through evaporative cooling) only if water supply is adequate.

Although the effect of rising temperatures on plant cells is to hasten metabolic activity in general, excessively high temperatures may cause enzymatic damage (Fitter and Hay, 1987). Up to a certain point, faster reaction rates are beneficial; beyond it, some plant processes are disrupted. The balance of the two effects determines the plant's overall response to increasing temperature. At lower ambient temperatures, the former tends to dominate, so the response to warming is generally favorable. At higher ambient temperatures, the latter effect dominates, so that a further warming produces a deleterious result. Above the optimum temperature range, denaturation of photoplasmic proteins, cellular enzymes, or membranes may cause rapid declines in or cessation of reaction rates and cytoplasmic streaming.

Respiration rates in temperate-region species tend to be low when temperature is under 20°C and to rise with temperature until a "compensation point" around 40°C. At that point, the rate of respiration equals the gross rate of photosynthesis so there is

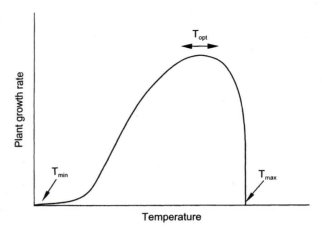

Figure 3.7 Schematic temperature response curve for plant growth (after Pisek et al., 1973).

no net carbon accumulation and, hence, generally no growth. When carbon consumption in respiration is subtracted from the gross carbon accumulation of photosynthesis, we obtain net photosynthesis. That value plotted against temperature is shown in Figure 3.9 (Fitter and Hay, 1987). Accelerated respiration reduces net accumulation of biomass, and this may in turn reduce economic yields. An example given by Paulsen (1994) showed that wheat kernels suffered increased respiratory loss when they were grown at high temperature. Such changes in respiration rate are likely to negate part of the advantage of an enhanced rate of photosynthesis in a CO_2-enriched and warmer environment. It is possible, however, that crops may adapt to warmer conditions gradually without a significant rise in respiration rate.

Vulnerability of crops to damage by high temperatures varies with development stage, but all stages of vegetative and reproductive development are affected to some extent. High temperatures during reproductive development are particularly injurious—for example, to maize at tasseling, to soybean at flowering, and to wheat at grain filling. Sterility may be induced in many species by high temperatures if such occur immediately before and during anthesis, although wheat appears to less sensitive than other species (Paulsen, 1994). In rice, the reproductive processes that occur within 1

Figure 3.8 The cardinal temperatures of net photosynthesis in members of the Gramineae from contrasting environments. (a) *Chionochloa* spp. tussock grasses (alpine, C3 photosynthesis), (b) wheat (temperate, C3), and (c) maize (subtropical, C4). The horizontal bars indicate the optimum range (Wardlaw, 1979).

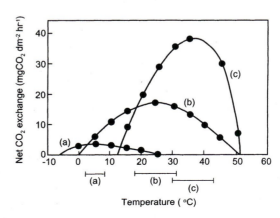

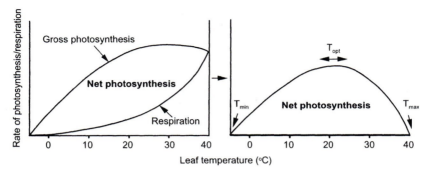

Figure 3.9   Schematic representation of the influence of temperature on plant photosynthesis and respiration in a typical plant (Pisek et al., 1973).

hour after anthesis—dehiscence of the anther, shedding of pollen, germination of pollen grains, and elongation of pollen tubes—are disrupted above 35°C (Paulsen, 1994).

Low yields are attributed to low assimilation rates, accelerated plant development, and diminished diurnal temperature range. High temperatures can initiate senescence and affect the functional properties of wheat grain for breadmaking. The negative effects of higher temperatures on crop yield are likely to be greatest in warm locations where antecedent temperatures were already close to the optimal for principal crops. If the mean warming reaches the upper end of the predicted range for doubled CO_2 concentration (i.e., a temperature rise of ~4°C), then developing heat-tolerant varieties of major crops will become a vital task for crop breeders.

Growing Season and Development

One consequence of the induced warming that will clearly benefit agriculture in the mid and higher latitudes is the lengthening of the potential growing season, usually defined as the period from the last frost in spring to the first frost in autumn. A longer potential growing season will allow earlier planting of crops in spring, hastened growth, and earlier maturation and harvesting. Consequently, multiple cropping (i.e., the planting of two or more crops in succession during the same season) may become possible. Multiple cropping is contingent on having sufficient water for two entire crops. Perennial crops such as alfalfa would also benefit from an extended growing season.

Temperature is the major environmental factor controlling crop development. Higher temperatures, in general, accelerate the phenology of plants, resulting in quicker maturation, except for extremely (i.e., excessively) hot conditions (Ellis et al., 1990). The shortened growth cycle, in turn, may reduce the yield potential of annual crops (Rosenzweig, 1990; Butterfield and Morison, 1992). In this connection, the duration of the vegetative growth stage and the integral of light interception during the reproductive stage are among the main determinants of total dry matter and yield (Monteith, 1981). Early plant growth in warm environments strongly influences grain yield potential at the end of the season (Paulsen, 1994).

Duration of all development stages is reduced by high temperature, but the period from double ridge to anthesis is reduced most and has the greatest impact on number of grains per spike and grain yield; higher temperatures frequently reduce the final leaf and spikelet numbers due to faster development (Paulsen, 1994). While the optimum temperature for photosynthesis in wheat ranges between 20 and 30°C, the optimum for grain yield is only 15°C. This difference highlights the importance of both duration and photosynthetic activity for maximum grain yield of temperate species (Paulsen, 1994). High temperature also reduces the number and weight of harvestable organs. Figure 3.10 shows the contrasting effects of high and low temperatures on crop development and yield (Acock and Acock, 1993).

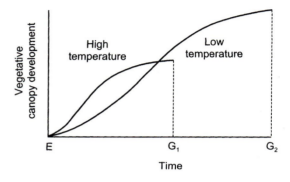

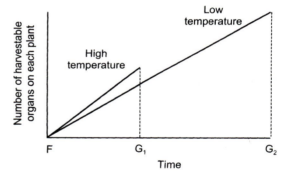

Figure 3.10 Schematic illustration of low and high temperature effects on crop development and yield through time. E = emergence; F = initiation of plant organs to be harvested; G = start of dry matter accumulation in plant organs to be harvested; H = end of dry matter accumulation in plant organs to be harvested. (Acock and Acock, 1993).

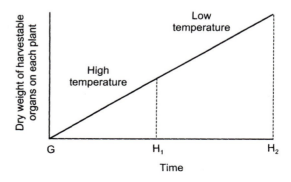

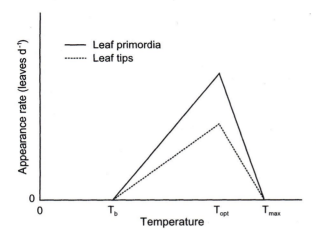

Figure 3.11 Idealized leaf-development rate response to temperature. The three cardinal temperature variables are base temperature (T_b), optimal temperature (T_{opt}), and maximum temperature (T_{max}) (Kiniry et al. 1991).

Many crop phenological and growth processes—including seed germination, leaf appearance and extension, and spikelet initiation—proceed in direct relation to cumulative heat, that is, the "thermal time" experienced by the crop (Figure 3.11). Below a base temperature, no effective thermal time accumulates and crop development cannot proceed. The rate of thermal time accumulation and the crop development rate rise with increasing temperature up to an optimum temperature value or range of values (plateau). Above that temperature value or plateau, the rate of thermal time accumulation and the crop response decrease with further increases in temperature until no further accumulation occurs and crop development ceases.

Agronomists have developed various linear and nonlinear indices to calculate the accumulated thermal time that a crop experiences. The thermal indices are generally characterized in terms of "growing degree days" based on the cumulative daily or monthly maximum and minimum temperatures.[4] Some of these have been modified to include the effects of daylength, solar radiation, and water stress. These indices may be used to calculate the expectable effects of higher growing season temperature on crop development, yield, and geographic region.

In a warmer climate, the greater and earlier accumulation of thermal units may induce shifts in the optimal zonation of crops (Rosenzweig, 1985). The resulting poleward shift of crop growing and timber growing could expand the potential production areas for some countries (notably Canada and Russia), though yields may be lower where the new lands brought into production consist of poorer soils. Other constraints on shifts in crop zonation include availability of water and technology, willingness of farmers to change crops, and sufficiency of product demand (i.e., marketing opportunities). Higher temperatures in mountainous regions will allow crop growth at higher altitudes, though soil resources also tend to be poor in such locations.

Minimum and Maximum Temperatures

Greenhouse theory, climate modeling results, and observations suggest that daily temperature minima may rise under greenhouse warming more than daily maximum temperatures (Karl et al., 1991), though this effect may diminish with time (Hansen et al., 1995). The reason for the expected change in the amplitude of the diurnal tempera-ture range is that the additional CO_2, water vapor in the atmosphere, and possibly clouds, would reduce the radiative heat loss from the ground at night while leaving unaffected the absorption of solar radiation during the day. Similarly, warming is pre-dicted to be greater in winter than in summer.

These differential effects can alter such phenomena as winterkill, vernalization, nighttime respiration rates, and high-temperature stress. High nighttime temperatures can suppress carbohydrate supply to reproductive organs (Hall and Allen, 1993) and, hence, interfere with the development of flowers, fruits, and seeds.

In certain crop plants, tissues are "hardened" by exposure to cool temperatures in the autumn, thus becoming more resistant to freezing injury in winter. While warmer winters should lead to fewer days with killing low temperature, warmer autumns may reduce winter hardening and thus increase crop vulnerability to occasional cold spells. Some studies have shown that winterkill may actually increase in certain locations with warming scenarios (Mearns et al., 1992).

In some crops derived from winter grasses (e.g., winter wheat), full flowering does not occur unless the plant experiences a period of cold temperature. This process is known as *vernalization*. As winter temperatures rise, vernalization may not take place. Farmers may adapt to this change by switching from winter to spring wheat types that do not require vernalization (Rosenzweig, 1985). This may entail changes in the qual-ity, processing, and marketing of wheat.

Warmer temperatures at night are likely to increase respiration and thus cause the partial depletion of plant reserves of carbohydrates and the lowering of yields. This ef-fect currently limits yields in some tropical agricultural settings.

In a simulation study of the effects of changes in minimum and maximum tem-peratures on wheat yields in the central United States, Rosenzweig and Tubiello (1996) found that negative effects on wheat yields are reduced when minima increase more than maxima. Responses varied, however, across a north-to-south transect, with larger negative effects occurring at the southernmost site studied due to declines in vernalization (Figure 3.12). A recent multiple-regression analysis of maize yields in the southern United States over the last 50 years highlighted the importance of asym-metries in minimum/maximum temperature changes in determining interannual yield variations (Stooksbury and Michaels, 1994).

High-Temperature Stress

High-temperature injury commonly can reduce productivity in crops grown in tropi-cal and temperate regimes. Such injuries, however, are difficult to measure and are often not perceived, even though high temperatures in many agricultural regions fre-quently exceed the temperature optima for crops. For example, wheat has an optimum temperature for grain growth of 15°C, but it is widely grown across the U.S. Great Plains where daily temperature maxima during the grain-filling period routinely ex-ceed 25°C (Paulsen, 1994).

Plants dissipate excess heat by emitting longwave radiation, convecting heat en-ergy, and releasing latent heat by transpiration (Gates, 1980) (Figure 3.13). When

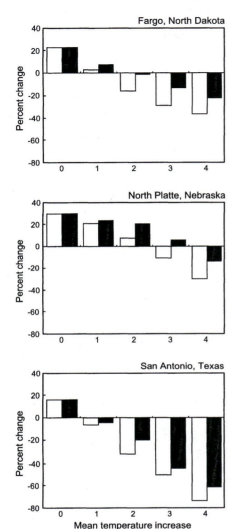

Figure 3.12 Simulated yield response to temperature change and elevated CO_2 concentration (550 ppm CO_2) for a transect of sites in the central United States. The percentage change is relative to baseline-simulated yields (1951–1980 observed weather, 330 ppm CO_2). Scenario A: Daily temperature minima and maxima change equally. Scenario B: Minima rise 3× maxima. (Rosenzweig and Tubiello, 1996).

plants are subjected to drought stress, their stomates close. Such closure reduces transpiration and, consequently, raises plant temperatures. Thus water stress and heat stress often occur simultaneously, the one contributing to the other. They are often accompanied by high solar irradiance and high winds as well. Hot dry winds such as the *sirocco* of North Africa or the *khamsin* of the Middle East exacerbate damage from high temperatures.

Air temperatures between 45 and 55°C that occur for at least 30 minutes damage plant leaves in most environments; even lower temperatures (35 to 40°C) can be damaging if they persist longer (Fitter and Hay, 1987). Plant temperatures above 40°C are

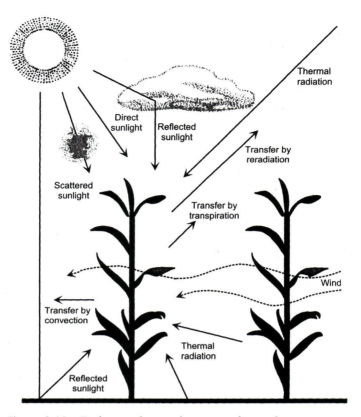

Figure 3.13 Exchange of energy between a plant and its environment by radiation, convection, and transpiration (redrawn from Gates, 1980).

associated with stomatal closure. Soybean is known for its ability to recover from short-term heat stress (Shibles et al., 1975). The effects of high temperature on the major agronomic crops are shown in Table 3.4 (Acock and Acock, 1993).

Photosynthesis appears to be more sensitive to temperature than is respiration. The Photosystem II (PSII) reaction center in the thylakoid membranes of leaf chloroplasts is highly sensitive, reacting rapidly to combined high temperature and high solar irradiance stresses, especially during reproductive growth (Paulsen, 1994).

Root temperature may be even more critical than aboveground temperature for crops because roots have lower temperature optima and are less adapted to extreme and/or sudden fluctuations (Paulsen, 1994). Root sink strength declines relative to other physiological sinks, at the same time that respiratory rates increase, until metabolic substrates are depleted (Paulsen, 1994). Thus, root growth and function are strongly inhibited when the plant is subject to heat stress. Temperature affects uptake of nutrients and water, as well as translocation of carbohydrates in roots. Prolonged exposure to high temperatures eventually stops root growth entirely.

Table 3.4 High temperature effects on growth stages of major crops (from Acock and Acock, 1993)

Crop	Effects
Wheat	Temperature >30°C for >8 h, can reverse vernalization
Rice	Temperature >35°C for >1 h at anthesis causes spikelet sterility
Maize	Temperature >36°C causes pollen to lose viability
Soybean	Great ability to recover from stress. No especially critical period in its development
Potato	Temperature >20°C depresses tuber initiation and bulking
Cotton	Temperature >40°C for >6 h causes bolls to abort

High-Temperature Tolerance

There appears to be a wide range of resistance to high-temperature stress both within and among crop species. This suggests a potential for genetic improvement of the trait. To promote adaptation to an environment of high CO_2 and high temperature, plant breeders may select cultivars that exhibit heat tolerance during reproductive development, high harvest index, high photosynthetic capacity per unit leaf area, small leaves, and low leaf area per unit ground area (to reduce heat load) (Hall and Allen, 1993). Selecting varieties characterized by high kernel number and large kernels may also be effective. Genes for heat tolerance should be tested for their performance over a range of increasing temperatures, although the heritability of resistance to high temperatures has not been clearly ascertained. So-called indeterminant species that flower repeatedly during the season may be able to recover from heat spells and to take advantage of longer growing seasons (Hall and Allen, 1993). According to Fitter and Hay (1987), physiological adaptation to high temperatures appears to be related, at least in part, to changes in the lipid composition of chloroplast membranes that tolerate higher temperature before disruption of the electron transport systems involved in photosythesis.

Sudden exposure to high temperature induces the creation of "heat shock" proteins, a subject of active experimentation at the current time. Such proteins may help crops to acquire tolerance to temperature stress, maintain cell integrity, prevent protein denaturation, and protect the PSII center, but their exact role is still unknown (Paulsen, 1994). Hardening[5] of plants to high temperature is highly complex, involving acclimation of photosynthetic processes and stability of mitochondrial electron transport.

Hydrological Regimes

Precipitation, being the primary supplier of soil moisture, is probably the most important factor determining the productivity of crops. While global climate models predict an overall increase in mean global precipitation, their results also show the potential for changed hydrological regimes (either drier or wetter) almost everywhere. A change in climate can cause changes in total seasonal precipitation, its within-season pattern, and its between-season variability. For crop productivity, a change in the pattern of precipitation events may be even more important than a change in the annual total per se. The water regime of crops is also vulnerable to a rise in the daily rate and seasonal pattern of potential evapotranspiration.

Precipitation and Evaporative Demand

In principle, a warmer climate can be expected to induce a greater evaporative demand. That demand is commonly characterized in terms of a region's potential evapotranspiration. A well-watered plant community is generally able to meet the climatically imposed evaporative demand, whereas a community insufficiently watered is likely to suffer moisture stress and curtailed growth. Both the intensity and duration of water stress are critical for crop growth response.

In the global hydrological cycle, water evaporated must be precipitated; hence, more evaporation implies more rainfall overall. However, the concurrent increases in potential evapotranspiration and in rainfall may not be commensurate in all locations. Climate model results suggest that potential evapotranspiration tends to increase most where the temperature is already high (i.e., in low to mid latitudes), while precipitation increases most where the air is cooler and more readily saturated by the additional moisture (i.e., in higher latitudes) (Rind et al., 1990). Thus, drier conditions may occur in many of the world's most important agricultural regions. This effect may be compensated only in part by the closure of stomates due to higher levels of CO_2, as described earlier in this chapter.

Where the rise in precipitation exceeds the rise in potential evapotranspiration, improved soil moisture conditions will promote growth, greater density of vegetative cover, protection of the land surface against erosion, and increased agricultural production under rainfed conditions.

Effects of Water Stress on Crop Growth and Yield

Water stress in plants is associated with reduced energy potential and activity of cellular water, lower cell turgor pressure, increasing concentration of solutes, shrinking of cell volume, and diminished hydration of tissues (Hale and Orcutt, 1987). Consequently, cell expansion and division, cell wall formation, protein and chlorophyll synthesis, and photosynthesis are all inhibited. When severe stress occurs, the respiration rate rises, sugars and proline accumulate, and metabolism is disrupted (Fitter and Hay, 1987).

As water stress develops, plants tend to lower the osmotic potential of their cells, a process of adjustment that helps maintain turgor. Osmotic adjustment allows cell enlargement and growth to continue at water potentials that would otherwise be inhibitory (Kramer, 1983). Characteristics that help maintain turgor include the ability to lower osmotic potential, a capacity to accumulate solutes, and elasticity of cells and tissues (Hale and Orcutt, 1987). In the initial stages of drought stress, turgor may be maintained by osmotic adjustment; if the stress persists, however, plants lose the capacity to adjust.

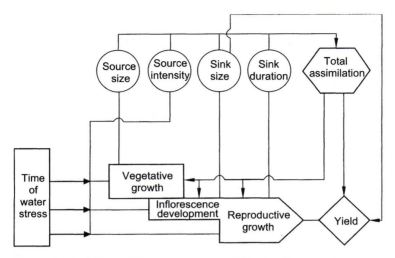

Figure 3.14 Effects of drought stress on yield by development stage (Hsaio et al., 1976).

Crop yields are most likely to suffer if dry periods occur during critical development stages such as reproduction (Figure 3.14). Water stress in rapidly transpiring leaves during reproductive development may draw water out of fruits and grains. Drought hastens the senescence of older leaves and induces premature abscission. Moisture stress during the flowering, pollination, and grain-filling stages is specially harmful to maize, soybean, wheat, and sorghum. Reduction of wheat yield due to water stress during crop development is shown in Figure 3.15. In wheat, leaf elongation, apex elongation, and spikelet initiation are all affected.

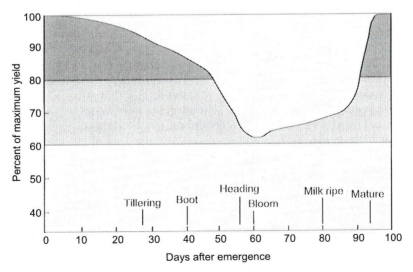

Figure 3.15 Effect of water stress at various growth stages on wheat yield (Bauer, 1972).

Drought Resistance

Breeding crop varieties with greater resistance to drought is a useful strategy for today's conditions—and all the more so for the possible future "greenhouse" climate. Different characteristics or combinations of traits may be selected, to enhance either dehydration avoidance (or postponement) or tolerance (Kramer, 1983). Dehydration may be avoided or postponed by morphological or physiological characteristics. The ability to roll their leaves when subjected to water stress helps plants reduce leaf area and thus the amount of surface exposed to solar radiation. Large, well-developed root systems with low resistance to water uptake increase the volume of stored soil moisture available to crops and serve to maintain water supply during periods of drought (Hale and Orcutt, 1987). A high level of osmotic adjustment is an example of dehydration tolerance (Kramer, 1983). Breeding programs must differentiate between cultivars that yield well only in favorable environments, those that yield well only in unfavorable environments, and those that yield well under a range of environmental conditions.

Management practices offer many strategies for growing crops in water-scarce conditions. One is drought escape, which can be achieved, for example, by the early planting of cultivars with rapid rates of development. However, seedlings may fail to germinate and late frosts may be damaging in these conditions. Fallowing and weed control help to conserve moisture in the soil. Another management strategy for responding to drought, of course, is the efficient use of irrigation, described in Chapter 6.

Effects of Excess Precipitation

Climate change is predicted to bring higher precipitation to some regions, a situation that may ameliorate crop-growing conditions in some cases and worsen them in others. Too much rain can cause leaching and waterlogging of agricultural soils, impeded aeration, crop lodging, and increased pest infestations. Flooded soil creates an anaerobic environment from which most crops cannot obtain the oxygen needed for root respiration. Flooding injury is characterized by yellowing leaves from the base to the top of the crop, drooping of petioles while the plant is still turgid, leaf epinasty (a condition in which the upper parts of an organ or part grow more rapidly, causing them to bend downward), hypertrophy (swelling of cells), new root formation from stems, and wilting under severe flooding conditions (Hale and Orcutt, 1987). Excess soil moisture in humid areas can also inhibit field operations and exacerbate soil erosion. High precipitation may prohibit the growing of certain crops, such as wheat, that are particularly prone to lodging and susceptible to insects and diseases (especially fungal diseases) under rainy conditions.[6]

Climatic Variability

Extreme meteorological events—such as brief spells of high temperature, torrential storms, or droughts—can have strongly detrimental effects on crop yields. Considerations of the potential impacts of climate change on agriculture should, therefore, be based not only on the mean values of expectable climatic parameters but also on the probability, frequency, and severity of possible extreme events (Katz and Brown, 1992; Mearns et al., 1992).

The relationship between changes in mean temperature and the corresponding changes in the probabilities of extreme temperature occurrences tends to be nonlinear: relatively small changes in mean temperature can result in disproportionately large changes in the frequency of extreme events (Mearns et al., 1984). Even if the variance of maximum daily temperatures remains unchanged, the probability of strings of successive days with high temperatures (i.e., of prolonged hot spells) increases substantially. In regions where crops are grown under conditions that are near their maximum temperature tolerance limits, such hot spells can, in effect, push a crop "over the brink" and thus have a significantly deleterious effect on crop yields. If temperature variability increases, crops growing at both low and high mean temperatures could be adversely affected since diurnal and seasonal canopy temperature fluctuations often exceed the optimum range. If temperature variability diminishes, however, crops growing near their optimum ranges might benefit. In a modeling study of wheat production in Kansas, Mearns et al. (1995) found that increases in daily temperature variability can reduce wheat yields due to lack of cold hardening and to subsequent winterkill.[7]

Interannual variability of precipitation is a major cause of variation in crop yields. During the 1930s, severe droughts reduced U.S. Great Plains yields of wheat and maize by as much as 50%. A study testing the possible impacts on Great Plains wheat production of a recurrence of the 1930s drought suggested that such a drought could still inflict widespread, heavy damage, even with today's agricultural management and technology (Warrick, 1984). Failure of the monsoon in 1987 caused yield shortfalls in Pakistan, Bangladesh, and India. By reducing vegetative cover, droughts exacerbate wind and water erosion, thus affecting future crop productivity. Excessively wet years, on the other hand, may cause yield declines due to waterlogging and lodging.

The relative increase in the probability of drought may be greater than the relative change in rainfall amount (Waggoner, 1986). This is especially so because a decrease in rainfall is generally accompanied by an increase in potential evapotranspiration, thus subjecting crops to the double jeopardy of a greater demand imposed on a reduced supply. Hence, the relative increase in the probability of low yield (or of crop failure) tends to be greater than the relative decrease in rainfall.

Precipitation events can vary in frequency, intensity, and persistence. Thus, conducting sensitivity tests to analyze potential impacts of changed precipitation variability is a complex task. Mearns et al. (1995) found that changes in daily precipitation variability resulted in substantial changes in mean and variability of simulated wheat yield. At Topeka, Kansas, where soil moisture is not limiting, mean yield decreased with increasing precipitation variability; at Goodland, Kansas, where soil moisture is limiting, mean yields increased.

Drought conditions may also be brought on by lower amounts of precipitation falling as snow and earlier snowmelt. In arid regions, such as the Sacramento River basin, these effects may reduce subsequent river discharge and irrigation water supplies during the growing season (Gleick, 1987).

Increased rainfall, caused by stronger atmospheric convection cells and greater humidity, may result in more intense (though not necessarily more frequent) rainfall.

Intense bursts of rainfall may cause damage to young plants, lodging of standing crops with ripening grain, and soil erosion.

CO_2, Climate Change, and Crop Yields: An Assessment

Changes in CO_2, temperature, and precipitation will alter the environment in which crops grow. The physiological effects of higher carbon dioxide on crop plants may well benefit agriculture. The enhancement of photosynthesis and water-use efficiency has been shown to occur in both optimal and resource-limited experimental growing conditions at least in the short term. However, agricultural yields in farmers' fields may not necessarily improve, especially where crop growth is limited by competition and lack of nutrients.

More effort is needed to investigate and assess the responses of crops grown in the developing world, in order to extend our knowledge beyond the few temperate crop species that are reasonably well studied. In a high-CO_2 world, enhanced synthesis of protein may require more nitrogen inputs to realize the full potential of CO_2 enrichment. Poorer farmers in developing countries may be at a disadvantage unless fertilizers become more readily available.

Physiological research on the responses of plants to changing levels of key factors of production is complicated by the fact that these variables are likely to change simultaneously. Previous experiments and models in which single factors were changed while holding all else constant are incomplete and their results may not apply to the new circumstances. To project impacts realistically, much more research is needed on the interactions of CO_2 and high temperature. We also need to improve our understanding of the interactions of CO_2, temperature, and precipitation with other environmental factors, such as tropospheric ozone, acid rain, and ultraviolet-B radiation.

4

Effects on Weeds, Insects, and Diseases

Climate change and rising atmospheric CO_2 will affect not just agricultural crops but their associated pests as well. The distribution and proliferation of weeds, fungi, and insects is determined to a large extent by climate. Such organisms become pests when they compete with or prey upon crop plants to an extent that reduces productivity.

Much more research has been done on potential changes in weed growth than on changes in the spread of insects and diseases. Like crops, weeds are primary producers and will be directly affected by both changes in climate and in CO_2 levels. Insects and diseases will probably not be affected directly by CO_2 changes, but may well be affected indirectly because altered host plant metabolism, development, and morphology will lead to changes in plant–pest interactions. The expected changes may result in some new, previously unobserved combinations of climate, atmospheric constituents, soil conditions—and, therefore, in new infestations of various pests. This chapter summarizes the research that has been done to date on potential changes in crop-pest interactions due to predicted changes of CO_2 concentration and climate.

Weeds

Competition, the struggle among plants for limited resources, occurs both at the individual plant level within species (intraspecific) and between species (interspecific). The latter affects the array and distribution of species that share the same habitat. Competition tends to be greatest when plants are crowded in habitats with otherwise adequate nutrients and soil conditions; when growth conditions are poor and highly disturbed, plant density may be too low for competition to occur (Patterson and Flint, 1990). In crop monocultures, the optimum planting density is that which maximizes final yield for the crop stand, though not necessarily for the individual plant.

But crops in farmers' fields rarely grow solely with intraspecific competition—that is, in true monocultures. Because weed seeds are ever present in agricultural soils, weeds grow repeatedly and must be controlled, lest they deprive crop plants of essential resources and lower yields below potential. From the standpoint of crop growers, the classic definition of a weed is "a plant out of place" (i.e., one that grows spontaneously in a cropped field). A weed in a crop field shares the same trophic level and competes with the crop for resources (water, nutrients, and light), thus reducing their availability to the crop and limiting its growth and productivity.

Management practices often promote competition between weeds and crops, albeit unintentionally, as soil preparation and field operations create regular disturbances and synchronize the germination and establishment of weeds together with crop plants. Fertilization and irrigation benefit weeds as well as crops. Most weeds are adapted to take advantage of these conditions by growing rapidly in environments that were recently disturbed and that are rich in nutrients (Patterson, 1993). These competitive interactions are likely to be altered by changing climatic and atmospheric CO_2 conditions.

In addition to direct competition, damage due to weeds includes interference with harvesting and the lowering of yield quality by contamination. The geographic dispersion of weed species is promoted by the widespread dissemination of seeds and other agricultural products.

Worldwide, weeds have been estimated to cause crop production losses of about 12% (Parker and Fryer, 1975). In the United States, annual losses in crop production due to weeds have been valued at approximately $12 billion, amounting to some 10% of potential production (Patterson and Flint, 1990). Losses may be even higher in subsistence agricultural systems — on the order of 25% (Parker and Fryer, 1975).

Massive efforts are being made to control weeds. More human labor is expended in hand weeding than in any other agricultural task, and most cultivation and tillage practices are designed to aid in weed control. A huge chemical industry manufactures herbicides, which next to fertilizers account for the largest volume of chemicals applied to crops (Furtick, 1978). In the United States, over $6 billion are spent on weed control every year (Patterson and Flint, 1990).

Differential Responses to Elevated CO_2

Experiments have tested the effects of higher atmospheric CO_2 on weeds with C3 and C4 photosynthetic pathways. While there is large variability in response and considerable overlap, generally the response of C3 weeds to high CO_2 tends to be greater than that of C4 weeds (Table 4.1) (Patterson, 1993). Variation in response to CO_2 can be high both within and among species, depending on experimental conditions (such as temperature, light, and the availability of water and nutrients) and on innate biochemical factors.

Most experimental work to date has focused on the interspecific competition among plants with the C3 and C4 pathways. The typical finding of such experiments — namely that C3 plants tend to respond more favorably than C4 plants to elevated CO_2 — has in general been replicated in competition experiments with weed species. In controlled environment studies with plants grown in pots under high CO_2 conditions, C3 species generally produced greater biomass increments than did C4 species, both individually and with even greater differences in performance when

Table 4.1 Effects of doubled atmospheric CO_2 concentration on biomass and leaf area of C3 and C4 weeds (based on literature review by Patterson, 1993)

Species	Range of response ($\times$ growth at ambient)		Species	Range of response ($\times$ growth at ambient)	
	Biomass	Leaf area		Biomass	Leaf area
C3 species			*C4 species*		
Abutilon theophrasti	1.00–1.52	0.87–1.17	*Amaranthus retroflexus*	0.96–1.41	0.94–1.25
Agropyron smithii	1.31	0.96			
Ambrosia artemissifolia	1.10–1.33	—	*Andropogon virginicus*	0.81–1.17	0.88–1.29
Anoda cristata	1.40	0.90	*Cyperus rotundus*	1.02	0.92
Arrhenatherum elatius	1.18	—	*Digitaria ciliaris*	1.06–1.61	1.04–1.66
Aster pilosus	0.95–1.20	0.81–1.15	*Echinochloa crus-galli*	0.95–1.59	0.95–1.77
Brachypodium pinnatum	1.00	—	*Eleusine indica*	1.02–1.21	0.95–1.32
			Paspalum plicatum	1.08	1.02
Bromus mollis	1.37	1.04	*Rottboellia cochin-chinensis*	1.21	1.13
Bromus sterilis	1.04	—			
Bromus tectorum	1.54	1.46	*Setaria faberii*	0.93–1.35	1.01–1.40
Cassia obtusifolia	1.38–1.60	1.04–1.34	*Setaria lutescens*	1.59	1.48
Chenopodium album	1.00–1.55	1.22	*Sorghum halepense*	0.56–1.10	0.99–1.30
Cirsium arvense	1.21	0.92			
Crotalaria spectabilis	1.67	1.54			
Dactylis glomerata	1.24	—			
Datura stramonium	1.74–2.72	1.46			
Deschampsia fexuosa	1.22	—			
Digitalis purpurea	1.16	—			
Elytrigia repens	1.64	1.30			
Epilobium hirsutum	1.11	—			
Festuca ovina	1.23	—			
Festuca rubra	1.00	—			
Holcus lanatus	1.60	—			
Lolium perenne	1.34–1.43	—			
Oryzopsis hymenoides	1.10	0.68			
Phalaris aquatica	1.43	1.31			
Plantago lanceolata	1.00–1.33	1.33			
Plantago major	1.55	—			
Poa annua	1.00	—			
Poa trivialis	1.03	—			
Polygonum pensylvanicum	1.53	1.15			
Rumex acetosella	1.31	—			
Rumex crispus	1.18	0.96			
Urtica dioica	1.30	—			

grown in competition (Table 4.2) (Patterson and Flint, 1990). Under high CO_2, tall meadow fescue (*Festuca elatior* L.), a C3 grass, improved its competitive advantage over Johnsongrass (*Sorghum halepense*), a C4 grass (Carter and Peterson, 1983). Bazzaz and Garbutt (1988) demonstrated that, while C3 annuals depress the biomass of C4 species, the overall productivity of competing C3 and C4 species is not always improved due to the canceling out of both positive and negative responses to high CO_2. Furthermore, the responses of competing species at intermediate CO_2 levels (500 ppmv) might be different from responses at 700 ppmv CO_2, leading to different predictions about future associations of competing plants.

Elevated CO_2 also affects competition among species of the same photosynthetic pathway, but in ways that depend on individual species' responses and environmental

Table 4.2 Competition effects of increased atmospheric CO_2. (Based on literature review by Patterson and Flint, 1990)

Species	Photosynthetic pathway	CO_2 conc. (ppmv)	Species favored by high CO_2	Effect on total production	Other responses
Festuca elatior	C3	350, 600	C3	0	Increased competitiveness of C3 species; community productivity unchanged.
Sorghum halepense	C4				
Glycine max	C3	350, 675	C3	+	Increased competitiveness of C3, species; RYT unchanged.
Sorghum halepense	C4				
Abutilon theophrasti	C3	350, 500, 700	C4(growth); C3(P_s)	+	Increased competitiveness of C3 species; greater P_s enhancement in C3 but greater RGR increase in C4 due to changes in C allocation.
Amaranthus retroflexus	C4				
Lolium perenne	C3	300, 620	Variable	+	Community production up; *Trifolium* favored before mowing, but *Lolium* more competitive after mowing.
Trifolium repens	C3				

0 = no response; + = increase; P_s = photosynthesis; RYT = relative yield total; RGR = relative growth rate.

conditions—for example, light, water, nutrients, and management. In a 2-year study, two C3 pasture species, perennial ryegrass (*Lolium perenne* L.) and white clover (*Trifolium repens* L.) at first benefited greatly from CO_2 enrichment. In the longer term, overall benefits were lower. In a follow-up study, clover benefited more when the plants were not cut, while ryegrass benefited more when plants were cut (simulating mowing or grazing) four times a year (Overdieck et al., 1984; Overdieck and Reining, 1986).

Climate Effects on Weed Ecology and Distribution

Other manifestations of the predicted climate change may favor plants of the C4 pathway, in contrast to favoring the C3 pathway by increases in CO_2. Climate change may exert a strong influence on the geographical distribution of weeds as well as crops, so entire agroecosystems will tend to shift together. However, warmer temperatures and drier hydrological regimes, which may occur in continental interiors, tend to favor C4 plants. Since C4 plants evolved originally in semiarid, subtropical regions, they are fundamentally better adapted to such conditions. Experimental studies have shown that C4 weed species compete favorably with C3 species in conditions of increased temperature and water stress. Flint and Patterson (1983) observed that smooth pigweed (*Amaranthus hybridus* L.) (a common C4 weed that can reduce soybean yields in the southern U.S.A. by as much as 50%) is likely to cause substantial growth reduction especially in late-planted soybeans because it is more competitive at higher temperatures.

Many of the worst weeds in temperate regions originated from tropical or warm temperature regions, and in the current climate their distribution is limited by low temperature. Such geographical constraints will be obviated under warmer conditions. Some leguminous C3 weeds and C4 weeds respond favorably to elevated temperature and may therefore expand their geographical range (Figures 4.1 and 4.2). Sicklepod (*Cassia obtusifolia* L.), hemp sesbania [*Sesbania exaltata* (Raf.) Cory], and showy crotalaria (*Crotalaria spectabilis* Roth.) are Leguminosae weeds now restricted to the southeastern United States by low temperature that may expand their ranges (Flint et al., 1984). An example from the C4 weeds is cogongrass [*Imperata cylindrica* (L.) Beauv.], a particularly strong perennial weed now confined to the Mexican Gulf States (Patterson, 1993). Other examples include itchgrass [*Rottboellia cochinchinensis* (Lour.) W. Clayton] and witchweed [*Striga asiatica* (L.) Ktze.], important weeds infesting corn and soybean fields in the southern United States. If they were no longer constrained by low temperature, these weeds could become problematic in the Corn Belt (Patterson and Flint, 1990). The spread of kudzu [*Pueraria lobata* (Willd.) Ohwi] and Japanese honeysuckle (*Lonicera japonica* Thunb.) is also limited by low winter temperature. The ranges of these weeds could extend northward and westward by several hundred kilometers with a mean and minimum winter temperature warming of 3°C and with improved chilling tolerance resulting from higher CO_2 levels (Sasek and Strain, 1990). Some weeds, on the other hand, could lose their present competitive ability under the changed conditions.

While warmer temperatures may increase the relative competitiveness of C4 weeds over C3 crops in many cases, certain heat-tolerant C3 crops, such as cotton, may still do well. CO_2 enrichment has also been shown to increase the tolerance of

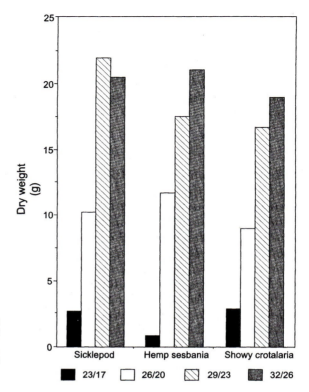

Figure 4.1   Effects of day/night temperature on dry weight of three leguminous weeds (Patterson, 1993 from data of Flint et al., 1984).

cotton to low temperature, thus enhancing its competitive ability at both ends of the temperature range (Patterson, 1993).

Weed/Crop Competition

Weed/crop competitive interactions can be expected to change in complex ways with the simultaneous rise of CO_2 and change of climate. Higher atmospheric CO_2 will stimulate the growth of both weeds and crops, thus negating some of the otherwise beneficial effects of CO_2 "fertilization" on crop yields. However, since higher CO_2 should preferentially favor plants with the C3 photosynthetic pathway and since a majority of major food crops are of the C3 pathway, whereas many of the worst weeds are C4 species, a net beneficial effect is likely to take place in many cases (Table 4.3) (USDA, 1972; Holm et al., 1977; Prescott-Allen and Prescott-Allen, 1990). Many weeds associated with the major crops having the C4 pathway (maize, sorghum, sugarcane, and millet) are C3 species; so here competition may indeed favor the weeds.

Apart from the specific effects of elevated CO_2, however, global warming and increasing aridity will tend to favor C4 crops and weeds. Since species responses to environmental factors also vary between C3 and C4 plants, experiments with both high CO_2 and increased temperature have shown mixed results in weed/crop combinations. Thus, the relative prevalence and vigor of different weed species will probably change and may, in turn, alter the composition of agroecosystems.

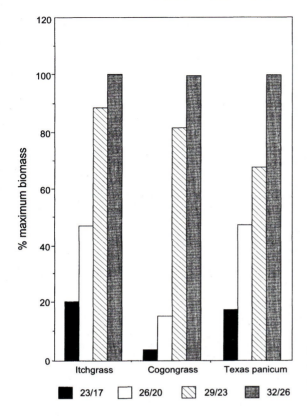

Figure 4.2 Effects of day/night temperature on dry weight of three grass weeds (Patterson, 1993).

Implications for Weed Management

In crop monocultures, undesirable competition is controlled through a variety of means, including crop rotations, mechanical manipulation (e.g., hoeing), and chemical treatment (e.g., herbicides). Selection for crop varieties that are adapted to changed weed dynamics and adjustment of management practices may reduce some of the potentially negative impacts caused by greater weed growth.

Climate change will certainly affect the timing of application and the efficacy of herbicides. Temperature, precipitation, wind, soil moisture, and humidity are known

Table 4.3 Biomass responses of C3 and C4 crops and weeds to doubled atmospheric CO_2 concentration (Patterson, 1993)

Category	Range of response ($\times$ growth at ambient)
C3 crops	1.10 to 2.43
C4 crops	0.98 to 1.24
C3 weeds	0.95 to 2.72
C4 weeds	0.56 to 1.61

to modify chemical toxicities affecting the metabolisms of both weeds and crops. Some predicted effects of climate change and high CO_2 on weed/crop interactions are speculative, including the possibility of reduced uptake of chemicals from soils due to the transpiration-inhibiting effects of CO_2, changes in the action of foliage-applied chemicals with changes in leaf anatomy and surface characteristics, and greater resistance to chemical control of perennial C3 weeds due to increased rhizome and tuber growth induced by high CO_2 (Patterson and Flint, 1990). Greater amounts of starch in plant leaves could possibly hinder herbicide or pesticide action, while higher temperatures could speed plant metabolism and therefore the uptake, translocation, and effectiveness of chemicals. Direct CO_2 effects will probably have the most influence on enhancing weed competitiveness, though that enhancement will probably be restrained by changes in foliar uptake, chemical activity, leaf structure, and starch content.

There is currently a growing interest in the biological (rather than chemical) control of weeds. For example, some plants produce compounds that are natural herbicides known as allelopathic chemicals, some of which are being replicated for commercial use. These and other techniques depend on complex interactions of biological agents and targets that may themselves change in a changing climate and higher atmospheric CO_2 concentrations.

Insects

An agroecosystem is a complex functioning community of numerous organisms operating on at least three trophic levels: crop, pest, and pest predator. These trophic levels interact to determine the severity of infestation of insect pests in the current climate. Humans intervene in agroecosystems in an effort to maximize productivity of the crop itself. Humans also intervene inadvertently by introducing exotic insect species to new environments and by extending the geographical range of host crops—for example, with the implementation of large-scale irrigation projects for production of export crops.

Agroecosystems have evolved over long periods of time and will continue to evolve under changing environmental conditions. Climate change will alter these interactions in multifarious ways. New interactions will result from shifts in the geographical ranges of crops, pests, and predators, and from changes in their development. The potential effect of climate modification on insect pests of various crops have been reviewed by Stinner et al. (1989), Cammel and Knight (1991), Porter et al. (1991), Sutherst et al. (1995), and the IPCC (1996).

Insect pests in agricultural systems are the second major cause of damage to yield quantity and quality, after weeds. Most analyses concur that in a changing climate with increasing levels of atmospheric CO_2, as projected for the enhanced greenhouse effect, insects may become even more active than they are currently, thus posing the threat of greater economic losses to farmers.

CO_2 Effects on Insects and Host Plants

Increasing atmospheric CO_2 at the levels predicted for the coming century will probably have little direct effect on insects. A higher level of CO_2 will more probably affect agricultural insect pests indirectly, through its effect on the host plants. Such indirect influences may be either enhancing or inhibiting. Bees and other pollinating insects may respond positively if more nectar is produced at flowering (Kimball, 1985), thus promoting fruiting.

On the other hand, as mentioned in Chapter 3, higher CO_2 will probably tend to increase the carbon:nitrogen (C:N) ratio in crop leaves, which stimulates the feeding of some insects. Lincoln et al. (1984) found that soybean looper (*Pseudoplusia includens*) larvae fed more on leaves from soybeans grown in higher CO_2 concentrations than from soybeans grown in ambient CO_2 levels. They hypothesized that this occurred because the larvae needed to eat more C-enriched leaves to gain adequate nutrition. Results of a recent study on leaf miners (*Pegomya nigritarsis* Zetterstedt (Diptera: Anthomyiidae) and their damage to two common species of dock (*Rumex crispus* L. and *Rumex obtusifolius* L.) were consistent with the hypothesis that insect herbivores compensate for increased C:N ratios by increased food consumption (Salt et al., 1995). C3 and C4 crops will be affected differently since C3 crops are more responsive to elevated CO_2, and thus C:N ratios should increase more in C3 crops; this means that insect feeding may become more damaging in these crops. Greater insect damage, in turn, may partially offset the predicted increases in biomass due to higher levels of atmospheric CO_2.

According to Stinner et al. (1989), the greater growth of crop plants due to high CO_2 will probably not affect insect populations per se because insects are rarely limited by food supply. Overall, the direct effects of rising CO_2 should be less important for insects than the concomitant climate effects (Kimball, 1985).

Climate Effects on Insect Pests

Climate strongly affects insect pests directly through abiotic processes and stresses (particularly high and low temperatures; excess and lack of precipitation, humidity, and wind speed) and indirectly through effects on food quantity and quality, as well as the abundance and activity of predators, parasites, and pathogens. Climate plays a determining role in creating insect habitats, and year-to-year climate and weather variability causes variations in insect populations. Thus, absolute changes in climate variables, the time rate at which climate changes occur, and the frequency and severity of extreme events—all can alter insect ecology. Organisms may adapt, change geographical distribution, or become extinct.

Insect habitats and survival strategies are strongly dependent on patterns of climate. Insect habitats are characterized as either stable or unstable. A stable habitat is long-lasting relative to an insect's life span, and engenders stable populations determined by density-dependent factors such as food supply. In an unstable habitat, the longevity of the habitat and an insect's life span are approximately equal. For an insect species to survive when the habitat changes, it must develop a strategy either to alter its life cycle or to migrate to another, more suitable, habitat. Such strategies often entail high mortality, high rates of reproduction to compensate for the high mortality, and, consequently, more destructive feeding rates. Agricultural systems generally constitute unstable habitats. That is one reason agricultural insect pests are characterized by high reproductive rates, mobility, and relative insensitivity to density-dependent constraints.

An environmental change that involves climatic instability will naturally tend to destabilize habitats and therefore to stimulate crop pests and migratory pest species. Thus, climate change is likely to exacerbate agricultural pest infestations. Evidence of fossil beetles from the late Quaternary period reveals that insect species are extremely mobile — capable of moving great distances to satisfy their ecological requirements in the face of changing climate (Elias, 1991).

Temperature Responses. Insects are particularly influenced by temperature because they are poikilothermic (cold-blooded), meaning that their body temperature follows ambient temperature. Many insects have relatively high thermal optima associated with their enzyme and membrane systems, so higher temperatures permit them to grow faster and larger, to suffer less mortality, and to lay more eggs per unit quantity of ingested food (Mattson and Haack, 1987). Insects respond to higher temperature with increased rates of development and, for species with life cycles consisting of more than one generation per year, with less time between generations. Warmer winters will reduce winterkill and therefore promote greater overwintering survival for both migratory and nonmigratory insect species. Consequently, there will be increased insect populations in the next growing season. High-latitude ranges will be extended and lower-latitude species will be able to migrate more freely. With warmer temperatures occurring sooner in the spring, pest populations will become established and thrive during earlier and more vulnerable crop growth stages. If climate becomes warmer and drier as well, the population growth rates of small, sap-feeding pests may be favored (Stinner et al., 1989).

Aphids are insects that cause direct damage to crops through sap feeding and indirect damage due to transmission of viruses and growth of fungi on their excretion. Mild winters allow aphids to survive — and continue reproduction — and to show greater and earlier spring migration. Warm spring weather further encourages migration because of temperature thresholds for flight. Unusually warm winter and spring conditions in England in 1988/1989 caused the year to be designated the "year of the aphid," with cereal aphid damage and widespread viral infections of grain and sugarbeet (Morison and Spence, 1989).

Since warmer temperature will bring longer growing seasons in temperate regions, this should provide opportunity for increased insect pest damage. Since insect pests and crops are linked ecologically, the effect of a lengthened growing season depends in part on whether crop breeders develop and farmers grow crops that can take advantage of the extended warm seasons. A longer growth period may allow additional generations of insect pests and higher insect populations. The Mexican bean beetle and bean leaf beetle, both major pests of soybeans, presently have two generations in the U.S. Midwest and three in the Southeast. An additional generation may be possible in the Midwest if the growing season there lengthens (Stinner et al., 1989).

Extended growing seasons will be important for species with the capability of multiple generations per year. Voltinism, the number of broods typically produced per year by a population of a particular insect species, depends both on thermoperiod (generally expressed in terms of accumulated daily temperatures) and photoperiod (daylength). Additional insect generations and greater populations encouraged by higher temperatures and longer growing seasons will require greater efforts at pest management. While the higher latitudes are predicted to become warmer, their photoperiod will, of course, remain the same. This contrast may lead to complex changes in pest–environment interactions. A longer growing season should not have much of an effect on insect species that produce only one generation per year.

In tropical regions, diurnal temperature variation separates daytime and night-time insect activity. If diurnal temperature ranges change, this separation may be affected and daily activity patterns and species composition could ultimately change (Porter et al., 1991).

Hydrological Responses. Precipitation—whether optimal, excessive, or insufficient—is a key variable that affects crop–pest interactions. Since regional changes in hydrological regimes are still uncertain, consequent changes in insect damages to crops are difficult to characterize for a given location. It is well known that drought changes the physiology of host crops, leading to changes in the insects that feed on them (i.e., phytophagous insects) (Figure 4.3) (Mattson and Haack, 1987). Because drought stress tends to bring increased insect pest outbreaks, insect damage may increase in regions destined to become more arid. However, severe and prolonged drought can be debilitating to insects, just as it can be to crop plants (Mattson and Haack, 1987). Abnormally cool, wet conditions can also bring on severe insect infestations, although excessive soil moisture may drown out soil-residing insects.

Plants normally contain nutrients desirable to insects, as well as defensive compounds that ward insects off. The balance between insect-attracting and insect-repelling compounds is affected by climate. For instance, drought tends to exacerbate insect damage because it changes plant nutrient composition. Drought tends to enhance the concentration of nutrients, especially nitrogen, in leaves and to reduce the plants' defensive systems, thus leading to improved conditions for pests and increased pest infestations. Drought-stressed plants are often yellower and warmer and exhibit greater infrared reflectance. Such traits evidently make plants more attractive to insects. Plant defense mechanisms may also be lowered. Genes conferring resistance of wheat to stem and leaf rusts and to the Hessian fly (*Mayetiola destructor*) appear to be sensitive to high temperature (Mattson and Haack, 1987). Drought may also enhance the ability of insects to detoxify plant compounds, induce genetic changes, and reduce their natural enemies. On the other hand, in some regions where snow cover is reduced, insect populations may decline because of increased winterkill.

Extreme Events. Heat spells, droughts, and persistent patterns of atmospheric circulation are often critical factors in pest infestations. There are complex interactions between the timing of insect life cycles and host crop development under climate extremes. These interactions will determine the intensity of pest outbreaks. While a single year of unusual weather may not bring a change in pest outbreaks, a run of extreme conditions, such as may occur in a warming climate, may trigger severe infestations. The widespread infestation of desert locust (*Schistocerca gregaria* (Forsk)) in Africa during 1986–1988 is attributed to a prolonged sequence of good rains and warm, favorable winds (Pedgley, 1989). By the end of that period, the desert locust reached new northern limits in Europe.

Ecological Mechanisms

Ecological mechanisms in insects that are sensitive to climate include reproduction and fecundity, phenology and life cycle duration, diapause (a period of retarded or sus-

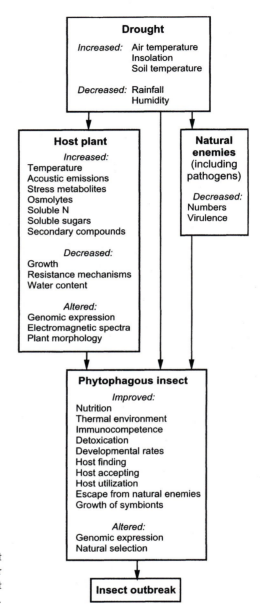

Figure 4.3   Drought influences on host plants, phytophagous insects, and their natural enemies, which lead to insect outbreaks (Mattson and Haack, 1987).

pended development) and overwintering, dispersal and speciation, population growth rate, and mortality. Indirect effects of changing climate and CO_2 on the synchrony of crop plant and pest development stages can also be important.

Reproduction and Fecundity.   Stressful climate conditions directly affect arthropod reproduction, fecundity, and hatching success. Studies have documented that temperature and humidity values that are either too high or too low can be harmful to reproductive processes (Stinner et al., 1989). An example of an important pest that is favored by warmer temperature is the spider mite (*Tetranychus* spp.), a pest of soybean, maize, wheat, sorghum, tomatoes, and apples. The severe drought of 1988 in the U.S. Midwest started early in the spring and continued throughout most of the summer, accompanied by higher than normal temperatures. Damaging outbreaks of two-spotted spider mites (*T. urticae*) occurred on soybeans throughout the entire region. The damage occurred during the critical flowering, pod development, and pod filling growth stages. An estimated 3.2 million ha were sprayed with insecticides to control the mites across the region and losses to Ohio farmers were estimated to be $15 to 20 million (Stinner et al., 1989).

Since insect fecundity also depends in part on how nutritious the food supply is, climate change and CO_2 effects on the nutrient status of the plant hosts are also likely to affect insect reproduction rates. Drought tends to concentrate nutrients in plant leaves and to encourage insect fecundity, whereas excess precipitation seems to deter reproduction (IPCC, 1995).

Phenology and Life Cycle Duration. Insect development rate depends on temperature and humidity, as well as on food quality and quantity. Insect species display high rates of development within optimum ranges of temperature and humidity, which vary with physiological stage and among species. Changes in temperature and hydrological regimes can either spur or repress development rates, depending on whether antecedent conditions were above or below the optimum for the insect species. Spring and growing season temperatures near latitudinal or altitudinal boundaries will rise with global warming and will allow earlier physiological development. This could hasten and shorten life cycles, thereby increasing insect populations. The occurrence of extreme climate events, however, will have different effects, depending on the phenological stage of insect development.

Smaller insect pests, such as aphids and mites, tend to be more vulnerable to adverse climate conditions than larger insects. They are more easily destroyed by either desiccation or heavy precipitation events. However, they also tend to have greater fecundity, shorter life cycles, and more generations per year; therefore, they may be able to shift genetically more readily in response to climate fluctuations (Stinner et al., 1989).

Diapause and Overwintering. Climate warming will affect agricultural insect pests especially in the winter (and perhaps during other seasons when crops are not growing) because low-temperature limits are important in determining the distribution of insect species. Diapause, being a dormant phase in an insect's life cycle, allows survival through periods of unfavorable climatic conditions. For example, many insects survive low winter temperatures while in a pupal stage in the soil. Diapause stages vary among and within species, and are evidently triggered by temperature and photoperiod.

In temperate latitudes, warmer temperatures in winter will promote overwintering by insect pests, raising the survival rate and contributing to greater infestations during the following seasons of both migratory and nonmigratory species. Nonmigratory pests normally overwinter in the same location as the crops they infest, but they may be held in check by severe cold temperature. For example, a cold spell in the U.S. Midwest in 1983–1984 evidently constrained the populations of both the Mexican bean beetle and the bean leaf beetle in soybean fields during the following summer (Stinner et al., 1989).

Warmer winter temperature will also affect those pests that currently cannot over-winter in high-latitude crop regions but do overwinter in lower-latitude regions and then migrate to the crops in the following spring and summer. For example, the potato leafhopper (*Empoasca fabae*)—a pest of soybeans, alfalfa, and other crops—may expand its overwintering range (now limited to a narrow band along the Gulf of Mexico) and thus be better positioned to travel to the U.S. Midwest earlier and in greater numbers during the cropping season (Figure 4.4). Such insect pests may also cause greater damage in their new overwintering areas.

Some species are pests in the America's South but not in the Midwest, because they do not migrate to the Midwest early enough or in significant numbers. Corn earworm [(*Heleliothis zea* (Hubner)] is an example of a current pest of corn and soybean in the South that is not a serious pest in field corn and soybean in the Midwest. With climate change, extension of overwintering range may bring the corn earworm to field corn and soybean crops in the U.S. Midwest (Stinner et al., 1989).

Since climate projections show that the rise of winter temperatures may exceed the rise of summer temperatures, at least in the first stages of climate warming, change in overwintering ability may be one of the most important impacts of climate change on agricultural insects. Greater overwintering and earlier spring activity will be especially important for pests in high latitudes.

Dispersal and Speciation.   Many insect species exhibit characteristic dispersal strategies in seeking new habitats to provide additional food sources and reproductive opportunities. These strategies often take advantage of climatic mechanisms (such as atmospheric circulations) to travel from older, depleted habitats to newer and richer ones. Warmer temperature and stressful environmental conditions are associated with

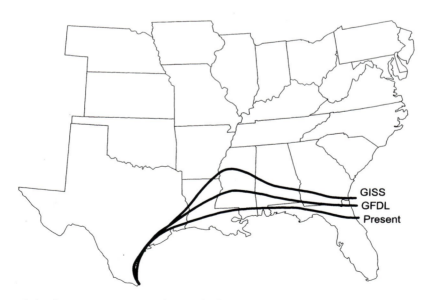

Figure 4.4   Overwintering range of potato leafhopper (*Empoasca fabae*) under current and two doubled-CO_2 climate change scenarios (Stinner et al., 1989).

greater movement of pests, both local and migratory. Large-scale climate change is projected to promote redistribution of pests and intermixing of gene pools.

Population Growth Rate.   Climate, the availability and quality of food, and the presence of natural enemies determine the favorableness of an environment for the proliferation of insect pests, and therefore the amount of economic damage sustained by farmers. Climate can affect the population growth rate of insects in at least six ways: initial population size (number of overwintering survivors), speed of development (time to reproductive maturity), per capita lifetime fecundity, survivorship of immature and adult stages, sex ratio, and dispersal rates (Stinner et al., 1989). Higher temperature, as long as it is below the lethal threshold, should hasten development and thus raise population growth rates of most pest species, especially where these are currently limited by early cool season temperatures in high-latitude locations. The ratio of male and female insects in a population is affected by climate conditions, with consequences to population growth rates. Temperature has been shown to influence sex ratio, with higher temperatures favoring an increased percentage of males in some species (Stinner et al., 1989).

Mortality.   Climate affects insect mortality in several ways. Many insect species cannot survive cold temperature. High rates of precipitation may drown insects that live in the soil. Wet conditions often tend to favor organisms that prey on insects (e.g., nematodes), thereby causing insect mortality indirectly.

Synchrony.   The severity of crop damage is often linked to the occurrence at the same time of vulnerable crop development stages and feeding stages of insects. A change of climate that induces shifts in crop–pest synchrony could have either positive or negative effects. If crops and pest development go out of phase so that crop plants are in a less vulnerable stage at the time of infestation, insect development may be slowed. In general, crops are more vulnerable in the early part of the season. Crop damage may worsen if insects are more numerous and larger—and, hence, consume more vegetative matter—during vulnerable crop stages. If warming should spur insect activity more than it does crop growth, insects may become established while crops are most vulnerable and cause greater damage than otherwise. However, since farmers are likely to respond to warming by setting earlier planting dates, the hazard of greater insect damage may be evaded in part as crops may thus be able to outgrow the pest damage. Drought and moisture regimes also influence (either enhance or disrupt) crop–pest synchrony.

Interactions among Species

Arthropod communities may be destabilized by the differential effects of weather on pests and their natural enemies (which include pathogens, parasites, and predators). The effect of this destabilization on agriculture will depend on the type and development stage of the crop; the possible occurrence, intensity, and duration of plant stress; the types and development stages of the pests and their predators; and the specific differential responses of the species involved to the various manifestations of climate changes. Any potential change in crop plant quality is also likely to alter interactions among pests and their enemies. For example, aphids (frequent pests of maize, cotton, alfalfa, hops, potatoes, and peas) seem to be better adapted to cool weather than some of their natural enemies; hence, warmer temperatures should benefit the enemies and help in the biological control of aphid infestation in some areas (Stinner et al., 1989). However, pest outbreaks may occur subsequently if natural predators of the pests die out due to their own loss of food supply. While some current pest species may be better controlled by natural enemies in a warmer climate, other insect species not currently damaging may become pests.

Insect Distributions

The potential geographical distribution of insects depends primarily on temperature thresholds (minimal and maximal temperatures for given species). Low temperature thresholds are often the more critical ones. As climate warms, the boundaries of insect habitation are predicted to extend into currently cooler areas, while regions of present infestation will probably not be affected (Kimball, 1985). A listing of temperature thresholds and response functions for some important crop pest species is given in Table 4.4. Computer models based on temperature and soil moisture requirements and stress levels have been devised to assess the potential establishment of exotic pests under the current climate (e.g., Worner, 1988); such sets of algorithms are useful in testing the potential spread of insects under projected climate change scenarios.

The actual (as opposed to the potential) distribution of pests depends on the prevalence of host crops, which is also affected by moisture availability (precipitation or irrigation). For example, grasshopper densities in southern Idaho depend on the abundance of native grasses, which, in turn, are correlated with the amount of precipitation occurring in November and with the temperature prevailing in April (Fielding and Brusven, 1990).

Other pest distribution studies predict that under warmer conditions earworm infestation will probably increase in midwestern soybeans, with potentially significant economic losses to grain farmers. Potato leafhopper, black cutworm, sunflower moth, and green cloverworm—all important migratory pest species in the United States— may extend their ranges northward due to increased overwintering capability (Stinner et al., 1989). This northward migration is likely to be accompanied by an earlier invasion of crops in the growing season. If the southern states become appreciably more tropical, insect pests that now primarily overwinter in Central America (such as the velvetbean caterpillar pest of soybean) may extend overwintering ranges to greater areas of North America.

The European corn borer (*Ostrinia nubilalis*) goes into diapause under conditions of shortening daylengths and cooling temperatures to prepare for winter. Under one climate change scenario, the European corn borer may extend its range by 1000 km or more and may produce an additional generation in its current regions (Porter et al., 1991). Another potentially serious migrant in the United Kingdom is the Colorado beetle [*Leptinotarsa decemlineata* (Say)], now found in potato fields of the northern coasts of France and Belgium (Porter et al., 1991). A climate change study on butterfly species in Britain projected mainly disruptive effects on butterflies, with range expansions and upland colonization predicted for the north and reduced population and population extinctions for the south (Dennis and Shreeve, 1991).

Table 4.4   Climatic thresholds and response functions for development stages of selected agricultural insect pests (from literature review by Westbrook and Lindgren, personal communication).

Species	Common name	Thresholds	Response functions
Helicoverpa zea	Corn earworm	12.5°C	185 DD larval development on sweet corn
	Cotton bollworm	12.6°C	557 DD egg to adult development on cotton
			Overwintering confined to 39°N latitude
Heliothis virescens	Tobacco budworm	12.6°C	448 DD egg to adult development on cotton
Spodoptera exigua	Beet armyworm	12.2°C	517 DD per generation
Spodoptera frugiperda	Fall armyworm	13.8°C	340 DD per generation
Pieridae	Butterflies	28°C[1]	
Dasyneura laricis	Larch gall midge	0°C	1600 DD
Lymantria monacha	Nun moth	0°C and 45°C	More frequent outbreaks in low-precipitation sites
Coleophora laricella	Larch casebearer	No lower limit	
Trypodendron lineatum	Striped ambrosia beetle	15.5°C dispersal flight initiated	19–26°C optimal air temperature for flight; 30°C maximum air temperature for flight
Scolytus ventralis	Fir engraver beetle	24°C flight initiated	
Dendroctonus ponderosae	Mountain pine beetle	16–30°C emergence	21–38°C flight initiation; −18°C lethal to eggs; −40°C lethal to larvae; single generation may require 2 years; multiple generations may develop in the southern part of the range
Ips typographus	Spruce bark beetle	47°C lethal	

DD = 5 degree days.

1. Body temperature required for flight.

As climate change modifies the geographical zonation of crops and weeds, alternative hosts, refugia, and "green bridges" may form. These are sites where pests can survive during interludes between infestations of agricultural crops (Porter et al., 1991). New infestations of incidentally introduced, or exotic, insect pests may also occur. The proliferation of introduced insects will depend, of course, on the presence or absence of natural enemies. There may well be time lags between the geographical shifts of crops and those of pests; but if the warming trend is slow, crops and pests may move together.

Insect Vulnerability

The survival of insect species depends on the availability of alternate resources and their ability to respond to environmental change by dispersal or adaptation. Dennis and Shreeve (1991) have delineated seven criteria for vulnerability to climate change and have ranked butterfly species in Britain according to these criteria (Table 4.5). The criteria include latitudinal range of the species (insects with narrowest ranges are most vulnerable), distribution (insects with lowest distribution within their range are most vulnerable), host plant type (insects that eat only from one type of host plant are most vulnerable), host plant abundance (insects whose host plants are substrate dependent are most vulnerable), major habitat (insects of climax vegetation are most vulnerable), range of habitat types (insects whose habitat types are limited are most vulnerable), and dispersal ability (insects with closed populations with little evidence of movement outside colonies are most vulnerable). Climax communities are more vulnerable to climate change than are early successional stages, since the former lack mechanisms to facilitate rapid colonization of new environments. The species least vulnerable to climate change are likely to be migrant, polyphagous (feeding on various kinds of food) species with extensive ranges, continuous distributions, ubiquitous host plants, and varied and extensive habitats (Dennis and Shreeve, 1991).

Table 4.5 Climate change vulnerability criteria for insects in Britain (adapted from Dennis and Shreeve, 1991)

Attribute	Criteria
A	Range, based on the latitudinal extent: 1, <25%; 2, <50%; 3, <75%; 4, < 100%.
B	Distribution, based on the proportion of 10-km squares occupied within the range of the species: 1, <24.5%; 2, <49%; 3, <73.5%; 4, <98%. Maximum extent of any species is 98%.
C	Host plant type: 1, Monophagous; 2, Oligophagous—1 species per habitat; 3, Oligophagous using more than one species per habitat; 4, Polyphagous.
D	Host plant abundance; 1, Substrate-dependent; 2, Patchy within habitats; 3, Ubiquitous within habitat types; 4, Ubiquitous and cosmopolitan.
E	Vulnerability of major habitat seral stage occupied: 1, Climax woodland or plagioclimactic bog; 2, Preclimax forest; 3, Shrubs, tall herbs, and grasses; 4, Bare ground, short herbs, and grasses.
F	Range of seminatural habitat types occupied: 1, <5; 2, <9; 3, <14; 4 <18. Maximum number of habitat classes occupied is 18.
G	Dispersal ability: 1, Closed populations with little evidence of any movement outside colonies; 2, Colonial species with evidence of dispersal outside colonies; 3, Open population structures with evidence of dispersal outside colonies; 4, Migrants and species that are vagrant and known to engage in long-distance movements.

Coding of attribute states: 1–4, most to least susceptible, respectively. Low total scores may indicate species under the greatest threat from climatic change. Equal weight is given to all variables, but this may be unrealistic since critical variables are not identifiable with current knowledge, and different variables will be of unequal importance. Ant associations in some Lycaenidae may interact with attributes C, D, E, and F, but the precise nature of such interactions is not comprehensively known and is probably accommodated in factor F. There are other potential important variables (e.g., voltinism, flight period), but too little information is available to assess their full significance.

Implications for Insect Pest Management

The projections for increased insect pest infestations with climate change imply a substantial rise in pesticide use and in associated costs to farmers, as well as greater environmental hazards. Principal concerns are for species that can increase their population size by producing an extra generation each year in warmer climates or for those species that may expand their geographical distributions (IPCC, 1996). Pest problems and environmental risks due to use of more agricultural chemicals are not only likely to intensify in the agricultural regions where they now occur but also to extend into new areas as cropping zones expand to higher latitudes and elevations. Pests now confined to greenhouses may infest open fields (Porter et al., 1991). Both the timing and the effectiveness of insecticide application will probably change. In some instances, pest control could become more difficult if small sap-feeding insects proliferate, since such insects tend to develop resistance to pesticides quite rapidly.

There is a growing awareness that crop system management must consider entire pest populations and life cycles rather than focus sporadic attention ad hoc on specific infestations in farmers' fields. This is particularly the case for migratory insects that depend on air currents for long-range transport from decaying to viable habitats (Stinner et al., 1982; Sparks, 1986). Examples of such migratory pests of economic importance to U.S. crops, now restricted in their southern overwintering ranges, are the sunflower moth [*Homoeosoma electellum* (Hulst)], black cutworm [*Agrotis ipsilon* (Hufnagel)], fall armyworm [*Spodoptera frugiperda* (J. E. Smith)], tobacco budworm [*Heliothis virescens* (F.)], and corn earworm [*Heliothis zea* (Boddie)]. Boundary-layer and regional synoptic weather models may be used in conjunction with known insect growth, movement, and mortality responses to environmental variables in order to calculate *potential* insect infestations (Stinner et al., 1982), both in the current climate and for possible future climates. These models often rely on minimum and maximum temperature thresholds, cumulative degree day requirements, and effects of dry or wet conditions on insect development and migration. The prospect of climate change and changing atmospheric circulation patterns requires such a comprehensive approach to projecting large-scale crop–pest interactions, although current limitations of general circulation models (GCMs) make specific predictions difficult. Better knowledge of how nocturnal wind jets in the United States and in the intertropical convergence zone (ITCZ) system may change in strength, timing, and extent will aid in more precise prediction.

The prospect of climate change should also spur the development and dissemination of integrated pest management (IPM) techniques, which aim to reduce the wanton application of chemicals by regularly monitoring pest conditions, establishing economic thresholds or other criteria for tolerable injury levels, and relying as much as possible on biological controls.

Increasing CO_2 and climate change will affect the development of IPM strategies, including the use of biological agents, genetic resistance, and new cropping systems in preference to exclusive reliance on agricultural chemicals (pesticides). Synchronized timing between the growth and development of the biological control agents and their pest targets is essential to successful IPM. Current means of biological control may lose some of their effectiveness if insect growth rates are speeded up, since natural enemies are more effective against insects that have slower growth rates. However, some natural enemies of pests may benefit from climate change.

Diseases

Little research has been conducted on how atmospheric CO_2 rise and consequent climate change may affect crop diseases. Plant pathogens include fungi, bacteria, viruses, and nematodes. Current understanding of the major factors that affect plant pathogens suggests that climate change will indeed influence the occurrence, severity, and geographical distribution of crop diseases.

CO_2 Effects on Pathogens and Host Plants

Atmospheric carbon dioxide has little direct effect on crop pathogens, at least within the concentration range of 0.03 to 0.06% considered in connection with the enhanced greenhouse effect. The possibility exists that, by changing the biochemistry or the structure of host crops, rising CO_2 may have an indirect effect on the susceptibility of crops to attack by pathogens, but this has yet to be defined. In one study, higher CO_2 did not appear to simulate the activity of a fungal pathogen on two grasses (Marks and Clay, 1990). If higher CO_2 causes greater amounts of carbohydrate root exudates, both disease-causing and disease-preventing soil bacteria may flourish. While plants grown under high CO_2 and those that are diseased sometimes show similar responses (e.g., increased starch accumulation in leaves, enhanced photosynthetic rates, and reduced stomatal aperture), we do not yet know how these responses might change for diseased crops in high CO_2 environments.

Climate Effects on Crop Diseases

Climate factors that influence the growth, spread, and survival of crop diseases include temperature, precipitation, humidity, dew, radiation, wind speed, circulation patterns, and the occurrence of extreme events. Crop pathogens are often carried by insect or nematode vectors that will also be affected by climate. Warming should intensify problems caused by both virus-vector nematodes in northern Europe through an increase in existing populations and by the spread of nematodes from the south (Boag et al., 1991). Higher temperature and humidity and greater precipitation are likely to result in the spread of plant diseases, as wet vegetation promotes the germination of spores and the proliferation of bacteria and fungi.

In regions that suffer greater aridity, however, disease infestation may lessen, although some diseases (such as the powdery mildews) can thrive even in hot, dry conditions as long as there is dew formation at night (Patterson, 1993). Fungi such as root and stalk rots and some wilts and foliar diseases colonize water-stressed plants more successfully than they do normal plants (Mattson and Haack, 1987). The geographic distribution of pathogens will change their ranges if host crops ranges shift. As with insects, natural controls may be disrupted and exotic diseases may invade new regions. As new varieties and crop species are introduced, new combinations of diseases and crop hosts may arise. Management and control measures will need to adapt to these changing conditions, and such adaptation may include the adoption of new exclusion and quarantine efforts (Patterson, 1993).

Climate Change and Crop Pests: A Summary

Rising temperature and CO_2 levels are predicted to affect pest–crop relationships, probably exacerbating their negative impacts on crops. Impacts are likely to be severe in many different environments and involve numerous species of weeds, insects, and diseases. For farmers and the economics of farming, negative pest impacts are felt both in terms of reduced crop productivity and of the higher costs of such additional control measures as will become necessary. When yield reductions occur and prices rise, the costs of pest infestation are shared by consumers. Reductions in quality of produce also affect both farmers and consumers alike. Agricultural pest and disease management must adapt to these changing conditions, not the least of which will be the changing effectiveness of agricultural chemicals in a modified environment. Chemical uptake and metabolism, as well as injury to crops and nontarget organisms, may also be affected. Weed specialists, entomologists, and pathologists will face the challenges of developing beneficial crop adaptations to minimize the greater hazard by new or more vigorous pests and diseases.

The impacts, potential economic consequences, and management of specific pests will depend on the nature of climate/crop/pest interactions in each case, and these, in turn, will be affected by future cropping systems. Since there are literally millions of potential insect and pathogen species, warming is likely to reduce the threat of damage from some and magnify the threat from others. When environmental stresses intensify, as they well might under a warmer climate, crops may become less competitive with weeds and more vulnerable to insects and plant diseases. As climate change will alter the geographical distributions of crops, weeds, and insects, it will also alter the infestation levels of plant pathogens. Regions of possible insect and disease infestations will shift poleward, while infestation levels in current regions will probably not be reduced. Because increasing concentrations of atmospheric CO_2 will benefit weeds as well as crops, the competitive balance between crops and weeds is likely to shift—perhaps to the detriment of the former.

Improved predictions of potential changes in climate variability and transient climate changes are needed before climate change effects on insect pests can be better defined. Projections of the potential agricultural impacts of climate change and increasing CO_2 are hindered by lack of knowledge about the specific climate changes likely to take place on a regional or local scale.

Much more must also be learned about the potential effects of climate change on pests and associated yield losses. Potential pest migrants should be identified, as well as possible new combinations of crops and pests. If new nonresistant crops or cultivars are introduced in response to climatic changes, new pest problems could occur. Development of fundamental knowledge on the biology of weeds, insects, and diseases (e.g., CO_2 and temperature responses) is also essential for determining the best crop management systems for pest control likely to cause the least environmental damage under climate change conditions. Because climate change will most probably affect natural pest enemies and, hence, the arsenal of biological controls available to farmers, acquiring the needed information becomes an especially important task. If climate change takes place rapidly and extreme events occur more frequently, severe disruptions of agroecosystems will ensue.

Simulation models that can incorporate crop interactions with weeds, insects, and diseases are in their initial stages of development; their use will allow more realistic projections of actual yield changes due to climate change. Basic research in controlled environments and in the field on the comparative ecophysiology of crops and pests should provide some of the data required to create such models. Knowledge of critical temperature thresholds for many more insect species is especially important for identifying potential migrants and invaders.

Long-term monitoring of pest populations may provide an early indicator of climate change effects on agricultural systems. This is because insects are very adaptive to changing environmental conditions by virtue of their short regeneration times. Sensitive regions that may be particularly vulnerable to future pest infestations should be identified, in an effort to target research efforts and prepare for early action. The relatively few studies that have been done thus far regarding climate change effects on agricultural pests have pertained to mid-latitude crops and regions, mostly in the Northern Hemisphere. Even though temperature changes are predicted to be lower in tropical regions, studies are needed there as well, since insect species diversity is greater and insect pest problems are inherently more acute in the tropics than in the mid-latitudes.

5

The Role of Soil Resources

The soil is a complex and dynamic system, consisting of a solid phase (both mineral and organic, particulate and amorphous), a liquid phase (water and solutes), and a gaseous phase (air with associated water vapor, often enriched with carbon dioxide and sometimes with methane as well). The soil system responds to short-term events such as the episodic infiltration of rainfall and also undergoes long-term processes such as physical and chemical weathering. The quantitative evaluation of the predicted climate change on soil conditions is difficult, due not only to the uncertainties in the forecasts but also to the complex, interactive influences of hydrological regime, vegetation, and land use. Therefore only rough, qualitative estimations may be made and only general conclusions may be drawn at present.

Higher temperatures, along with changes in soil moisture, will lead to a wide range of soil and plant responses to global climate change. The physiological effects of increased atmospheric CO_2 on plants may also have significant consequences on soil organic matter, which is in itself a major sink in the global carbon cycle. Thus, we perceive that soils will respond to climate change in complex ways. The overall outcome will depend on numerous processes with often opposing effects. Changes in the soil are also likely to depend on topography, specific soil composition and properties, and crop or vegetation cover, all of which vary from place to place. Changing climate is likely to create climate–soil patterns that have not previously been observed in particular locales.

Soil Properties and Thermal Regimes

Soil properties change on different time scales. Properties such as bulk density, porosity, moisture content, infiltration rate, permeability, composition of soil air, and nitrate content can vary on a daily to monthly basis, while many soil properties tend to change on yearly, decadal, and century time scales (Table 5.1). Even though the rate of cli-

Table 5.1 Time scales of changes in soil properties (Varallyay, 1990).

Time	Soil parameter	Properties and characteristics	Horizons and phases	Regimes
$<10^{-1}$ yr	Bulk density; total porosity; moisture content; infiltration rate; permeability; composition of soil air; nitrate content	Compaction		Aeration; heat regime
$<10^{-1}$ yr–10^{-0} yr	Total water capacity; field capacity; hydraulic conductivity; pH; nutrient status; composition of soil solution	Microbiota		Microbial activity; human-controlled plant-nutrient regime
$<10^{0}$ yr–10^{1} yr	Wilting percentage; soil acidity; cation exchange capacity; exchangeable cations; ion composition of extracts	Type of soil structure; annual roots biota; meso-fauna; litter, fluvic, gleyic, stagnic properties; slickensides	Sulfuric horizon; gelundic, inundic, salic, yermic phases (fine earth properties only)	Moisture; natural fertility; salinity-alkalinity; permafrost
$<10^{1}$ yr–10^{2} yr	Specific surface; clay mineral association; organic matter content	Soil biota, tree roots; salic; calcareous, sodic, vertic properties	Histic (<20 cm), ochric, gypsic, albic, and immature natric and spodic horizons (Podsols); gilgai, placic, sodic, takyric phase	
$<10^{2}$ yr–10^{3} yr	Primary mineral composition; chemical composition of mineral part	Tree roots; color (yellowish/ reddish); iron concretions; soil depth; cracking; soft powdered lime; indurated subsoil	Histic, mollic, umbric, calcic, albic, natric, cambic, spodic, and nitic horizons; plinthite, placic, yermic phases (stone surfaces)	
$>10^{3}$ yr	Texture; particle-size distribution; particle density	Parent material; depth; abrupt textural change	Argic, oxic, petrocalcic, petrogypsic horizons; duripan, fragipan, skeletic, petroferric, lithic, rudic phases	

mate change is expected to be relatively rapid (0.1 to 0.8°C per decade), many effects of climate change on soils will take decades or centuries to be manifested. High chroma colors; iron concretions; histic, mollic, and umbric epipedons; calcic, natric, cambic, and spodic horizons; petrocalcic and petrogypsic horizons; duripans and fragipans—all take more than 100 years to form (Arnold, 1990). Younger, less weathered, soils will probably change more rapidly and to a greater extent than will mature, fully developed soils (Varallyay, 1990).

Texture and Structure

Soil texture—the relative contents of sand, silt, and clay particles—changes slowly with time, because physical and chemical weathering are slow processes. The characteristic response time for soil texture is on the order of a millennium. Maximum clay illuviation (migration within the soil profile) occurs in warm, wet climate regimes with acid forest litter, whereas dry/cold conditions limit clay formation (Figure 5.1). As climatic zones shift, these textural processes will slowly change in response.

The development of soil structure is even more complex than that of soil texture. The type, spatial arrangement, and stability of soil aggregates is affected by the intensity of precipitation, amount of surface runoff and infiltration, root distribution, earthworms and other soil fauna, and compaction by agricultural machinery.

Thermal Regimes

Climate is a major factor in the long-term process leading to soil formation. As expressed by Jenny (1941), the character of a soil is a function of the temperature gov-

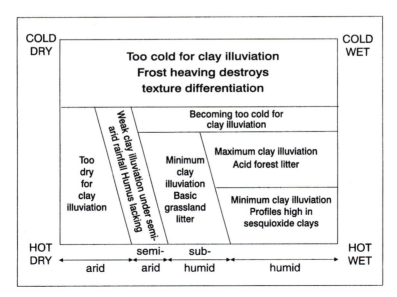

Figure 5.1   Effect of climate on texture differentiation (IGBP Report No. 5, 1989).

erning its formation (provided other affecting variables such as moisture, organisms, parent materials, relief, and time are unconstraining): $S = f(T)_{m,o,p,r,t,}$ where $T =$ temperature, $m =$ moisture, $o =$ organisms, $p =$ parent material, $r =$ relief, and $t =$ time.

The anticipated climate change from the enhanced greenhouse effect can be expected to influence soils by increasing the energy input to the surface. The additional input of energy will tend to raise both air and soil temperatures. The soil's thermal regime is governed by the gains and losses of radiation at the surface, heat conduction through the soil profile, convective transfer via the movement of gas and water, and transformation of sensible into latent heat in the process of evaporation (Hillel, 1980). Characteristics of the soil that affect its thermal regime include surface albedo, bulk density and porosity, and moisture content. The moisture content affects the soil's thermal regime because water has a high heat capacity and its presence increases the soil's thermal conductivity. High bulk density also increases thermal conductivity.

Soil-Warming Experiments

Several long-term soil-warming experiments are now under way (and more are being planned) in a variety of ecosystems, and these are complemented by controlled laboratory experiments (Workshop, 1991). Experimental sites in tundra and boreal forest ecosystems are important because soils there contain large amounts of organic matter and because global warming at high latitudes is predicted to be higher than the global average. Studies in agricultural systems are needed, for example, in the central U.S. Great Plains, so that the effects of soil warming on agricultural production can be predicted. Such experiments are planned with a 10-year time horizon because determining effects on some soil properties (e.g., the storage of soil organic matter) requires long-term monitoring over many growing seasons. Other properties—such as changes in microbial biomass and nitrogen mineralization—may be noticeable within a year, yet continue to react over long time periods. Experimental results of CO_2 flux from soils in response to increases in temperature are needed to validate models of soil carbon dynamics.

The experiments either increase the mean annual soil temperature of a site by a specific increment relative to a control plot or alter the energy input to the soil surface, based on the change in thermal forcing predicted by general circulation models (GCMs). The latter is predicted to range between 2.2 and 7.2 Watts m^{-2} by the year 2050 (Dickinson and Cicerone, 1986). The temperature response of different soils to the thermal forcing will differ from one soil or location to another, depending on the moisture content and other soil properties. For example, the surface zone of a wet soil generally shows a smaller temperature rise than that of a dry soil (both the heat capacity and thermal conductivity of the former would naturally be higher than of the latter).

The measurements made in these experiments include soil temperature and moisture; fluxes of CO_2, N_2O, and CH_4 into and out of the soil; soil microbial biomass; the dynamics of nitrogen and phosphorus mineralization; the biomass and distribution of plant roots; CO_2 concentration in the pore spaces of the soil; and the ionic balance of cations and anions in the solutions draining from the soil profile (Workshop, 1991). The last of these measurements should indicate changes in the rates of chemical weathering and of cation depletion in the soil.

Trace metals also are being monitored in an effort to understand the activity of organic chelates and to ascertain whether heavy metals are released in greater amounts as a result of enhanced decomposition of soil organic matter at higher temperatures. Finally, among the important physiological processes to be monitored are such processes as the proliferation of fine roots and the net primary production of plants growing in the soil.

Nutrients

Crops require a wide array of nutrient elements for healthy growth. Among these are elements that are normally taken up by plants and incorporated into their biomass in relatively large amounts, and that are therefore called "macronutrients" (e.g., C, H, O, N, P, K, S, Ca, and Mg). Just as essential are elements required in relatively small amounts and therefore called "trace nutrients" or "micronutrients" (e.g., Mo, Cu, Zn, Mn, Bo, Fe, and Cl). Such elements are normally provided by the environment in which plants grow. However, many nutrients can be added to the crop-growing medium as fertilizer, thus freeing crops from one or another natural environmental constraints. The inherent fertility of a soil, while composed of many different attributes, generally depends on its nutrient storing and supplying power, which, in turn, is often related to its cation-exchange capacity and the relative degree of base saturation. These characteristics are measurable and are known for many soils and may therefore be used to classify the efficacy of soils for agricultural production. The processes by which nutrient elements become actually available to plants are affected by climate variables (Russell, 1973). Nutrient dynamics typically take place in the topsoil (within a few tenths of a meter from the surface), where microbiological activity is concentrated.

Temperature Effects on Soil Nutrients

In general, warmer temperatures tend to hasten the chemical processes that affect soil fertility. The most important process is probably the accelerated decomposition of organic matter, which releases nutrients in the short run but may reduce soil fertility in the long run. In some cases, these decomposition losses may be balanced by increased carbon fixation of crops and vegetation and, hence, in a greater accumulation of organic matter residues. The net effect—the long-term depletion or accretion of carbon storage of the soil—is hard to predict. The enhanced potential growth in a CO_2-enriched atmosphere may require additional applications of fertilizers, since cycling of nutrients is likely to be accelerated with warmer temperatures. The fixation of atmospheric nitrogen by symbiotic bacteria will tend to increase with the greater root biomass of the CO_2-enriched crop. However, both biomass accumulation of crop roots and decomposition of organic matter will be suppressed if temperature risk is excessive or if soil moisture is limiting.

Soil temperature affects the rates at which organic matter decomposes, nutrients are released and taken up, and plant metabolic processes (including growth) proceed. Some of the major effects of temperature on soil properties are shown schematically in Figure 5.2 (Buol et al., 1990). Chemical reactions that affect soil minerals and organic matter are strongly influenced by higher soil and water temperatures. Both organic matter and the carbon:nitrogen ratio tend to diminish in warmer conditions due

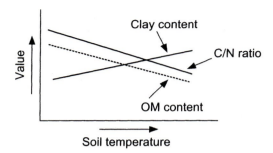

Figure 5.2   Temperature effects on
soil properties (Buol et al., 1990).

to accelerated decomposition by microbial action, while clay content tends to increase due to accelerated weathering of primary minerals. The biomass and the activity rate of soil microbes also increase. Nitrification—the process by which bacteria oxidize organic forms of nitrogen to nitrates—is inhibited in cool soils and accelerated in warm ones. Denitrification also increases with soil temperature. The rate of phosphate uptake is also enhanced as soil temperatures rise. However, high soil temperatures may have a depressing effect on symbiotic nitrogen-fixing bacteria (specifically those that attach themselves to the roots of legumes). Even if nutrient availability increases with temperature, it is difficult to make accurate predictions of how crops may respond since the rates at which nutrients are lost to the atmosphere and groundwater may also increase.

Higher soil temperatures should generally accelerate chemical reaction rates and diffusion-controlled reactions. The solubilities of solid and gaseous components may either increase or decrease by a small amount (Buol et al., 1990). The solubility of potassium and sodium salts rises with temperature, while that of calcite diminishes as temperatures rise. Carbon dioxide, nitrogen, and oxygen gases all exhibit reduced solubility in warmer conditions. Higher temperatures could also increase mineralization rates—that is, the decomposition of organic materials and their transformation into mineral salts—improving the availability of phosphorus and potassium and speeding colloid formation. The consequences of these changes may take years to become significant, however, given the generally long reaction times in the soil compared to experimental conditions in pure solutions (Buol et al., 1990). Higher temperatures should also contribute to increased evaporation and thus drier soil moisture regimes.

An interesting possibility suggested by Buol et al. (1990) is that the soils of higher latitudes may become redder, due to the altered equilibrium states of goethite and hematite (iron oxide minearls) in favor of the latter.

Hydrological Effects on Soil Nutrients

Some of the major effects of precipitation on soil properties are shown schematically in Figure 5.3 (Buol et al., 1990). As the amount of precipitation rises, calcium carbonate tends to fall steeply, while soil acidity and clay content tend to increase. The process of nitrification is inhibited in wet soils. The contrary process of denitrification is enhanced where high precipitation raises the water table in poorly drained soils (Buol et al., 1990). In well-drained soils, increased precipitation also promotes leaching of nitrates.

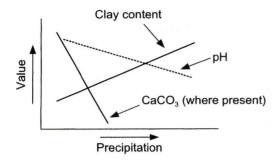

Figure 5.3   Precipitation effects on soil properties (Buol et al., 1990).

Carbon Dioxide Effects on Soils

As discussed in Chapter 3, experiments have shown that relative carbon assimilation, biomass accumulation, and positive yield responses to higher atmospheric CO_2 are suppressed when soil nutrients are lacking (Cure and Acock, 1986). If nutrient uptake is insufficient to satisfy the full requirements for photosynthetic carbon fixation, at some point further carbon assimilation becomes limited. The internal allocation of the products of photosynthesis (carbohydrates) may also be affected. Thus, increased applications of fertilizer will evidently be needed to take full advantage of the enriched CO_2 atmosphere. If, following the summaries of Kimball (1983) and Cure and Acock (1986), we assume that yields may increase by a third, then approximately a third more nutrients will be removed from the soil at each harvest. Unless the additional nutrients are added, the soil's own reserves will be depleted and soil fertility will decline (Kimball, 1985).

Of course, the yield versus nutrient requirement relationship will not be exactly 1:1. One reason is that an increased root:shoot ratio in the crops grown in higher CO_2 may allow the root system to draw nutrients from a larger volume of soil. However, higher root:shoot ratio values have not been found consistently in CO_2-enrichment experiments (Cure and Acock, 1986). Conditions for growth of mycorrhizal fungi that enhance phosphorus uptake may also improve with greater root carbohydrate exudation (Lamborg et al., 1983). High CO_2 will probably not have direct effects on soil microbes because the CO_2 concentration in the soil's air phase during the growing season is in any case generally 10 to 50 times higher than atmospheric CO_2 (Lamborg et al., 1983).

Greater plant growth should contribute more crop residues to the soil, thereby helping to reduce erosion, maintain or even increase soil organic matter, and provide additional nitrogen (Kimball, 1985). However, the nutritional quality of the residues may be lowered and may inhibit the action of microbial decomposers since the residues will tend to have higher carbon:nitrogen and carbon:phosphorus ratios (Kimball, 1985). Overall, however, improvements in soil characteristics—thanks to the added organic matter—including greater nutrient availability and moisture retention, better tilth, and reduced erodibility may prove to be additional benefits from higher atmospheric CO_2.

Nitrogen is also made available by symbiotic nitrogen-fixing bacteria associated with legumes. Experiments have shown that when legumes are grown in higher atmospheric CO_2, nitrogen fixation tends to increase (e.g., Williams et al., 1981), so legumes may have adequate supplies of nitrogen under future higher CO_2 conditions and, therefore, may continue to be a good source of nitrogen in crop rotations in the future.

Soil Carbon

The net effect (the long-term depletion or accretion of carbon storage of the soil), is hard to predict, although such prediction is an important task since soil organic matter is a major reservoir in the global carbon cycle. While the physiological effects of increased atmospheric CO_2 on plants may add organic matter to the soil, warmer temperatures may accelerate its decomposition. Agricultural activities that alter the carbon (and nitrogen) held in the soil include deforestation and afforestation, biomass burning, cultivation, rice paddy residue management, and fertilizer application. Alterations in these practices offer means to augment the amount of carbon held in organic matter in agricultural soils and to mitigate the enhanced greenhouse effect.

Global Carbon Cycle

Soils are a major reservoir of carbon, altogether holding about 1.5×10^{12} t C. The amount of organic carbon held in the top meter of the soil (SOC) is about twice the amount in the atmosphere (7.5×10^{11} t C) (Post et al., 1982). An additional 1.7×10^{12} t of carbon is held in inorganic forms contained in the deeper layers below 1 meter depth (SIC). This pool is composed primarily of calcium carbonate ($CaCO_3$) (Lal et al., 1995), but the material is hardly labile and participates in the carbon dynamics in the atmosphere only at very long time scales. Different soils store varying amounts of organic carbon, mostly near the surface, depending on climate regime (Figure 5.4). The organic soils of the tundra and boreal forests and the caliche ($CaCO_3$-containing) desert soils hold high amounts of carbon (Post et al., 1982). A significant portion of soil carbon is relatively labile (i.e., readily released to the atmosphere as CO_2 following decomposition processes), but that lability depends on how much carbon is contributed annually by plant residues and the rate at which those residues are oxidized by microbes. Raich and Schlesinger (1992) estimated that the global mean residence time (mass/output) of soil organic matter is as long as 500 years in the tundra and peaty wetlands and only about 10 years in the tropical savannas. Soil respiration is defined as the annual flux of CO_2 from the soil to the atmosphere, and it is currently estimated to total about 6.8×10^{10} t C yr^{-1} (Raich and Schlesinger, 1992). This amount is about 14 times the annual release from the burning of fossil fuels (about 5 $\times 10^9$ t C yr^{-1}). The soil produces CO_2 in two ways: through the decomposition of organic matter by microbes and through the respiration of live roots and mycorrhizal fungi.

Higher temperature should extend the depth of the soil's active layer, with accompanying enhancements in aeration, organic matter decomposition rates, and CO_2 efflux. Higher temperature is associated with greater annual releases of CO_2 from soils (Figure 5.5) (Raich and Schlesinger, 1992), although there is large variation in the measurements due to differences in such factors as soil moisture regime and length of the growing season. In principle, global warming should accelerate soil respiration be-

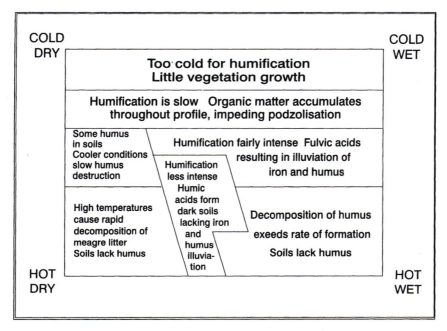

Figure 5.4   Effect of climate on soil organic matter (IGBP Report No. 5, 1989).

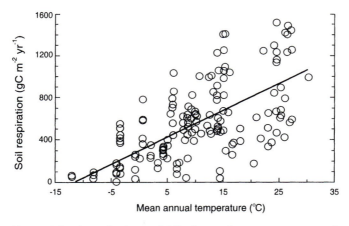

Figure 5.5   Annual release of CO_2 from soils versus mean annual air temperature for various ecosystems (Raich and Schlesinger, 1992).

cause of both the hastened decay of soil organic matter and the greater rate of root respiration. C:N ratios might thereby be narrower than otherwise expected, and CO_2 concentration in soil pores should increase (Figure 5.6). These effects should be offset somewhat by the greater root biomass and crop residues expected to result from positive plant responses to higher CO_2. The tendency of a rising temperature to hasten decomposition may also be offset in part by the negative impact of increased C:N ratios on decomposition, as well as the negative impact of drought on decomposition, where droughts become more frequent and prolonged.

Might the accelerated release of the large pool of soil carbon induce a runaway greenhouse effect? While warmer temperatures should, in general, cause a significant positive feedback on CO_2 release, Jenkinson et al. (1991) believe that a runaway greenhouse effect is unlikely because of the cold temperatures that will continue to prevail in northern latitudes where much of the carbon is stored. At high latitudes, the soil is so cold (on the order of $-15°C$) that even an increase of $10°C$ will have little effect on biological activity.

Several studies have attempted to calculate the positive feedback caused by soil warming and greater CO_2 emissions to the atmosphere. Jenkinson et al. (1991) have estimated that if world temperatures rise at the rate of $0.03°C$ yr^{-1} (the Intergovernmental Panel on Climate Change "best estimate"), the additional release of CO_2 from soil organic matter will be 6.1×10^{10} t C over the period of the next 60 years. This would be equivalent to roughly 20% of the projected CO_2 flux from fossil fuels over the same period.

In another study, Buol et al. (1990) classified soils according to their temperature regimes in an effort to estimate potential changes in soil organic carbon. Soil carbon content tends to decrease with increasing mean temperature (Figure 5.7) (Buol et al., 1990). They estimated that a $3°C$ warming would cause an 11% decrease in soil or-

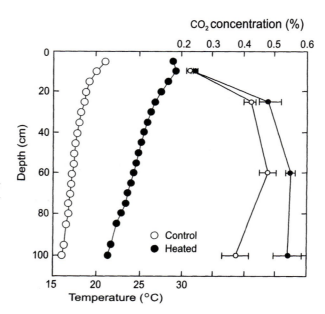

Figure 5.6   Distribution of CO_2 in the soil profiles of control and heated plots (Cape Cod, Massachusetts). Each value is the mean from three replicate access tubes in each plot on August 7, 1991 (Schlesinger et al., 1997).

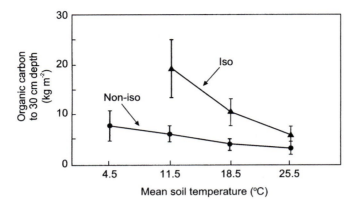

Figure 5.7 Organic carbon content in the top 30 cm of soils. Iso = soils where mean summer and mean winter soil temperatures differ by less than 5°C; Non-iso = soils where mean summer and mean winter soil temperatures differ by 5°C or more (Buol et al., 1990).

ganic matter in the upper 30 cm of "average" soils in the temperate zone. This could contribute to an 8% increase in atmospheric CO_2 (compared to the 1990 levels) over a 50-year period. These calculations were done without consideration of possible increases in plant biomass residues caused by physiological effects of higher atmospheric CO_2. Buol et al. then calculated that an average increase in total (above- and below-ground) biomass production of 568 kg ha^{-1} yr^{-1} would be needed to offset the soil carbon losses due to accelerated decomposition in warmer soils. Higher temperatures would cause C:N ratios to narrow slightly, and soil nitrogen content would rise by some 10% in temperate zones over a 50-year period.

Other potential effects of accelerating soil organic matter decomposition include a boost in the production of organic acids (such that may intensify rock weathering), a greater concentration of dissolved organic carbon in streams, and an overall rise in the productivity of riverine ecosystems and fisheries (Workshop, 1991). Friedland and Johnson (1985) have suggested that depletion of soil organic matter may result in the release of heavy metals (e.g., lead, mercury, and cadmium) from soils exposed to atmospheric pollution and acid rain. Agricultural conversion of peatlands, a source of carbon emissions, may be accelerated if warming extends crop regions into higher latitudes (Hartig et al., 1997).

Other agricultural processes that affect the global carbon balance include accelerated soil erosion, biomass burning, and depletion of soil fertility. Soil erosion due to water may cause about 1 Gt of carbon to be lost to the atmosphere each year (Figure 5.8) (Lal, 1995; Lal et al., 1995). Biomass burning, an important practice in shifting cultivation and tropical grasslands, releases carbon directly from the vegetation and indirectly from the soil; this releases nitrous oxide and methane as well, and erosion is usually accelerated. The loss of soil carbon by shifting cultivation is estimated to be 6.25×10^6 t C yr^{-1}; the loss due to natural fires in tropical savannas may be as much as 1.88×10^8 t C yr^{-1} (Lal et al., 1995).

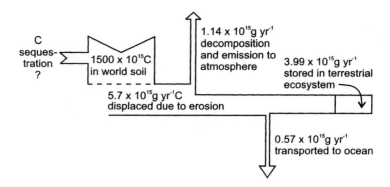

Figure 5.8
Global soil
erosion and
dynamics of soil
organic carbon
(Lal, 1995).

Agricultural practices that contribute to soil degradation include mechanized de-forestation, conventional tillage farming, continuous cropping on marginal lands, low-input and resource-based shifting cultivation, subsistence farming that leads to fertil-ity depletion, and overstocking and overgrazing of livestock (Lal et al., 1995). Tropical ecosystems, especially in dry regions, are more prone to degradation and carbon loss than temperate ones (Houghton and Skole, 1990). There are many agricultural prac-tices that can build soil organic matter, restore fertility, improve net primary produc-tivity, and restore soil structure, among them afforestation, conservation tillage, mulching, planted fallows and cover crops, pasture management and low stocking rates, and agroforestry. Chemical fertilizers, if properly applied, can aid in the process of restoration.

Soil Organic Carbon Sequestration

Increasing the amount of carbon held in organic matter in agricultural soils has been proposed as a means of mitigating the enhanced greenhouse effect. For example, the U.S. National Action Plan for Global Climate Change (USDS, 1992) emphasizes conservation reserve and conservation tillage programs. At present, about 32 million hectares are included in the conservation tillage programs sponsored by the 1990 Farm Bill. Since cultivation of soils over time results in a loss of organic carbon of about 30%, there seems to be a considerable potential for building back the levels of carbon stored in agricultural soils (Schlesinger, 1986). That potential, however, varies widely, depending on soil characteristics and management practices.

Using a model of soil carbon and nitrogen dynamics with crop growth and crop-ping practices, Li et al. (1992a, 1992b, 1994) reported that the simulated equilibrium soil organic carbon tends to rise with lower temperatures, increasing clay content, en-hanced nitrogen fertilization, greater manure application, and crops with higher residues. The largest carbon sequestration occurred with manure additions to soils that had a high clay content. Increased nitrogen fertilization generally enhanced carbon sequestration, but these results were dependent on soil texture, initial soil carbon con-tent, and annual precipitation. Reduced cultivation, as in zero-tillage or minimum-tillage practices, also tended to augment soil organic carbon, but the magnitude of the response was relatively small. When simulated precipitation was low, reducing tillage had little effect on modeled soil organic carbon in clayey soils and in soils with a low initial carbon content. (Reduced tillage, however, has other effects beneficial to the environment, such as lowered rates of wind and water erosion, energy consumption, and leaching of nitrates.)

The Li et al. model predicted, however, that it would take up to several hundred years for soil organic carbon to come into equilibrium under these practices, with climate being the dominating factor in determining the long-term equilibrium state. Crop rotation can also have a significant impact. Continuous corn provides large amounts of crop residues and, thus, high equilibrium soil organic carbon, while continuous wheat provides less residue and consequently lower soil organic matter in the long term. Li et al. concluded that for carbon sequestration in agricultural soils to be effective for greenhouse mitigation, efforts should focus on the specific climatic regions and agricultural practices offering the highest potential for increasing soil carbon content.

Most agricultural soils are currently not in equilibrium in regard to carbon storage; those soils furthest from equilibrium have the greatest potential for short-term carbon gains or losses. For carbon sequestration to be significant, substantial additions of organic matter to the soil are needed, in the form of either manure or crop residues. Reduced tillage and/or improved efficiency of nitrogen fertilization may be effective in some locations. Land set aside in conservation reserve programs should be monitored for carbon storage dynamics to determine if net carbon sequestration is indeed taking place, and to what extent. Other environmentally important processes, such as nitrate leaching to groundwater, should be monitored simultaneously.

Soil Resources for Agriculture

The basic soil characteristics that affect crop plants include temperature, moisture, aeration, fertility, depth, texture and stoniness, acidity, salinity and toxicity, tilth, slope, and flooding. Ranges of these characteristics may be defined for given crops with respect to (1) optimal production, (2) marginal growth threshold, where yield begins to diminish, and (3) unsuitability, where the crop cannot be grown under present technology. The ranges for these characteristics have been estimated for 12 major crops under rainfed cultivation (FAO, 1978). This type of tabulation is useful for evaluating soil-resource suitability for projected shifts in geographical crop regions. According to the U.S. Soil Taxonomy (Soil Survey Staff, 1992), soils now classified as having an ustic moisture regime may change to udic due to increased precipitation. Similarly, frigid soils may change to mesic soils with warmer temperatures.

Higher air temperatures will be accompanied by higher soil temperatures. In general, soil temperatures average about 1°C higher than air temperatures in the mid-latitudes and about 3°C higher than air temperatures in colder regions under the present climate (Buol et al., 1990). Soil temperature plays an important role in determining the adaptation and productivity of crops (Rykbost et al., 1975). Regions now limited by cold temperature at high altitudes and in the high latitudes may become suitable for cultivation, thanks to the earlier snowmelt and longer growing seasons. However, early planting of some crops (maize, for example) may still be vulnerable to late spring frosts.

Buol et al. (1990) estimated the potential extension of maize and wheat production in the Northern Hemisphere that might result from a 3°C mean temperature in-

Table 5.2   Potential increases in the areas of maize and wheat belts assuming a 3°C increase in mean annual air temperature (Buol et al., 1990)

Country	Current area[1]	Million ha Potential increase[2]	New total	Increase (%)
Maize				
Former Soviet Union	5	30	35	600
Canada	1	31	32	3100
China	17	11	28	65
United States	30	8	38	26
Wheat				
Former Soviet Union	50	71	121	142
Canada	13	19	32	146
China	29	0	29	0
United States	26	0	26	0

1. FAO, 1985. *Production Yearbook.* Food and Agriculture Organization of the United Nations. Rome.

2. Assuming 50% of new potentially arable land is locally suitable for cultivation.

crease on the basis of the mean annual and summertime air temperature isotherms (as a proxy for soil temperature) and the FAO world soil map. Currently, maize production is limited below a mean annual air temperature of 5°C and wheat is restricted when mean summer temperatures are below 18°C. The FAO/Unesco Soil Map of the World (FAO, 1990) was used to determine suitability of terrain and soil type for commercial farming. The results showed that areas potentially suitable for maize production will expand dramatically in the former Soviet Union and Canada, with smaller but still significant increases in potential wheat regions in both countries. The quantitative predictions are shown in Table 5.2 (Buol et al., 1990). The potential for expanding maize- and wheat-growing areas is quite low, however, in the United States and China. The actual expansion of these crops into higher latitudes depends on the economics of production and the comparative advantage of growing the crops in the new regions relative to existing areas. Overall, expanded production of wheat in regions where it is now limited by cold may be balanced by reduced production in regions where high temperatures will limit its growth. The production of maize, however, will probably not decline, since it is a crop of tropical origin (Buol et al., 1990).

Soil characteristics from warmer areas may be used to infer eventual changes in what are now cooler areas. For example, soils from lower latitudes tend to be highly weathered and of low fertility; such characteristics may slowly spread to higher latitudes if the warming trend persists over many decades.

Erosion

Changes in precipitation and wind regimes can lead to changes in the rates of soil erosion by water and wind. Soil erosion depends on climate (primarily on the quantity and intensity of precipitation), relief, vegetation (type, patchiness, and density), and soil erodibility characteristics (Lal, 1995). In areas where climate change brings higher precipitation (and more precipitation falling as rain rather than snow), erosion should increase. Where precipitation becomes lower, rate of erosion should fall. However, soil surface desiccation might make cultivated land more vulnerable to water and wind erosion, especially where the surface is devoid of vegetative cover and is pulverized by cultivation. Such conditions could generate "dust bowl" effects in some regions. The hazard of water erosion might also grow worse, as sudden—albeit infrequent—rainstorms strike at the soil. Soil erosion has a small but significant effect on the global carbon cycle, as described by Lal (1995).

A study of potential climate change in relation to land degradation in New South Wales, Australia, predicted significant increases in soil erosion, stream and estuary sedimentation, and soil salinity if rainfall intensities increase (Aveyard, 1988). Another study was conducted on the erosion potential for cropland, pastureland, and rangeland in the United States (Phillips et al., 1991). The study, which used the "universal soil loss equation" (Wischmeier, 1976; Hudson, 1995) projected that average sheet and rill erosion rates may change by +2 to +16% in croplands, −2 to +10% in pasturelands, and −5 to +22% in rangelands, depending on the climate change scenario. The areas with erosion rates above the soil loss tolerance level, and areas classified as highly erodible increased slightly. The results depended on whether precipitation changes were characterized as changes in storm frequency or storm intensity. A recent study using the Erosion/Productivity Impact Calculator (EPIC) model found that total erosion increased with 2°C warming due to increased wind erosion (by 15−18%), even though water erosion declined slightly (by 3−5%) (Lee et al., 1996). If semiarid grasslands become more arid, wind erosion will cause heavier loading of the atmosphere with dust (Schlesinger et al., 1990).

Actual changes in soil erosion will depend on whether and how farmers respond with timely and effective changes in their cropping patterns and other management practices.

Fertilization and Soil Management Practices

Buol et al. (1990) foresee no major changes in fertilization practices where crops are already heavily fertilized, since modern rates of fertilizer applications are much higher than the rates of nutrient release due to the weathering of naturally occurring soil minerals. Alterations in the timing and method of fertilization (e.g., adjustment of topdressing or side-dressing applications of nitrogen during the vegetative phase of crop growth) are expected with changes in temperature and precipitation regimes. Lamborg et al. (1983), on the other hand, projected that greater amounts of nitrogen and phosphorus will be needed to realize the full potential productivity gain due to high CO_2.

Hypothetically, even without atmospheric CO_2 enrichment and consequent climate change, improved crop varieties with higher yield potential in the future will probably require greater applications of fertilizers. To take full advantage of the higher growth potential offered by the buildup of CO_2, fertilizer needs will rise still further, perhaps by as much as one-third for doubled levels of CO_2. At present, the efficiency of nitrogen fertilizer use (i.e., the fraction of applied fertilizer that is actually taken up by the crop) is generally quite low—for example, Kimball (1985) reported a figure of only 50% for maize. Hence, some (but not all) of the increased need for nitrogen by crops can probably be met by improving the efficiency of fertilizer use. While increases in fertilizer requirements may be readily met in developed countries, farmers in the poorer developing countries may be hard pressed to pay for additional amounts of fertilizer. Fertilizer demand in developing countries is already projected to fall short of amounts needed by 2020 to provide food security, resource conservation, and nutrient replenishment (Bumb and Baanante, 1996).

Local soil tests and crop fertilizer response trials will be needed to ensure optimal adaptation to the foreseen changes in climate and CO_2 conditions. Increased rainfall generally causes greater leaching of mineral solutes through the root zone and a greater tendency toward acidification. To counteract that tendency, greater applications of lime (as well as fertilizers) will be required. However, Buol et al. (1990) estimated that a 10% increase in precipitation would not change the lime requirement of most agricultural soils significantly.

Nitrogen is the most critical nutrient for crop growth. In most agricultural soils, it is usually not available in sufficient amounts to attain full yield potential. External sources will continue to be needed (perhaps in greater amounts) to augment natural supplies of nitrogen. Clearly, the adequate supply of nitrogen for crops in the future will be a crucial task, especially under changing climatic conditions.

The amount of fertilizer that farmers actually apply depends on the price of the fertilizer relative to the price of the fertilizer-enhanced crop—that is, on the economic advantage to be gained by the practice. The prices of fertilizers and crops depend on the supply and demand for each commodity. Global supplies of nitrogen, potassium, and phosphorus are projected to be adequate in the coming decades, but fertilizer supplies in developing regions may be more limited (Bumb and Baanante, 1996). The major inputs to nitrogen fertilizer production are hydrocarbons and fossil fuels (including natural gas and coal) as energy sources for which there is likely to be greater competition from other sectors of the economy. If policies to mitigate the greenhouse effect are put in place, restrictions on the use of coal for energy-intensive activities may be instituted that would further raise the price of nitrogen fertilizer.

Summary and Research Needs

Soils play a major role in agriculture and in the global carbon cycle. In some cases, modifications of soil processes may result in loss of fertility, while in other cases they may provide an opportunity for mitigating the greenhouse effect through sequestration of carbon. Much research is still needed to understand the contribution of soil to the global carbon cycle and the gamut of possible responses of soils to global climate change. Understanding is especially limited by the lack of standardized methods for measuring and monitoring the dynamics of soil organic carbon. Data are not available for all ecosystems. These lacunae may lead to substantial errors in the aggregated estimates of soil carbon pools and fluxes.

The processes to be studied include nutrient dynamics, soil carbon accumulation or depletion, and trace gas exchange. Processes that require improved understanding include (1) the dynamics of both carbon dioxide and methane in the upper soil layers and (2) potential geochemical changes (such as mineral weathering, dissolution, and precipitation) in deeper soil layers. The exact nature of the effects of management practices is still largely unknown. Research is particularly needed on the interactive effects of high CO_2, temperature, moisture, and nutrient availability in order to project future fertilizer requirements. In this regard, important processes include mycorrhizal activity, nitrogen fixation, biomass production, mineralization-immobilization, nitrification-denitrification, organic matter accumulation and decomposition, and phosphorus availability. Simulation models can be useful in the prediction of climate change effects on soils, accounting for the many interactive processes and feedbacks that occur in the soil profile. Such predictions must be validated by measurements and experiments in the field.

6

Water Resources and Sea-Level Rise

W ater, being essential to all life and a basic requirement for agricultural production, is a valuable resource. Its efficient utilization demands careful planning and meticulous management. This is so not only for farmers who practice irrigation, but also for farmers in rainfed regions, as well as urban, commercial, and recreational users. Since climate change, should it occur, is quite likely to change the hydrological regimes of entire regions, it must be factored into water-resource planning and policies for the future, at least on a contingency basis.

Crops growing in the field are subject to an evaporative demand imposed primarily by the ambient climate. The parameters that affect evaporative demand are temperature, net radiation, atmospheric humidity, and windiness. As global climate change will be manifested in these variables, changes in the water regimes of crops will ensue. In the global hydrological cycle, water evaporated must be precipitated; hence, more evaporation implies more rainfall overall. However, increases in potential evapotranspiration and in rainfall may not be commensurate or concurrent in all locations. Climate models suggest that potential evapotranspiration tends to rise most where the temperature is already high (i.e., in low to mid-latitudes), while precipitation tends to increase most where the air is cooler and more readily saturated by the additional moisture (i.e., in higher latitudes and near seacoasts). Thus, drier conditions may occur in many of the world's most important agricultural regions, a consequence that could have great practical importance.

Both the demand for and the supply of water for irrigation will be affected by changing hydrological regimes. Important social and economic effects, either beneficial or adverse, could accompany such changes. Irrigated production tends to be more highly capitalized and expensive to manage. Hence, it must be profitable enough to justify the initial capital investment and the subsequent running costs, both of which are higher than in the case of rainfed production. The demand for irrigation water is generally projected to rise in a warmer climate. This may exacerbate competition for water resources between agriculture and other economic activities, including urban and industrial uses. Where water supplies are limited, more efficient irrigation, such as trickle or drip systems, could be applied more widely, notwithstanding their high investment costs. Extra demand and rising costs might cause some land to be withdrawn from irrigation. For example, in the region supplied by the Ogallala aquifer in the United States (parts of Nebraska, Oklahoma, Texas, Colorado, and New Mexico), the curtailment of irrigation (already begun) may accelerate, as water table depths and energy demands for pumping make water supplies more expensive. If, in addition, climate change requires more water per unit area, the practice of irrigation may indeed become prohibitive for some crops in some regions.

The future availability of water resources for agriculture will depend on possible changes in precipitation, potential and actual evaporation, and runoff at the watershed and river basin scales. Warmer winters will induce loss of natural storage in mountain snowpacks and subsequent reduction of stream flows in late summer and fall (Gleick, 1987). Such changes will affect the entire management of water resources, including reservoir operation, hydropower production, urban water use, flood control, environmental protection, and irrigation systems.

Global warming is predicted to lead to thermal expansion of seawater, along with the partial melting of glaciers and ice packs. The resulting rise of sea level may pose a threat to agriculture in low-lying coastal areas, where impeded drainage of surface water and of groundwater, intrusion of seawater into estuaries and aquifers, and land inundation might take place.

Soil Moisture

Although major emphasis has been placed on temperature rise per se as the most important manifestation of the enhanced greenhouse effect, potential changes in soil moisture may be equally important for agriculture. Changes in soil moisture arise secondarily from changes in radiation and temperature, so they are difficult to simulate. In general circulation models (GCMs), precipitation and melting of snow add water to the amount stored in the upper layers of the soil, while evaporation and runoff remove it. Regional precipitation depends in part on the moisture content of the wet or dry surfaces over which the air masses pass, the moisture that remains in the air masses, large-scale convergence and divergence patterns, and local precipitation triggering mechanisms. While these processes are included in GCMs, they are often represented in simplified forms. Hence, conclusions drawn from GCM results about potential changes in soil moisture should be taken with caution.

Early analysis based on the GCM of the Geophysical Fluid Dynamics Laboratory (GFDL) projected that soil moisture would be reduced in summer over the midcontinental regions of the middle and high latitudes. The regions include the North American Great Plains (Figure 6.1), western Europe, northern Canada, and Siberia (Figure 6.1) (Manabe and Wetherald, 1986, 1987). The anticipated reduction of soil moisture is on the order of 20%. The broadscale mechanisms proposed for the midcontinental summer drying were earlier snowmelt followed by a period of intense evaporation; lesser precipitation; and reduced cloud cover leading to more intense solar energy reaching the surface and, hence, greater evaporation (Manabe and Wetherald, 1987). The reduced cloud cover sets up a positive feedback that enhances

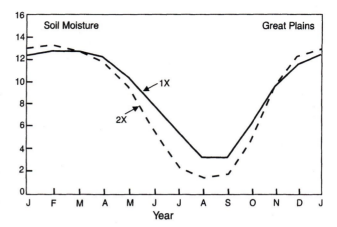

Figure 6.1 Seasonal variation in simulated soil moisture in the Great Plains between high CO_2 and normal CO_2 cases for the GFDL GCM with variable clouds (Manabe and Wetherald, 1987).

the drying process. An opposite effect was found in winter (Manabe and Wetherald, 1987).

An analysis of soil moisture results obtained from five GCMs for a hypothetical doubling of atmospheric CO_2 (to ~600 ppmv) reveals a tendency toward drier conditions in the inland sections of the North American continent (Kellogg and Zhao, 1988), although the soil moisture predictions from the different GCMs were not uniform. In a companion analysis of soil moisture conditions for Asia, the five GCMs all predicted that the monsoonal circulation will intensify in southern Asia, resulting in a trend toward drier winters and wetter summers. The same study predicted the reverse for northern Asia, with summertime drying likely to be most pronounced in the region of Manchuria (Zhao and Kellogg, 1988).

Rind et al. (1990) have argued that the prospective drier soil conditions are equivalent to an increased frequency of drought due to the greater atmospheric demand for water (i.e., the potential evaporation) relative to the atmospheric supply of water (i.e., precipitation). In another study, soil moisture changes were projected for Dodge City, Kansas, as calculated on the basis of two GCMs (Figure 6.2) (Rosenzweig and Hillel, 1993). These projections tend to be more severe than the Dust Bowl decade of the 1930s and the extreme drought year of 1934.

There is also general agreement among GCMs that wetter conditions may prevail in northern high latitudes during winter. Wetter conditions in the climate models are caused by increases in winter precipitation, more precipitation falling as rain rather than snow, hastened snowmelt, and smaller changes in potential evaporation at lower winter temperatures (IPCC, 1990a). Coastal areas along the Gulf Coast and the West Coast of the United States are also projected to become wetter in most of the GCM simulations.

How might a drying of mid-continental Northern Hemisphere regions in summer affect crop production? In principle, lower soil moisture can be expected to lower productivity because of the greater likelihood of crop water stress. There are several mitigating factors that should tend to dampen, although probably not completely overcome, negative effects on yields. First, the earlier thaw should lengthen the growing season, allowing earlier planting and avoidance of severe summer drying. Second, if

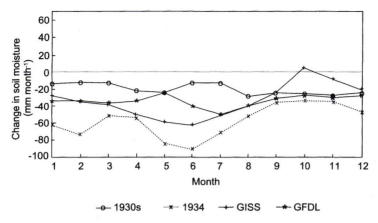

Figure 6.2   Soil moisture anomalies calculated by the Palmer Drought Severity Index method relative to 1951–1980 in Dodge City, Kansas, for the 1930s, 1934, and the GISS and GFDL climate change scenarios (Rosenzweig and Hillel, 1993).

crop growing periods are further shortened by faster physiological development (caused by higher temperature), less moisture will be removed from the soil. Consequently, a greater opportunity may exist for soil moisture recharge, with positive prospects for the following crop growing season. Finally, if crop growth is limited by nutrient supply or by shortened duration of development, increases in biomass due to high CO_2 may not occur and decreased transpiration rates may lead to increases in the amount of soil water remaining after the crop growing period (Goudriaan and Unsworth, 1990). These biological factors would tend to reduce but not eliminate the negative effects of summer drying on crop yields.

Irrigation

Irrigation is the artificial enhancement of soil moisture aimed to promote crop productivity. Various techniques exist for supplying water to crops. In arid regions, irrigation generally provides most of the water required for crop growth. In more humid regions, supplemental irrigation is provided periodically to prevent yield losses caused by seasonal moisture stress. About 17% of the world's cropland is under irrigation, and irrigation has been increasing steadily since the 1970s, particularly in developing nations in Asia (Figure 6.3). In Korea and Pakistan, for example, over 50% of the cropland is irrigated. Water for irrigation is taken either from surface water resources (lakes, streams, and rivers) or from groundwater (aquifers). Due to higher quantity and quality of yields, irrigated crops tend to be more economically valuable than rainfed crops.

An important task is to project the combined impacts on irrigation of both the general warming and the specific physiological responses of crops to higher atmospheric CO_2 content. Such projections can contribute to assessments of future water requirements for agriculture, since the future welfare of farmers and rural communities in regions that depend on irrigation may be critically affected by climate change.

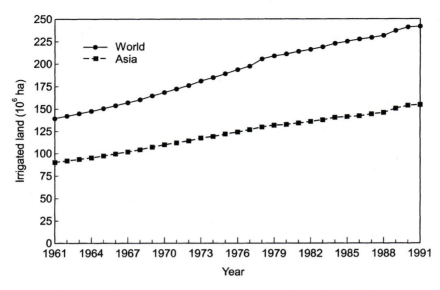

Figure 6.3   Irrigated land (million hectares) in the world and Asia (World Resources Institute Database, 1994).

The current intense competition for water and the growing economic importance of irrigated agriculture should provide stimuli for conducting such assessments. Irrigated agriculture also has significant secondary effects on runoff and downstream water quality, with associated human and environmental consequences.

Irrigation Requirements

The amount of water needed to irrigate a crop (the so-called irrigation requirement) depends principally on crop evapotranspiration. Rising air temperatures generally intensify vapor pressure deficits, reduce the transfer of sensible heat from crop surfaces in humid areas, stimulate plant development, and lower the latent heat required to evaporate water. The overall effect is to increase crop evapotranspiration (Allen et al., 1991).

Rosenberg et al. (1990) conducted a detailed sensitivity study of changing climatic and biophysical effects on a grassland in Kansas, based on the Penman-Monteith equation (Monteith, 1965). They estimated that a 1°C rise in mean air temperature would entail a 4 to 8% increase in evapotranspiration. They also projected for the same setting that:

1. A 1% increase in net radiation will raise grassland evapotranspiration by an average of 0.6%.
2. The combination of a 3°C rise in air temperature, a 10% increase in net radiation, and a 10% reduction in atmospheric vapor content will boost evapotranspiration by 20 to 40%.
3. An enlargement of the total leaf area of plant canopies by 15% will result in 5% more evapotranspiration.

4. A 40% gain of leaf stomatal resistance will reduce evapotranspiration by about 15%.

Peterson and Keller (1990) attempted to project the potential net irrigation requirement, defined as the volume per unit area of irrigation water needed to maximize crop production, for the United States. They calculated the required crop evapotranspiration less the effective rainfall for a number of climate change scenarios. The scenarios were as follows: (1) a 3°C rise in temperature without change of precipitation; (2) the same warming with a 10% rise in precipitation; and (3) the same warming with a 10% reduction in precipitation. In all cases, irrigation requirements increased under the warmer climate because of the lengthened growing season, multiple cropping, shifts in crops, greater evapotranspiration, and lower effective precipitation (Figure 6.4) (Peterson and Keller, 1990). The most severe impacts were predicted to take place in the Great Plains, and the least severe in the Pacific Northwest. Further, Peterson and Keller projected that irrigators in the western United States would have difficulty in maintaining agricultural productivity, even with improvements in irrigation efficiency. For the eastern United States, they predicted an expansion of the irrigated area. If CO_2 enrichment induces closure of stomates, thereby reducing evapotranspiration, these same results would occur with a temperature rise of about 4°C instead of 3°C.

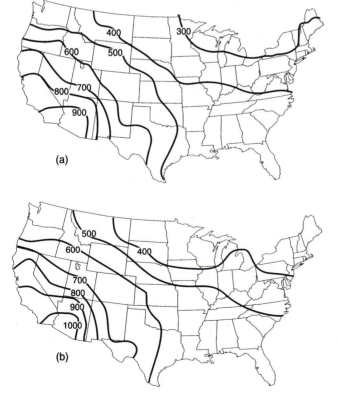

Figure 6.4 Potential net irrigation requirement (mm) over the United States for present (a) and (b) +3°C (Peterson and Keller, 1990).

If irrigation increases both in amount and in area under warmer conditions, watershed runoff is predicted to diminish, particularly in arid areas (Peterson and Keller, 1990). In drier regions, irrigation tends to reduce runoff because of higher crop transpiration. More water is also exposed to the air during storage, conveyance, and application, causing yet more evaporation. Greater seepage also reduces surface runoff. In the western part of the United States such changes may be dramatic. Depletion of runoff due to irrigation is expected to increase by about 25% with a 3°C warming (Peterson and Keller, 1990).

The utilization of water for agriculture may, incidentally, affect sea-level rise, since some of the water so used would otherwise flow into the sea. A quantitative assessment of this effect has recently been made by Gornitz et al. (1994; 1997).

Allen et al. (1991) also assessed the potential impacts of CO_2-induced changes in climate and in plant physiological mechanisms on irrigation-water requirements of the southern and central Great Plains. They utilized a model with crop phenology and with evapotranspiration that responds to varying canopy and aerodynamic resistances. Alfalfa, maize, and wheat were the three crops tested. Hypothetical increases in canopy resistances (the quotient of stomatal resistance of single leaves and leaf area index (LAI)) of 20 to 80% over current values were tested. The climate factors were surface air temperature, precipitation, solar radiation, wind speed, and humidity. In general, the climate change scenarios utilized in the study (based on the GCMs of the Goddard Institute for Space Studies [GISS] and GFDL) projected for the Great Plains region increases in temperature, precipitation, humidity, and solar radiation, as well as widely fluctuating changes in wind speed.

The Allen et al. (1991) study projected that seasonal evapotranspiration of alfalfa will increase due to increases in length of the crop life cycles, while seasonal evapotranspiration of maize and winter wheat will decline due to decreases in the length of the growing period and to changes in crop calendars. Peak monthly evapotranspiration was projected to increase for the doubled CO_2 scenarios for all crops and for all sites except one, because of the greater evaporative demand imposed by the warmer atmosphere, greater wind speed, and higher solar radiation. Of these three variables, higher air temperature had the greatest effect. Changes in irrigation requirements varied by crop.

According to Allen et al. (1991), seasonal irrigation requirements for alfalfa can be expected to rise with the predicted extension of the frost-free growing seasons and the higher evaporative demands. Only modest changes of seasonal irrigation requirements were estimated under the doubled CO_2 scenarios for maize and winter wheat because of the hastened maturation and the consequent reduction in length of the crop life cycles. Enhanced canopy resistance moderated the rise in irrigation requirements, thus manifesting a possibly beneficial physiological CO_2 effect on crop production. Peak monthly irrigation requirements tended to increase in all crops however.

Cohen (1991) researched the possible impacts of climatic warming on water resources in the Saskatchewan River subbasin in Canada. Results with five GCM scenarios and ten hypothetical warming scenarios indicated decreases in summer soil moisture and increases in irrigation demand, but no consensus on changes in runoff or annual net basin supply.

In general, irrigation requirements are predicted to rise in response to greater atmospheric evaporative demands, lengthened cropping seasons, and reduced precipitation wherever they occur. The rise in irrigation requirements may be moderated somewhat by increasing canopy resistance, although this effect may itself be partially obviated by the possible rise of leaf surface temperatures. The rise in leaf surface temperatures could also have significant effects on crop metabolism and yields, and it may make crops more sensitive to moisture stress.

Adaptation

Under future climates, irrigators may change their cropping systems to take advantage of or adapt to new climatic regimes. Some of the changes may include growing longer season cultivars, increasing cropping intensities (i.e., the number of successive crops produced per unit area per year), or planting different types of crops. How farmer adaptation to the projected climatic conditions might affect future irrigation requirements was also included in the study by Allen et al. (1991), who considered the effects of switching to longer season crop varieties to counteract the compression of crop development and to take advantage of a longer potential growing season. The use of extended-season maize and wheat cultivars tended to increase seasonal irrigation requirements. Whether farmers will adopt such cultivars depends on future economics, which will be governed by the costs of applying extra irrigation water and the relative changes in yields. (High temperatures will likely depress yields of irrigated crops, relative to irrigated yields under current temperature regimes).

Increased length of the total potential growing season and compressed lengths of specific life cycles for annual crops may encourage farmers to grow two or more crops per year in regions with sufficient water supplies. Such increases in cropping intensity would almost certainly result in greater irrigation requirements. Improving the efficiency of water use will aid in adapting to such greater demands.[1]

If hydrological patterns change markedly and irrigated agriculture is required to relocate in response, prior investments may be lost as existing infrastructures become obsolete, and new investments will be needed for the new installations. The economic and social costs of relocating irrigation systems could be considerable. Regional water-resource managers may need to factor in future construction of irrigation-supply facilities such as dams, reservoirs, wells, pumps, canals, and pipes. Irrigation also tends to be highly energy-intensive, a factor to be considered in view of the foreseeable need to reduce the overall use of CO_2-emitting fossil fuels.

Water Resources for Agriculture

The hydrological changes likely to result from the global greenhouse effect will influence the supply of water available for agriculture, as well as the demand for water as described in the preceding section. Farmers who practice irrigation should be less vulnerable to climate change than dryland farmers, provided, of course, that the former are assured of a continuing and adequate supply of water. However, in the United States, for example, water resources are already limited in terms of supply, demand, requirements of hydroelectricity, groundwater withdrawal, and current climate variability in several regions (Figure 6.5) (Gleick, 1990). Competition between hydropower and irrigation may intensify if fossil fuel use is curtailed in an effort to mit-

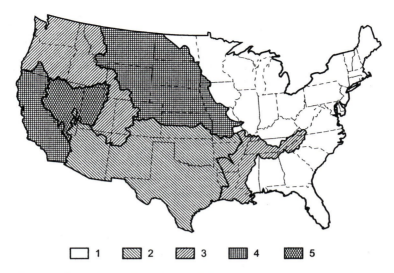

Figure 6.5 Vulnerability index for U.S. water-resource regions (redrawn from Gleick, 1990). Numbers 1–5 indicate lower to higher vulnerability, based on the number of vulnerability categories present.

igate greenhouse gas emissions (Rosenberg, 1996). Development of new supply systems (including storage, pumping, and conveyance) will be expensive.

Preparation for a change in climate variability should include the provision for the possibility of increased flooding as well as the incidence of drought. Flooding can threaten storage dams, diversion structures, and canals. Under changing climate conditions, records of past climate variability will no longer be reliable predictors of future events, and thus the likelihood of damage by unexpected extreme events will rise.

Regional Examples: The Ogallala Aquifer and Midwestern River Basins

The improvement and spread of well drilling and pumping technology after World War II permitted the extraction of water from the immense Ogallala aquifer (Figure 6.6). As of 1982, 19 million acres, or 12% of the cropland in the Great Plains, mostly in the southern Plains, were under irrigation. Groundwater has supplied most of the water for irrigation: 61 to 86% of the water applied to crops in Nebraska, Oklahoma, and Kansas, compared with only 20% nationally. In 1982, this aquifer supplied irrigation for approximately 14 million acres in the Great Plains states of Colorado, Nebraska, Kansas, Oklahoma, New Mexico, and Texas (High Plains Associates, 1982). The aquifer allows the irrigation of terrain too far from surface supplies, and it provides water for municipal and industrial needs as well. The section of the aquifer that is in the southern reaches, however, is already seriously depleted.

The Ogallala aquifer varies spatially in depth of the water table, rate of natural recharge, and thickness of the saturated water-bearing strata (Frederick and Hanson, 1982). In the Texas panhandle and its neighboring areas of Oklahoma and New Mexico, where the Ogallala has long been tapped chiefly for cotton (and to a lesser extent for corn, wheat, and sugarbeets), the depletion has been most serious. Here, the

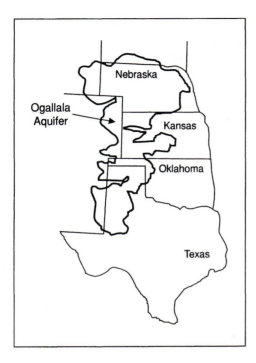

Figure 6.6 Ogallala aquifer (Powers, 1987)

high withdrawal and low recharge rates resulted in "mining" of the resource and have recently forced the abandonment of thousands of formerly irrigated acres (Wilhite, 1988). In Nebraska, where the aquifer has a higher recharge rate than in the southern areas, significant drawdown problems have not yet occurred. Farmers in Nebraska recently began to use the aquifer to irrigate corn, which is grown mostly for livestock feed.

Since 1980, water levels in some parts of the Central and Southern High Plains have continued to decline, but at a slower annual rate (Dugan et al., 1994). The slower rate of decline is attributed to a decrease in groundwater application for irrigation, reduction in irrigated acreage, water-conserving practices, and improved technology. Greater-than-normal precipitation may have contributed to slowing declines or even increases in water levels in some locations of the aquifer.

Glantz and Ausubel (1984) have argued, in any case, that projections of the region's future must include consideration of its diminishing water resources as well as of its susceptibility to future droughts such as are projected by GCM simulations, since both factors are critical to the future of agriculture in the area.

Frederick (1993) compared current (1951–1980) and historical (1930s) streamflow in Midwestern rivers as part of a coordinated study on the effects of climate change (Rosenberg, 1993). The Dust Bowl climate reduced the long-term means by 28%, 28%, and 7% in the Missouri, Upper Mississippi, and Arkansas-White-Red River basins, respectively. Projecting to a future with similar climate conditions, Frederick estimated tht it would be necessary to curtail hydropower production by about 50% if the large reservoirs on the upper Missouri continued to operate under current practices. The study also projected that demand for irrigation water would be increasing at the same time. With no change in irrigated acreage (an unrealistic assumption), the study projected an increased demand for irrigation water of 39% in Nebraska and 12% in Kansas (Rosenberg, 1993). Thus, significant shortfalls in supply may occur based on Frederick's streamflow calculations.

Social and Economic Factors

Communities dependent on irrigation have in the past learned to adapt to climatic variability, changing economic conditions, receding groundwater, and transfers of water rights. The resiliency with which communities have responded to varying circumstances can offer analogues or models by which to prepare for potential climate change in the future (Peterson and Keller, 1990). Government programs have often aided such communal restructuring. In less fortunate cases, however, overdevelopment of regional irrigation systems—often subsidized by government programs—has actually increased the vulnerability of irrigation-based farming communities.

Many social and economic factors must enter into a comprehensive assessment of future regional conditions for irrigated agriculture under changing climate conditions. These include farmland values, crop prices, costs of irrigation (including pumping energy costs), costs of production in addition to irrigation, government subsidy programs for idling land, and the economic situation of both prosperous and marginal farmers in a region. Studies of responses to past long-term droughts have shown that financial and social effects are operative by the end of the second year of a drought, mainly impelled by lowered farm income (Peterson and Keller, 1990). If deficits are severe relative to capital investments and running costs, long-term droughts can cause widespread bankruptcy. The effects of short-term droughts may be alleviated by drilling additional wells and by drawing on groundwater with wells already in place, but such measures tend to diminish groundwater reserves and raise the cost of pumping.

In the United States, social services are now in place in some rural communities to survive short-term droughts and return to normalcy when a drought ends. However, with projections of prolonged drying in mid-continental regions comes the possibility that some rural communities will be unable to return to rainfed farming at all (Peterson and Keller, 1990). The maintenance of social services and infrastructure such as schools, hospitals, and roads for rural communities will then be more difficult.

International River Basins

An international study recently examined how climate change might affect the development of five major international river basins: the Nile, Zambezi, Indus, Mekong, and Uruguay (Riebsame et al., 1995). A wide range of hydrological responses was found, based largely on the different water regimes of the basins in the current climate. Figure 6.7 shows the relative dryness of the five basins by graphing the proportion of basin precipitation that runs off (the runoff:precipitation, R:P, ratio) against the evaporation regime (the ratio of potential evapotranspiration:precipitation, PET:P) of each basin (Riebsame et al., 1995). In general, sensitivity to climate change increases with lower R:P ratios and larger PET:P ratios. Schaake (1990) has defined an index of "hydrological elasticity," defined as the ratio of percentage change in runoff to percent-

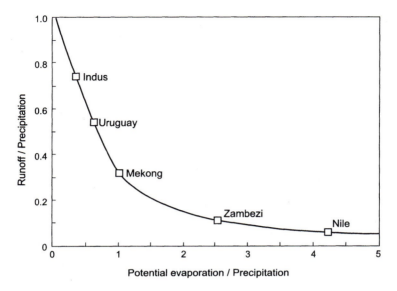

Figure 6.7 Generalized relationship between runoff as fraction of precipitation (R:P) and potential evapotranspiration as fraction of precipitation (PET:P). Sensitivity to climate change increases with lower R:P ratios and larger PET:P ratios (Riebsame et al., 1995).

age change in precipitation, shown in Table 6.1 for three river basins (Riebsame et al., 1995). The drier basins, the Nile and Zambesi, show the greatest hydrological sensitivity and indeed are projected to suffer the greatest impacts for the climate change scenarios tested. Runoff is seen to diminish in these river basins even when precipitation increases because warmer temperatures cause even greater evaporation.

The wetter basins (namely, the Mekong, Uruguay, and Indus) show less sensitivity to temperature change but more sensitivity to precipitation change (Riebsame et al., 1995). Although increased runoff could augment water supply and hydropower production, it might also cause deleterious effects, such as greater flooding, waterlog-

Table 6.1 Hydrological elasticities[1] for three major river basins (Riebsame et al., 1995)

Parameter	Indus	Zambezi	Nile
Precipitation	0.950	1.88	3.6
Temperature	−0.125[2]	−1.68[3]	−4.5[4]

1. Ratio of percentage change in runoff to percentage change in precipitation (Schaake, 1990).

2. Based on a 2°C warming on a 25°C basin average.

3. Based on a 2°C warming on a 21°C basin average.

4. Based on a 4°C warming on a 20°C basin average.

ging, and salinity. Over 25% of the irrigated land in the Indus basin is already affected by waterlogging and salinization (Hillel, 1991).

The five river basins varied in the most climate-sensitive factors of management and development (Table 6.2) (Riebsame et al., 1995). Irrigation is a high priority in the Indus river basin, while hydropower production is more important in the Zambezi and Mekong basins. To some extent, the degree of development of river basins is also a measure of sensitivity to changing climate conditions. The ratio of available water storage to annual runoff gives an indication of how the water-resource managers of a basin may be able to moderate the effects of both high and low flows. Table 6.3 shows that the Indus, Mekong, and Uruguay have lesser storage ratios than the Zambezi and the Nile basins and therefore permit less control over water supply (Riebsame et al., 1995). The relative vulnerabilities to climate change of the five basins based on hydrological characteristics and development levels are shown in Table 6.4 (Riebsame et al., 1995).

Major water-development decisions may have greater relative hydrological, environmental, and social impacts than climate change per se in the shorter term. However, climate change will affect the benefits to be accrued by future water development and should therefore be taken into consideration in the context of water-resource policies and planning (Riebsame et al., 1995).

In some cases, climate change may reduce the benefits of investment in augmenting water supplies (Figure 6.8) (Riebsame et al., 1995). Increased efficiency of water use is to be encouraged, especially since new and larger projects may not always guarantee effective adaptation to climate change. Cooperation among the countries sharing a river basin is essential, so that the interrelated needs of all the users, as well as the potential impacts of climate change, can be integrated into water-resource development (Hillel, 1994).

Water Quality

As the recycling of waste water gains importance in the future, the problem of water quality, already serious in some areas, will become yet more acute. Water quality tends to deteriorate under conditions of low flows and higher water temperatures, which are predicted for arid areas. In such areas, the impact of climate change on water quality may be especially significant. A study of the potential effects of climate change on the water resources of the southern United States projected that surface water temperature may rise by as much as 7°C (Cooter and Cooter, 1990). The study also found that although advanced treatment technology (presently available) would be sufficient to maintain desirable levels of instream dissolved oxygen even under elevated water temperature conditions, the use of such expensive technology would become necessary on a much wider scale than at present.

Sea-Level Rise

Greenhouse warming on a global scale could raise sea level between 10 and 40 cm by the middle of the 21st century, and between ~20 and ~90 by 2100 (IPCC, 1996). While sea level is already slowly rising in many locations, due primarily to geologic processes and anthropogenic manipulation, the projected sea-level rise associated with

Table 6.2 Critical factors of water-resource development and management for five major river basins (Riebsame et al., 1995).

Basin	Sea-level rise	Salt-water intrusion	Hydro-electric power	Fisheries	Navigation transport	Irrigation	Municipal industry	Domestic water supply	Riverine flooding	Environmental quality
Uruguay			xx	x		x	x	xx	x	
Mekong	x	xx		x	xx	xx		xx	x	x
Indus	x	x	x	x		xx		xx	x	x
Zambezi	x		x	xx		x			x	
Nile	x	x	xx		xx	xx	x			

xx = Most important

x = Important

Table 6.3 Storage-to-runoff ratio for five major river basins (Riebsame et al., 1995)

Uruguay	Mekong	Indus	Zambezi	Nile
0.10	0.10	0.25	3.00	2.50

the enhanced greenhouse effect represents an increase of 2 to 5 times over present rates. Several mechanisms will contribute to this rise—namely, thermal expansion of seawater, melting of mountain glaciers, and changes in accumulation or depletion (melting) of polar ice sheets. The potential impacts of this accelerated sea-level rise include inundation of low-lying coastal areas and estuaries, retreat of shorelines, and changes in the water table. With such changes, coastal zones may be permanently inundated to an elevation equivalent to the vertical rise in sea level. Further, episodic flooding from high storm surges could penetrate much further inland. Increasing salinization of coastal aquifers and of streams as a consequence of sea-level rise could contaminate the soil itself and adversely affect agriculture.

Crops, Coastlines, and Countries

In saline conditions, crop plants are continuously under osmostic stress as a result of high solute content in the rooting medium. Furthermore, the high concentrations of specific ions may cause toxic reactions. The majority of crop plants are sensitive to saline conditions, responding with reduced yields or, in extreme conditions, in total crop failure. Some crop plants, especially forage grasses, do have a degree of salt tolerance. Economically important crops (rice, sorghum, and barley, among others) are being tested for genetic salt-tolerance potential (Hale and Orcutt, 1987). It is interesting to note, incidentally, that salt tolerance as a trait does not preclude a positive response to CO_2 enrichment. In fact, halophytes (naturally salt-tolerant plants) of both the C3 and C4 photosynthetic pathways have responded with increases in dry matter accumulation to enriched atmospheric CO_2 (Schwarz and Gale, 1984).

The impacts of global sea-level rise on coastal zones will vary from region to region because of local factors such as land subsidence, susceptibility to coastal erosion

Table 6.4 Sensitivity and adaptability of five major river basins to climate change (Riebsame et al., 1995)

Basin	Hydrological sensitivity	Structural robustness	Structural resiliency	Adaptive capacity
Uruguay	Moderate	High	High	Moderate
Mekong	Low	Low	High	High
Indus	Moderate	High	Moderate	High
Zambezi	High	Low	Low	Low
Nile	High	High	Low	Low

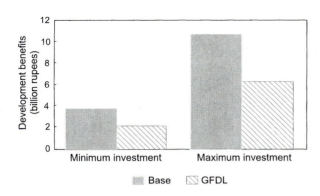

Figure 6.8　Changes in expected benefits from planned Indus Basin water developments caused by the GFDL climate change scenario. The base is equal to the total value added by irrigation agriculture in the year 2000 if no new projects are built. The "no new projects" scenario is unfeasible and was estimated. A 3.25% annual economic growth rate was assumed (Riebsame et al., 1995).

or sedimentation, varying tidal ranges, and cyclonicity. For example, local subsidence in the Mediterranean region (e.g., in the Po and Nile deltas) could result in sea-level rise greater than the global average, affecting highly populated areas (Milliman, 1992). Criteria for assessment of coastal vulnerability to rising seas have been developed, and they account for the hazards associated with permanent and episodic inundation and erosion (Table 6.5) (Gornitz et al., 1991, 1994). The Food and Agriculture Organization of the United Nations (FAO) has developed a more detailed database for major deltas and coastal plains in developing countries (Jegersma et al., 1993).

Sea-level rise will cause saltwater intrusion and rising water tables in agricultural soils located near coastlines. The deltas of major rivers are often extensively used for agriculture. In areas that already suffer from poor drainage (e.g., Bangladesh, China, Egypt, Indonesia, the Netherlands, and, in the United States, Louisiana and California), agriculture in coastal areas could become increasingly difficult to sustain.

The major low-lying river deltas of southeast Asia (Ganges-Brahmaputra in Bangladesh, Irrawaddy in Burma, Chan Praya in Thailand, Mekong and Hong in Vietnam) are among the regions most vulnerable to sea-level rise. These deltas support extensive rice cultivation areas that face permanent flooding by even a small rise in sea level. A substantial percentage of rice paddies in low-lying areas of southeast and east Asia will be vulnerable to anticipated sea-level rise (Milliman et al., 1989; Parry et al., 1992, Asian Development Bank, 1994). For example, around 16% of the rice production of Bangladesh might be lost under a 1-m sea-level rise scenario (Figure 6.9) (Huq et al., 1995). Both cultivated land and farmers' homestead lands would be affected (Karim et al., 1996). These losses could be even greater if the effects of salinization beyond the inundated areas are taken into account.

Hydrological Aspects

A coastal aquifer is bounded on at least one side by an extensive saltwater body—a lagoon, a sea, or an ocean. Due to the direct contact between the freshwater in the aquifer and the saltwater body, coastal aquifers are vulnerable to encroachment of saline water and consequent degradation of water quality, which may affect human welfare and agricultural production. Freshwater is rendered marginal for human con-

Table 6.5 Risk classification for sea-level rise (Gornitz et al., 1994)

Variable	Risk Level				
	Very low	Low	Moderate	High	Very high
Elevation (m)	≥30.0	20.1–30.0	10.1–20.0	5.1–10.0	0–5.0
Geology (relative resistance to erosion)	Plutonic Volcanic (lava) Metamorphic	Low-grade metamorphics Sandstone and conglomerate (well-cemented)	Most sedimentary rocks	Coarse and/or poorly sorted unconsolidated sediments	Fine unconsolidated sediments Volcanic ash
Landform (geomorphology)	Rocky, cliffs coasts fiords	Medium cliffs Indented coasts	Low cliffs Glacial drift Salt marsh Coral reefs Mangrove	Beaches (pebbles) Estuary Lagoon Alluvial plains	Barrier beaches Beaches (sand) Mud flats Deltas
Relative sea-level change (mm/year)	<−1.0 (Land rising)	−1.0–0.99	1.0–2.0 (Within range of eustatic rise)	2.1–4.0	>4.0 (Land sinking)
Shoreline erosion or accretion (m/year)	>2.0 (Accretion)	1.0–2.0	−1.0–+1.0 (Stability)	−1.1–−2.0	<−2.0 (Erosion)
Mean tide range (m)	<1.0 (Microtidal)	1.0–1.9	2.0–4.0 (Mesotidal)	4.1–6.0	>6.0 (Macrotidal)
Maximum wave height (m)	0–2.9	3.0–4.9	5.0–5.9	6.0–6.9	>6.9
Annual tropical storm probability (%)	0–8.0	8.1–12.0	12.1–16.0	16.1–20.0	>20.1
Annual hurricane probability (%)	0–4.0	4.1–8.0	8.1–12.0	12.1–16.0	16.0–20.0
Hurricane frequency-intensity index	0–20	21–40	41–80	81–120	>120
Mean forward velocity (m/sec)	>15	15.0–12.0	12.1–9.0	9.1–6.0	<6.0
Annual mean no. extra-tropical cyclones	0–10.0	10.1–20.0	20.1–30.0	30.1–40.0	>40.1
Mean hurricane surge (m)	0–2.0	2.1–4.0	4.1–6.0	6.1–7.0	>7.0

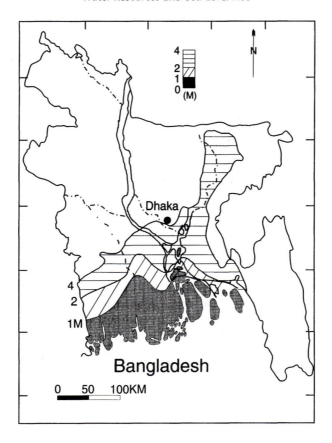

Figure 6.9 Projected effect of sea-level rise in Bangladesh. Land up to 1 m (gray) could be permanently flooded, according to the IPCC 1990 high sea-level rise scenario. Episodic flooding due to severe typhoons could affect the 2 to 4 m zone. (V. Gornitz, personal communication).

sumption if it is contaminated with seawater at the 2 to 3% level, and it becomes unfit for human use at the 5% level.

Groundwater is the main source of freshwater in many coastal areas, especially in arid and semiarid regions. The natural flow of the freshwater aquifers toward the sea creates a dynamic balance between the freshwater body and the saline water body, which normally prevents the landward intrusion of seawater into the aquifer. As population in coastal zones grows and increases the demand for freshwater (for agricultural, industrial, and domestic purposes), the rate of groundwater withdrawal by pumping rises correspondingly. This leads to a lowering of the water table and an upsetting of the freshwater-saltwater equilibrium. Saltwater intrusion then ensues. This phenomenon is already occurring in many coastal zones around the world. One relevant example is Egypt, where seawater has intruded as much as 35 km into the Nile delta.

Other locations where saltwater intrusion is now occurring include the Netherlands, Israel, Dakar, and Madras. In the United States, cases are found in Long Island and Miami and in the coastal regions of California.

The freshwater in the aquifer and the saline water of the sea are miscible fluids (Hillel, 1980). The two bodies of water meet in a transition zone in which the concentrations vary from entirely freshwater on the land side to seawater on the ocean side. At the seaward boundary, seawater enters the system and because of its greater density migrates to the bottom of the aquifer and tends to displace the freshwater there. The conformation and the degree of saltwater intrusion into a coastal aquifer depend on the type of aquifer (either confined, phreatic, leaky, or multilayered) and its characteristics (M. M. Sherif, personal communication). Other factors include the level of water table, peizometric head, seawater concentration, rate of flow, capacity and duration of water withdrawal or recharge, and tidal effects.

When pumping exceeds the rate of recharge, as is likely in areas of coastal development in the future, saltwater upconing into wells generally takes place. Productive wells often pump from a freshwater layer that is underlain by saline water strata. As pumping continues, the water table (or the piezometric head) falls, and the interface between the fresh and saline water rises in the form of a mound toward the pumping well. If the bottom of the well is close to the saline water strata or if the well discharge exceeds a certain critical value, the saltwater cone will enter the pumping well, and the withdrawn water will be brackish.

With climate change and sea-level rise, near-shore soils will likely be exposed to intermittent flooding, incremental salinity, and perhaps to acidification as well. In some regions, the colonization of mangrove or reed species will enhance accumulation of organic matter and precipitation of pyrite from sulfate in brackish water, and the weathering or oxidation of pyrite during seasonally dry periods will promote soil acidification. The saltwater–groundwater interface will tend to migrate landward, and the salinity front will begin to penetrate further upstream. Drainage will be impeded due to rising water table levels.

The combination of sea-level rise and an intensified hydrological regime and flooding could damage agricultural productivity and local fisheries in many coastal zones. Pumping programs must be carefully planned to include adequate water-quality monitoring. Changes in annual river floods should also be considered in projecting coastal zone impacts, as river floods help to hold back saltwater intrusion.

Preparing for the Future

A possible change in regional water resources is perhaps the most critical impact of climate change, not only for agriculture but also for other economic activities. As planners prepare for the future, the dynamic nature of water-resource management needs to be fully considered, both physically and socially. Lessons from past development (especially from past failures) should be remembered and applied judiciously. Approaches drawing on current water problems and political and cultural perspectives are also useful (Wescoat, 1991).

Given the possibility of coming climate change, improvement of irrigation efficiency is essential to better utilize limited and vulnerable water resources. Such improvement is imperative in any case as competition for water resources from nonagricultural uses grows. Technology is involved in the task, but so is institutional infrastructure. Good public management of water will be especially crucial should climate change and its hydrological effects exceed the range of current climate variability. If water transfers are proposed across basins, states, and countries, environmental constraints will doubtless come into play.

In the United States, the possibilities for expanding supplies of irrigation water from groundwater are limited. In fact, groundwater is declining in some of the major aquifers currently used for irrigation. Development of new irrigation regions involves high costs for building storage, conveyance, and pumping facilities. Developing additional storage will be more difficult in the future, as the most suitable sites are already taken. Some enhancement of supply will come from water recycling, particularly from reuse of urban and industrial wastewater. The potential of increased supplies from conservation and technological improvements is also promising. In general, it appears economical to improve the efficiency of existing technologies prior to investing in developing expensive new ones.

There are many opportunities in irrigation systems for improved efficiency. The efficiency of every part of the total irrigation system should be measured, and an effort should be made to effect improvements throughout the system. Water losses from reservoirs and conveyance may be reduced by lining canals, substituting pipes, and improving facilities and management. Efficiency can also be enhanced by proper maintenance and operation of pumping installations, by optimizing scheduling, and by capturing and reusing the runoff from surface irrigation. Land leveling of fields to ensure more even water distribution is often necessary. Sprinkle and trickle irrigation, if properly managed, can be more efficient than surface irrigation.

Farmers will probably try to adapt to changing hydrological regimes by changing crops. For example, farmers in Pakistan may grow more sugarcane if additional water becomes available, and they will grow less rice if water supplies dwindle (Riebsame et al., 1995). New crop varieties, alternative rotations, adjusted cropping calendars, and improved cultivation methods may be adopted. Crops with higher harvest index (usable biomass as a fraction of the total biomass) produce more marketable product per unit amount of water consumed in evapotranspiration. The continuing development and adoption of such crops therefore contribute to the efficiency of water use in irrigation. Irrigation may also help alleviate some of the stresses on crops that warmer temperatures may bring. Sprinkle irrigation is used to delay flowering until danger of frost is past, to reduce high temperatures, and to lengthen the dormant season to improve winter hardening (Peterson and Keller, 1990). Economically, farmers seeking to maximize returns per unit of water rather than to maximize yield per area will use water more efficiently. More effective cooperation among farmers, agricultural researchers, and water-resource planners is essential for sound irrigation management. Good management involves equitable and efficient distribution of irrigation water of high quality, while avoiding canal breaches, waterlogging, and salinization.

Water transfers from nearby basins or even across state lines are sometimes proposed as solutions to dwindling water supplies for irrigation. At the present time, economic, environmental, and political considerations constrain large projects of this type. Some small-scale interbasin transfers now occur through market mechanisms by purchase of water rights, but such transfers take place primarily from rural watersheds to urban users. Transactions of this sort will probably be more prevalent in the future.

Under changing climate conditions, the pressure toward small-scale and large-scale transfers both for agricultural and urban uses may well grow. In a warmer and more populous world, facilities already constructed for storage and conveyance of irrigation water may be allocated to (or purchased by) other sectors in response to changing hydrological and economic conditions.

Internationally, transnational cooperation may be needed to promote large-scale river basin development, including reservoir construction or enlargement and watershed protection (Hillel, 1994). Such actions generally require long time frames to plan and implement. Appropriate water pricing and international and interprovincial water accords may take decades just to negotiate. Environmental problems such as salinization of lands, rivers, and aquifers also require long-term solutions. Thus, climate change poses the challenge of enhanced cooperation to avoid situations that might otherwise exacerbate conflict over water resources.

7

Analysis of Climate Change Impacts

Any attempt to peer into the future of agriculture under the potential effect of global warming demands multifaceted analyses involving the study of both biophysical and socioeconomic processes. This chapter surveys the methods employed to study the various aspects of climate change impacts on agriculture. There are several approaches to this study. One approach is based on climate change scenarios—projections of what values climate parameters may assume in the future and how agriculture might fare in the new circumstances. Equilibrium climate change scenarios have been most often used in this approach, which may also include the study of responses of agricultural systems to past climatic variations.

Another approach is based on thresholds and attempts to define the limits of tolerance of an agricultural system as it is currently configured to changes in climatic variables. The first approach addresses the question, "What will agriculture be like in a given changed climate?" while the threshold approach asks, "What type, magnitude, and rate of climate change would significantly impact the agricultural system as we know it?" The latter approach typically assumes a transient (i.e., gradual rather than abrupt) climate change. Both approaches construct a chain of causality from the biophysical responses of crops and livestock at the farm level to the socioeconomic effects at the regional, national, and international levels.

Several different techniques from the field of economics have been used in analyses of the potential impacts of climate change. One technique uses economic data (such as land prices) to evaluate the impact of climate on farming in different regions through regression equations. Linear programming models of national agricultural sectors are also employed, as well as linked national and regional models that simulate the world food trade system.

Analysis of adaptive responses is an important aspect of research on the impacts of climate change. The biophysical approaches described in this chapter allow the explicit examination of farm-level adaptations, such as changes in planting practices, crop varieties, or species. The economic approach tends to deal with adaptation more implicitly through the evaluation of integrated measures such as land prices or the aggregate response of regional production to changes in comparative advantage.

Climate Change Scenarios

Climate change scenarios are the essential first step in an assessment of the impacts of climate change. Climate change scenarios are defined as plausible combinations of climatic conditions that may be used to test possible impacts and to evaluate responses to them. Scenarios may be used to determine how vulnerable agriculture (or any other sector) is to climate change and to identify thresholds at which impacts become negative or severe. They are also used to compare the relative vulnerability among sectors in the same region or among similar sectors in different regions. Thus, to be useful, climate change scenarios should be applied on a regional scale (Kellogg and Zhao, 1988).

While there is scientific consensus that increased atmospheric concentrations of greenhouse gases will likely raise global temperatures (with associated increases in global precipitation and sea level), there is no consensus on how fast and by how much the climate may change, on how different regions may experience the change, or on how the variability (as well as the mean values) of climatic parameters may change. To cope with these uncertainties, climate scenarios of different types have been developed for analysis of impacts on regional agriculture. These include scenarios based on arbitrary changes in climate variables, analog warming in previous times, and on general circulation model (GCM) and regional climate model simulations.

The design of an impact study often includes a set of scenarios consistent with the current state of knowledge regarding global climate change. Since current knowledge is admittedly incomplete and fraught with uncertainty, the scenarios are meant to span a range of possible climatic conditions. By analyzing multiple scenarios, the direction and relative magnitudes of potential responses may be assessed. Studies have been done with one, several, or many alternative scenarios. However, it is still difficult, if not impossible, to ascribe probabilities to any of the various climate change scenarios, owing to uncertainties regarding future emissions of radiatively active trace gases and tropospheric aerosols and the potential response of the climate system to those emissions. Therefore, impact studies based on climate change scenarios do not make actual predictions; rather, they describe hypothetical possibilities. Nonetheless, they are useful in defining, for critical biophysical and socioeconomic systems, directions and relative magnitudes of change, as well as potentially critical thresholds of climate-sensitive processes. By these means, researchers and resource managers are able to conduct "practice" exercises, which may help them anticipate future conditions and prepare possible adaptations to those conditions in a flexible manner.

Arbitrary Scenarios

The simplest type of scenario is the application of prescriptive changes, such as a given rise in temperature and/or a given (absolute or relative) reduction in precipitation, to observed climate. Waggoner (1983) used this approach to study yields for the major cropping regions of the United States using statistical regression and crop growth models. For a 1°C warming and a 10% decrease in rain, he predicted a decrease in yields ranging from 0.04 to 0.18 t ha^{-1}—that is, a loss of between 2 and 12%. Newman (1980) used growing-season thermal units to measure the sensitivity of the Corn Belt to +/− 1°C daily temperature changes. He found that the Corn Belt would shift 175 km per degree of temperature change, and that the shift would take place in a SSW or NNE direction. Newman did not consider precipitation changes per se, but he did include the impact of temperature change on potential evapotranspiration in the simulation.

Tests with such simple changes can help identify the sensitivities of systems to changes in defined variables. One may attempt to isolate the effects of a simple climate variable (e.g., temperature) while holding all other variables constant. However, such tests do not offer a comprehensive and consistent set of climate variables, since in reality evaporation, precipitation, wind and other variables are all likely to change concurrently and interactively with change in temperature. For example, a 2°C rise in temperature would also increase the evaporation rate, which is of importance to crop growth, but that change would not be reflected in the simplistic scenario. Arbitrary scenarios do provide an opportunity to define threshold sensitivities and a set of responses to which other types of scenarios may be compared, however. Response surface diagrams may be created from multiple simulations of such scenarios (Figure 7.1) (Rosenzweig et al., 1996).

Historical Analogs and Paleoclimatic Scenarios

Another type of climate change scenario is based on the historical record. Observations from cool or warm, wet or dry historical periods are used to construct scenarios for use in modeling studies of the impact of climate change. For example, years when the Arctic regions were anomalously warm in this century may be useful as analogs of the high-latitude amplification of greenhouse warming (Jäger and Kellogg, 1983).

Such periods are also useful for the insights they can provide into the responses of farmers and farming systems to periods of climatic extremes. The Dust Bowl of the 1930s in the Southern Great Plains is probably the best known example (see, e.g., Warrick, 1984; Rosenberg, 1993), but past events (such as an extreme cold spell or heat spell, aquifer depletion, and lake-level change) have also been used to study patterns of societal response to regional climate change (Glantz, 1988).

Lough et al. (1983) constructed climate scenarios for the Northern Hemisphere based on the instrumental observations of the warmest 5-year and 20-year periods in this century. The warmest period occurred from 1934 to 1953 (Figure 7.2). The effects of these climate scenarios on crop yields in England and Wales were calculated by means of regression models that employ the principal components technique. The results showed diminished yields for most crops due to warmer summer, drier spring, and wetter autumn seasons.

Scenarios have also been derived from paleoclimatic records of pollen, lake-level, or other climatic indicators for periods that are known to have been warmer than the current climate. Examples are the Altithermal (or Hypsithermal) period about 4,000 to 8,000 years ago and the Eamian interglacial period of 125,000 years before present (B.P.) (Pittock and Salinger, 1982; MacCracken et al., 1990). These periods were about 1–1.5°C and 2–2.5°C warmer, respectively, than our current climate. The Pliocene

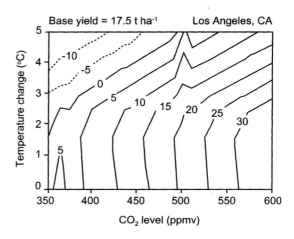

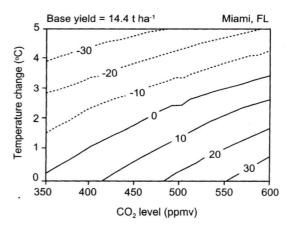

Figure 7.1   Valencia orange yield response to elevated temperature and CO_2 at currently producing sites in the United States. Solid lines indicate contours of the percentage of yield increases, and broken lines indicate the percentage of yield decreases, above and below base yields that were predicted at current temperature and CO_2 levels. Contours were interpolated from the means of 28 simulated years for each climate scenario (Rosenzweig et al., 1996).

Optimum 12,000,000–2,500,000 years B.P. was 3–4°C warmer (Budyko and Sedunov, 1990). Broad patterns of regional changes in temperature and precipitation have been determined from paleoclimatic data.

The use of historical and paleoclimate scenarios as analogs for future climate change caused by the enhanced greenhouse effect has been a subject for debate. On one hand, historical data based on station observations may provide details of daily, local weather during a warmer period. The use of specific historical or paleoclimatic analog scenarios for impact studies is also justified by the assumption that patterns of climate warming and their impacts are similar, regardless of atmospheric forcing mechanisms. One such important pattern may be a lower equator-to-pole temperature difference (Kellogg and Zhao, 1988).

On the other hand, a difficulty with either of these scenario approaches as proxies for the expected global warming due to increasing CO_2 and other trace gases is that the patterns of climate warming may indeed be different, depending on the nature of the atmospheric forcing mechanisms.

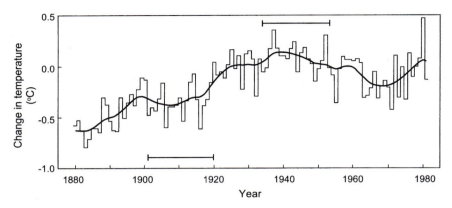

Figure 7.2   Northern Hemisphere temperature variations (°C) with selected warm and cool 20-year periods (Lough et al., 1983).

While warm periods within the historic climate record were probably not caused by buildup of radiatively active trace gases, warm paleoclimatic periods may well have been forced by a combination of CO_2 augmentation and shifts in the inclination of the earth's axis and other orbital changes. Orbital shifts are not projected to be part of the currently predicted future warming. The physical mechanisms of cause and effect between CO_2 and temperature change in the paleoclimate record are difficult to define, in part because changes in both records appear almost simultaneously in the data when proceeding from glacial to interglacial periods (i.e., during periods of warming), whereas the temperature signal apparently decreases before the CO_2 signal during transitions from warm to cold periods (Barnola et al., 1987).

Furthermore, there are no historical periods for which we have observations with climatic fluctuations as great as those predicted for the future "greenhouse" climate. The temperature rise of the warmest period in the 20th century is only 0.4°C (warm minus cold temperature change for the Northern Hemisphere) (Lough et al., 1983) compared to the rise of 1.5 to 4.5°C predicted for future greenhouse warming (IPCC, 1990, 1992). Even with the cooling effects of anthropogenic aerosols taken into account, the projected warming is still estimated to be between 1.0 to 3.5°C (IPCC, 1996). Therefore, both the forcing mechanisms and the magnitude of the historical scenarios lack complete correspondence to the predicted future changes.

Scenarios developed from paleoclimatic data are also hampered by lack of spatial and temporal detail. Regional specificity is often lacking, and construction of monthly time-series of changes in temperature and precipitation (such as are needed for comprehensive analyses) is difficult to derive from pollen or lake-level data. Even if the regional-scale mean values of a few variables (including temperature, precipitation, and solar radiation) can be derived from paleoclimate data, the local daily and inter-annual climate data can hardly be determined from such records.

Notwithstanding these limitations, the study of past climates can help us understand forcing mechanisms and patterns of atmospheric dynamics. Simulation experiments with GCMs that reconstruct past climates subjected to different forcing mechanisms can serve as important validation tests for models designed to predict future climates (e.g., Kutzbach and Street-Perrott, 1985).

GCM-Based Scenarios

Climate change scenarios are also derived from global climate model experiments with specified forcing mechanisms (e.g., doubled or quadrupled atmospheric CO_2 levels). The GCM model experiments are conducted to produce either equilibrium or transient climate projections. GCMs estimate how global and regional climates may change in response to increased concentrations of trace gases. Regional and global climate responses are mutually and physically consistent, as heat, moisture, and energy processes are calculated from the same set of equations representing physical processes. A full set of climate variables (including wind, solar radiation, cloud cover, and soil moisture) is provided by GCM output for use in a wide variety of impact models. Changes in both the means and the variances of these climate factors are available as well. Thus, GCM scenarios yield "example" climates for impact researchers and resource managers to test responses of important systems to simultaneously altered conditions in different regions. In agriculture, this is particularly important since a change in comparative advantage among regions can be a major driver of changes in national and international food systems. GCM climate change scenarios provide a global framework in which to embed more detailed regional case studies. A range of GCM scenarios should be included in the design of impact studies in order to incorporate a range of climate sensitivities to greenhouse gas forcing.

Disadvantages of GCM scenarios include (1) the present lack of realism of the current climate simulated by GCMs at regional scales; (2) their crudely modeled ocean, cloud, and land–surface processes; and (3) their coarse spatial resolution. GCMs represent current climate at global and zonal (latitudinal) scales, but do not simulate regional climate very accurately (Table 7.1) (Grotch, 1988; Kalkstein, 1991). Differences in climate projections among GCMs increase as the scale is reduced from the global to the regional and gridbox levels. GCM simulations of current temperature regimes are better than their simulation of current hydrological regimes. The lack of regional accuracy in simulating current climate conditions by GCMs is due to several factors. The characterizations of the physical processes in the oceans, in clouds, and at the land surface is often simplistic, due in part to incomplete understanding of the complex interactions involved. Topographical features (e.g., mountains and lakes) within the gridboxes are also rather crudely represented.

While GCM resolution is improving due to increasing computer power, the large size of the gridboxes (varying between about 2 by 4° latitude to about 3 by 7° longitude) continues to be a major disadvantage of current GCMs and hampers their use in impact analyses (Figure 7.3). GCMs do not account for variations in climate factors within each gridbox. Rather, they provide a single value for temperature, precipitation, and other predicted variables even though such variables may range widely within the encompassed area. To moderate the artificially abrupt differences from one gridbox to another, interpolations between gridbox values of climate variables are often calculated for use in GCM climate change scenarios. Some researchers recommend that regional climate variables be averaged over a minimum number of gridboxes in order to reduce the large variability. For example, the climate variables output

Table 7.1 Differences between observed and simulated temperatures by four GCMs (Grotch, 1988).

Variable and model	Global mean	Domain of comparison		
		North America	Contiguous U.S.	Midwestern U.S.
December–January–February				
Observed median temperature (°C)	8.5	−5.8	0.9	−1.5
Difference in median temperatures (GCM minus observation)				
CCM	−1.6	−0.3	−2.1	−0.5
GFDL	1.5	−1.8	−0.8	−1.3
GISS	0.8	−0.5	0.0	1.1
OSU	0.3	0.5	−0.6	−1.0
June–July–August				
Observed median temperature (°C)	13.9	18.9	23.0	23.0
Differences in median temperatures (GCM minus observation)				
CCM	1.3	6.0	6.3	6.8
GFDL	−0.2	0.6	0.1	3.7
GISS	0.4	−3.1	−4.5	−4.8
OSU	−0.6	−2.2	−2.2	−1.6

CCM = Community Climate Model (National Center for Atmospheric Research) (Washington and Meehl, 1984).
GFDL = Geophysical Fluid Dynamics Laboratory (Manabe and Wetherald, 1987).
GISS = Goddard Institute for Space Studies (Hansen et al., 1984).
OSU = Oregon State University (Schlesinger and Zhao, 1988).

from a GCM could be averaged over an area of ten $5 \times 5°$ gridboxes (an area of about 3 million km^2) for regional climate projection purposes (Robock et al., 1993).

For all of the above reasons, GCM regional climate change projections should be regarded as examples of possible future climates rather than as actual predictions.

Equilibrium Scenarios

A climate equilibrium is defined somewhat arbitrarily as a climate in which average conditions remain constant, as do the ranges within which year-to-year variations occur. In reality, we know that climate is always changing, the degree of instability depending on the time scale under consideration.

To create an equilibrium climate change scenario, a GCM is run long enough for the simulated climate to come into equilibrium with the radiative forcing. Boundary conditions and forcing factors are usually specified for a number of years (usually 10 to 30). This is called the control run. The last 10 years or so of the run are used to compare to observations of the present climate, in order to test how realistically the model simulates current conditions. Then an instantaneous doubling of atmospheric CO_2 is introduced in the climate model, and it is run again for a number

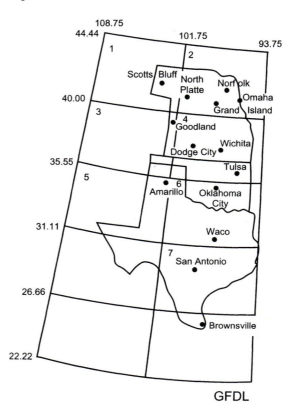

Figure 7.3 Gridboxes of the
GFDL GCM at 4.44° (lat.) × 7°
(long.) resolution (Manabe and
Wetherald, 1987) for the Southern
Great Plains (Rosenzweig, 1990).

of years. This is called the perturbed run. The model is run long enough with the perturbed conditions to reach a new equilibrium and the last period of ~10 years is then
used for creating statistics of the simulated climate. The difference between the two
runs indicates the sensitivity of the model to the perturbation under equilibrium conditions (Figure 7.4), but does not provide direct information about the time-
dependent response of the climate to the forcing (i.e., the period of transition from
one equilibrium state to another).

Doubled CO_2 in GCM equilibrium simulations is taken as a surrogate for the
projected change in forcing from all the greenhouse gases (including, in addition to
CO_2, methane, nitrous oxide, chlorofluorocarbons, and tropospheric ozone). The
character of the actual warming that occurs and the magnitude of the change are likely
to differ, however, due to differences in specific radiative properties among these gases
(Wang et al., 1992).

Because GCM simulation of current regional climates is often inaccurate, direct
projections of GCM-generated future climates are seldom used. Changes in climate
variables in the perturbed simulations relative to the control run are generally applied
to historically observed weather data, to create the climate change scenarios used in
impact studies. Absolute model biases are omitted by using the relative changes. Thirty
years of current climate data are often used to develop the baseline climate scenario

GISS

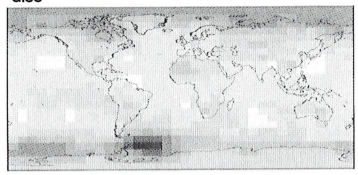

GFDL

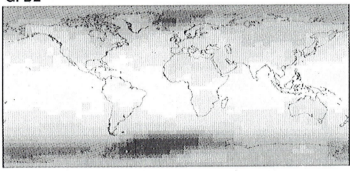

UKMO

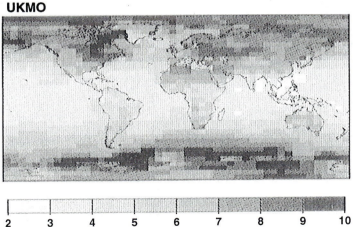

DEG (C)

2 3 4 5 6 7 8 9 10

Figure 7.4 Projected changes in annual temperature for three GCM doubled-CO_2 climate change scenarios. GISS = Goddard Inst. for Space Studies (Hansen et al., 1983); GFDL = Geophysical Fluid Dynamics Laboratory (Manabe and Wetherald, 1986); UKMO = United Kingdom Meteorological Office (Wilson and Mitchell, 1987).

to which the relative changes in GCM outputs are applied. A 30-year period is considered long enough to represent "normal" climate variability. Two recent periods, 1951–1980 and 1961–1990, are often selected, recent periods representing current climate and having accurate data most easily available. The latter period is somewhat less appropriate since the 1980s included some of the warmest years on record and since that warming may itself have been caused by the enhanced greenhouse effect.

An early assessment of climate change impacts in the United States (Smith and Tirpak, 1989) utilized scenarios of predicted changes in climatic parameters under conditions with doubled atmospheric CO_2 concentrations (660 ppmv versus 330 ppmv) developed from two GCMs, namely GISS and GFDL. Relative changes in the average monthly values of climatic parameters were estimated by the GCMs on a grid basis. Due to computation limitations, grid sizes in the GISS and GFDL models were relatively large, averaging 10° (longitude) by 7.8° (latitude) and 7.5 by 4.4°, respectively. Daily values of air temperature, humidity, solar radiation, wind speed, and precipitation for specific locations within the GCM grids were estimated by multiplying observed 1951–1980 values by average "change ratios" for each month of the year. These change ratios were calculated for each GCM grid element by dividing values of weather parameters from the $2\times CO_2$ simulations by values of weather parameters under control run simulations. For each site in the study, the results provided 30-year sets of daily weather data that are hypothesized to occur with doubled atmospheric CO_2 concentrations.

Most climate change impact studies have been conducted using equilibrium doubled-CO_2 GCM scenarios, simulating a step change in trace gas content and assuming an abrupt change to a new equilibrium climate. This group of studies has created, in effect, a set of "research blinders," which has focused attention on a rather unrealistic vision of the future. In reality, climate change is evidently being forced by a gradual increase in trace gases, and the responses of the climate system may well be nonlinear. Cline (1992) argues that effective CO_2 levels could increase eightfold in the next few centuries, and that these higher levels and longer time frames are essential to assess the full effects of continuing greenhouse gas emissions.

Transient Scenarios

A transient scenario is one in which climate change occurs gradually over time. It is therefore a more realistic representation than the equilibrium-doubled-CO_2 scenarios. GCMs have been used to simulate the response of the climate system to a continual rise (rather than an abrupt doubling) of greenhouse gas concentrations (IPCC, 1996). In a transient GCM experiment, the simulated climate is not in equilibrium with the radiative forcing, and thus the time-dependent response of the climate system may be examined. The GCM is forced with steadily increasing levels of CO_2 (recent runs have a rise of 1% yr^{-1}) to represent increases in all radiatively active trace gases. Some experiments include effects of anthropogenic sulfate aerosols as well as of CO_2. GCMs used recently for transient simulations include a more detailed treatment of the oceans than did older GCM versions, so the combined models are called coupled Atmosphere-Ocean General Circulation Models (AOGCMs). In the case of transient climate change simulations with CO_2 alone, the realized warming at the time of CO_2 doubling in the atmosphere is lower (by about 50 to 80%) than the equilibrium-doubled-CO_2 warming response. The attainment of equilibrium lags behind the change in atmospheric composition due to the slow uptake of heat by the oceans.

Besides the general uncertainties arising from the use of GCMs, transient GCM scenarios carry specific uncertainties regarding future greenhouse gas emissions and the roles of oceans and clouds in affecting the rate of climate change. The regional patterns of climate change are affected by sulfate aerosols from industrial and urban burning of fossil fuels. While CO_2 emissions are projected to increase over time, sulfate aerosol injections into the troposphere may decrease with time if regulations curbing urban pollution are enacted.

Another difficulty with transient GCM simulations is known as the "cold start problem." This refers to the start of the transient runs with the climate model in a state of equilibrium, even though the climate in reality is not, given the already increasing levels of CO_2 and other trace gases. The cold start problem makes it difficult to assign calendar years to GCM outputs, even though more accurate projections of the timing of global climate change are desired for the purposes of devising national and international policies to avoid or cope with its potential effects.

Various transient-GCM simulations show some agreement in broad regional patterns. Land areas warm faster than the oceans due to the greater thermal inertia of large bodies of water. As the transient runs progress, the early decades are characterized by interannual and interdecadal climate fluctuations ("noise"). It takes more than several decades for the clear temperature trend (the "signal") to emerge from the noise, and even longer for discernible changes in the hydrological cycle to become statistically significant.

Regional Climate Models and Downscaling

Limited-area regional climate models (RegCMs) nested within GCMs simulate climate at finer resolutions (up to a few tens of kilometers) over selected regions. The effects of complex topography, vegetation mixtures, coastlines, and large lakes that regulate local circulations and regional distribution of climate variables are represented in more physically realistic ways in these models (Giorgi and Mearns, 1991). For example, Giorgi et al. (1994) have nested a regional model with a resolution of 60 km for the continental United States within the GENESIS version of the National Center for Atmospheric Research (NCAR) GCM and have tested the sensitivity of the model to CO_2 (Thompson and Pollard, 1996a,b). Other major mesoscale models that have been tested for this type of application include NCAR/Pennsylvania State University MM4/5 (Dickinson et al., 1989; Giorgi, 1990); the Japan Meteorological Agency limited area model (Kida et al., 1991); and DARLAM of CSIRO (McGregor and Walsh, 1993).

At present, long runs with nested mesoscale models are computationally expensive, so their use in impacts research is just beginning. As computing systems improve, these nested models can serve as a basis for regional climate change scenarios for impacts research. An example of their use for agriculture is found in Mearns et al. (1996), who tested climate change impacts on wheat in Kansas and Washington based on scenarios developed from a regional climate model.

Regional climate detail for climate change scenarios is also provided by a technique known as downscaling. Here, GCMs are used to describe the atmospheric response to large-scale forcings and empirical techniques account for mesoscale forcings. Statistical climate inversion is used to derive relationships between large-scale and local surface climate variables. These techniques hae been reviewed by Giorgi and Mearns (1991). An example used for an agricultural study is Wilks (1989), who developed regression formulas to predict regional distributions of daily climatic variables in the Great Plains from time-series of large-scale averages. Other methods consist of developing empirical relationships between observed surface weather variables and observed or model-produced atmospheric and surface weather predictors. The predictors can include regional average surface air temperature and precipitation, mean sea-level pressure, height of the 700-mbar pressure surface, and the zonal and meridional pressure gradient near the center of the region of study (Wigley et al., 1990).

Downscaling techniques appear to improve regional climate projections more than does the direct or interpolated use of GCM data at gridbox scales (Giorgi and Mearns, 1991). However, the methods work less well when climate variables, such as summertime precipitation, are not spatially well correlated. They suffer, too, from lack of physical explanatory power and thus lack of ability to apply under different climate forcings and beyond the range of the data used to specify the relationships.

Weather Generators

Mathematical techniques for generating synthetic time series of weather are important tools for climate change impact studies. They are particularly useful in developing scenarios of changed climate variability. Stochastic parameters are developed from historical weather data;[1] these parameters may then be adjusted to produce long-term transient climate series with changes in variance of key variables as well. The site-specific parameters are adjusted to produce simulated daily meteorological values consistent with specified changes in the monthly means and variances of the relevant climate variables. The weather generator WGEN developed by Richardson (1981) and Richardson and Wright (1984) has been used as a basis for the generation of climate change scenarios. Examples of these techniques are provided by Mearns et al. (1992, 1996), Wilks (1992), and Semenov and Porter (1995).

Biophysical Modeling Techniques

Modeling techniques of several kinds are used to study the potential impact and response of agriculture to changing climate and atmospheric composition. The choice of technique depends on the type of analysis conducted and the research questions posed.

Spatial Analysis

A crucial element in analyzing the impact of climate change on agriculture is geographical distribution.

Crop Suitability. Spatial analysis of climate–agriculture interactions consists of identifying the critical environmental limits (primarily climate, soil, and water regimes) of specific crops or farming systems, applying climate change scenarios, and calculating the consequent spatial shifts of crop-production regions. This agroclimatic method provides an approximation of possible changes in crop areas from a biological perspective, but does not address explicitly any possible changes in either varieties or management methods (e.g., those arising from new technologies).

Rosenzweig (1985) specified the environmental requirements for North American wheat-growing regions and projected that wheat could still be grown in most parts of the United States under a doubled CO_2 climate (Figure 7.5). However, fall-sown spring wheat could replace hard winter wheat where warmer winter temperatures prohibit vernalization. This type of substitution has implications for protein quality and marketing, as fall-sown spring wheat tends to be of lower quality than hard winter wheat and is therefore used more for animal feed than for human consumption.

An interesting spatial analysis was conducted by Rosenberg (1982), detailing the expansion of the profitable growing range for hard red winter wheat in the North American Great Plains region from 1920 to 1980. Genetic and soil management improvements facilitated the move both northward to cooler climates and southward to warmer ones. Such previous adaptation of crops to regional variation in current climate may presage future responses to climate change.

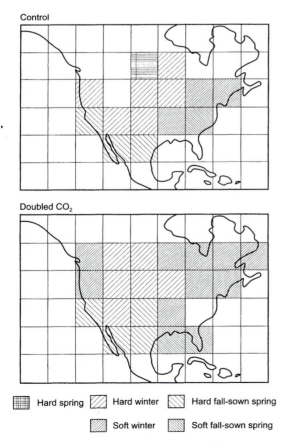

Figure 7.5 North American wheat regions assuming climatic conditions drawn from (*top*) the GISS GCM control run and (*bottom*) the doubled-CO_2 run (Rosenzweig, 1985).

Potential Production.   Potential production may be estimated from climatic variables or indices such as length of frost-free or rainy season, precipitation, evapotranspiration, solar radiation, and temperature. This approach is exemplified by the FAO Agro-Ecological Zone (AEZ) Project (FAO, 1978; Doorenbos and Kassam, 1979). One problem with this approach is that high temperature effects on crop biomass and yields depend on crop type. While the annual net primary productivity of perennial vegetation should rise with warmer temperatures, many agricultural crops are annuals that respond to higher temperatures by hastening their metabolic rate of development, thus shortening their growth cycle and reducing yields. The potential production method also tends to smooth the high degree of variability in production that is attributable to factors other than climate, such as management and economics. Dudek (1989) used the potential production method to estimate climate and CO_2 effects on agricultural crops in California and found that statewide average yields of most crops (except cotton and some vegetables) diminished slightly (1 to 10%) under conditions imposed by two climate change scenarios when the direct effects of high CO_2 on crop growth were taken into account (Figure 7.6).

The FAO Agro-Ecological Zone modeling technique simulates both crop zonation and potential production. A climate-crop model based on this methodology has been used to test potential change in global yield and crop distribution under one GCM scenario (Cramer and Solomon, 1993; Leemans and Solomon, 1993). These studies found large differences in regional response to climate change. High-latitude regions uniformly benefitted from projected longer growing periods and increased productivity. Regions in the mid- and low latitudes, however, did not benefit significantly or even tended to lose productivity. Declines in agricultural potential were caused primarily by reduced moisture availability.

Detailed country studies using AEZ methods have been carried out in Kenya and Bangladesh (Fischer and Van Velthuizen, 1996). The heterogeneity of biophysical resources and agricultural land use in these nations make regional responses to warming extremely diverse.

Statistical Regression Models

Multiple regression models have been developed from the historical relationships between past crop yields and climatic variables in specific locations. Among the many models of this type are those of Thompson for corn (1969a) and wheat (1969b) for parts of the central United States and those of Baier (1973) and Ramirez et al. (1975) for wheat in the Great Plains. Such relationships were used in early studies to predict future yields in climate change impact studies, usually with simple prescriptive scenarios (Table 7.2) (see, e.g., Thompson, 1975; Bach, 1979; Waggoner, 1983).

The use of regression models has been criticized (Katz, 1977; Biswas, 1980; Hayes et al., 1982; Rosenberg, 1982), especially in regard to the prediction of crop responses to future climate. Regression models are limited by their lack of explanatory power since the techniques rely on statistical coefficients alone and not on the fundamental biophysical mechanisms underlying crop responses to climate factors. Another problem associated with statistical models, besides their "black-box" nature,

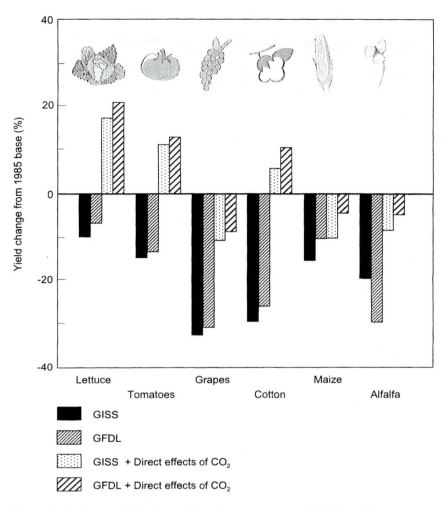

Figure 7.6 Average yield changes for various crops grown in California for two climate change scenarios, with and without direct CO_2 effects (from Dudek, 1989).

is the difficulty of separating changes in yield due to climate from changes due to differences in management or technology over time. For the most part, regression models have assumed linear relationships between crop yields and environmental variables and have considered no changes in crop varieties or production technology. Thus, regression models appear to be more appropriate for studying the effects of current climate *variability* than the potential effects of climate *change*, since the latter may involve extrapolation beyond the range of present-day climate and crop management used to specify the model. Nonetheless, regression models have been found to be useful, simple tools for testing the potential effects of relatively small changes in climate variables.

Table 7.2 Effect of weather on corn yield in three states based on regression; values are in quintals ha^{-1} (Waggoner, 1983)[1]

	State		
	Iowa	Illinois	Indiana
Variable			
Yield, average 1978–1980	72.7	68.8	65.3
Temperature, °C			
July	—	−1.56	—
Aug	—	−0.64	—
Oct	—	0.57	—
July to Aug Average	—	—	−2.34
Precipitation, mm			
May	—	—	−0.017
Sept to June	0.013	—	—
Sept to June SDFN[2]	−0.0001	−0.00006	—
July	—	—	0.045
Combined Variables			
Apr and May PET[3]	−0.12	—	—
May Prec/PET	—	−1.49	—
July Prec minus PET	0.076	0.025	—
(June ET/$\overline{ET}$[4] + July ET/$\overline{ET}$)/2	—	—	2.27
Calculated Estimated Change			
Yield, quintals/ha	−2.36	−1.72	−2.80
Change from 1978–1980 average	−3%	−3%	−4%

1. The variable effects are given as b coefficients in quintals ha^{-1} per unit of variable, i.e., mm of precipitation, °C of temperature or fraction of a ratio (after Leduc, 1980). The calculated estimates of the change in yield with a 1°C increase in temperature and a 10% decrease in precipitation from the historic average temperature and precipitation are recorded for the regions.

2. SDFN = departure from normal precipitation, squared.

3. PET = potential evapotranspiration in millimeters, a measure of the demand for water.

4. June ET/$\overline{ET}$ = June evapotranspiration divided by mean evapotranspiration.

Dynamic Crop Models

Dynamic plant growth simulation models track the growth of the crop quantitatively and progressively in terms of defined state variables (Rimmington and Charles-Edwards, 1987). In such models, the present state of a system at each stage depends on the initial conditions and on the cumulative and interactive influence of all the relevant inputs up to the present. Dynamic simulation models generally include *exogenous, endogenous, state,* and *rate variables* (Hillel, 1977).[2]

   Dynamic plant growth models formulate the principal physiological, morphological, and physical processes involving the transfers of energy and mass within the crop and between the crop and its environment. There exists a continuum of empirical and functional relationships in the structures of the many different crop models developed to date. From such relationships, these models derive predictions of integrated crop performance under various conditions (Loomis et al., 1979).

Dynamic crop models are now available for most of the major grain crops; a selected list is given in Table 7.3. In each case, the aim is to predict the response of a given crop to specified climatic, edaphic, and management factors governing production (Joyce and Kickert, 1987). Crop growth models capable of simulating the response of agricultural plants to climatic variables may be used in conjunction with GCM or other climate change scenarios to explore the consequences of atmospheric CO_2 enrichment and of temperature rise on yields and phenology, or to determine thresholds of crop sensitivity to changing conditions. Decision support systems utilizing the models allow the testing of possible adaptations to climate change, such as altered planting dates, irrigation scheduling, and crop selection (IBSNAT, 1990).

To accomplish these evaluations, the crop growth models must account for the primary responses to CO_2 enrichment of such processes as photosynthesis, photorespiration, dark respiration, and stomatal functioning (Strain and Cure, 1985), and of their interactions with temperature and moisture. Existing crop models have been modified to simulate the major physiological responses to a range of CO_2 levels (see, e.g., Peart et al., 1989; Stockle et al., 1992a; Tubiello et al., 1995). The dynamic crop models are run with baseline and changed atmosphere scenarios (including both climate variables and CO_2 levels) to investigate potential changes in yields, evapotranspiration, crop growing season, and irrigation requirements. Studies of this type include Peart et al. (1989), Ritchie et al. (1989), Rosenzweig (1990), Stockle et al. (1992b), Rosenzweig and Iglesias (1994), and Tubiello et al. (1995). However, more work still needs to be done to improve the use of crop models for this purpose. For instance, the effects of elevated CO_2 on dark respiration rates are still uncertain (Bazzaz, 1990).

Economic Analysis

Economic considerations inevitably enter into the evaluation of climate change issues. Economic analysis is concerned with the reciprocal relations between the biophysical conditions induced by climate change and the range of options available for individuals and institutions to respond to those altered conditions. To be appropriate, institutional response measures affecting human welfare must be both possible and economical. While biophysical analysis focuses primarily on the production of agricultural crops, economic analysis considers both producers and consumers of agricultural goods. Economic considerations include the likely effect of changing conditions upon input and output market prices and the opportunities available for individuals to minimize losses or maximize gains. As climate change affects the costs of production, it also affects the price and quality of products, which, in turn, can lead to further market-induced output changes. Even if prices remain constant, accurate indications of potential output changes are needed since production practices and types of outputs may change.

Research on the economics of agriculture subject to environmental stresses is based on several general principles (R. Adams, personal communication):

1. Both producers and consumers must be considered conjunctively.
2. Economic activities constitute a type of societal adaptation to environmental stresses, aimed at mitigation of negative effects.
3. Environmental stresses may alter the relative productivity of regions and countries.

These general principles guide more detailed economic analyses.

Table 7.3 Selected crop models and references[1] (adapted from Jones and Ritchie, 1990; U.S. Country Studies Program, 1994)

Crop	Model name	Reference
Alfalfa	ALSIM (Level 2)	Fick (1981)
	ALFALFA	Dennison and Loomis (1989)
Barley	CERES-Barley	Ritchie et al. (1989a)
Cotton	GOSSYM	Baker et al. (1983)
	COTCROP	Brown et al. (1985)
	COTTAM	Jackson et al. (1988)
Dry beans	BEANGRO	Hoogenboom et al. (1989)
Maize	CERES-Maize	Jones and Kiniry (1986); Ritchie et al. (1989b)
	(unnamed)	Stockle and Campbell (1985)
	CORNF	Stapper and Arkin (1980)
	SIMAIZ	Duncan (1975)
	CORNGRO	Childs et al. (1977)
	(unnamed)	Morgan et al. (1980)
	VT-Maize	Newkirk et al. (1989)
	GAPS	Buttler (1989)
	CUPID	Norman and Campbell (1983)
Peanuts	PNUTGRO	Boote et al. (1989)
	(unnamed)	Young et al. (1979)
Pearl millet	CERES-Millet	Ritchie and Alagarswamy (1989)
	RESCAP	Monteith et al. (1989)
Potatoes	(unnamed)	Ng and Loomis (1984)
	SUBSTOR	Griffin et al. (1993)
Rice	CERES-Rice	Godwin et al. (1990)
	RICEMOD	McMennamy and O'Toole (1983)
	(unnamed)	Horie (1988)
Sorghum	SORGF	Arkin et al. (1976)
	CERES-Sorghum	Ritchie and Alagarswamy (1989)
	SORKAM	Rosenthal et al. (1989)
	RESCAP	Monteith et al. (1989)
Soybeans	SOYGRO	Wilkerson et al. (1983); Jones et al. (1989)
	GLYCIM	Acock et al. (1983)
	REALSOY	Meyer (1985)
	SOYMOD	Curry et al. (1975)
Sugarcane	CANEMOD	Inman-Bamber (1991)
Wheat	AFRCWHEAT2	Porter (1993)
	CERES-Wheat	Ritchie (1985); Godwin and Vlek (1985)
	(unnamed)	Stockle and Campbell (1989)
	TAMW	Maas and Arkin (1980)
	(unnamed)	Aggarwal and Penning de Vries (1989)
	(unnamed)	van Keulen and Seligman (1987)
	SIMTAG	Stapper (1984)
General model	EPIC	Williams et al. (1984)

1. See crop model section in Chapter 7 bibliography for references.

Adams (personal communication) has summarized the general findings of previous economic studies of environmental stresses on agriculture. These findings include:

1. Increasing environmental stress causes accelerated rates of economic losses.
2. Growers may gain from yield losses because of environmental stress up to a point, due to price increases.
3. Consumer losses are a substantial portion of the total loss from environmental stress.
4. Economic losses (in terms of percentage changes) are less than the underlying biophysical yield changes because producers and consumers adjust activities.
5. Environmental stress affects both productivity and demand for inputs.
6. Environmental stress has differential effects on the comparative advantage of regions or countries.
7. Climate change alters international trade flows with attendant gainers and losers.

Economic models are designed to estimate the potential impacts of climate change on production, consumption, income, gross domestic product (GDP), employment, and farm value. These may be only partial indicators of social welfare, however. Not all social systems, households, and individuals (for instance, subsistence farmers) may be appropriately represented in models that are based on producer and consumer theory. Furthermore, many of the economic models used in impact analyses to date do not account for climate-induced alterations in the availability of land and water for irrigation, though such nonmarket aspects of a changing climate may be critical.

Studies and models based on market-oriented economies assume profit and utility maximizing behavior. These models are data intensive and are relatively expensive to construct since they depend on access to detailed data regarding time series of price, quantities, resource use, and other economic information. Several different types of economic models have been used for climate change studies, including mathematical programming models at farm, regional, and national levels and econometric models at regional, national, and international levels. Because of the expense of model development, many climate change impact studies have utilized currently available models, which are relatively accessible and inexpensive to use. Such models have already been calibrated and validated to economic conditions in the present or recent past, and have been subject to peer review. However, such models may not address the specifics of the climate change issue. Constructing new models to specifically address climate change issues is desirable, but is an exacting and time-consuming process.

Data of a variety of types are needed to conduct an initial economic assessment of climate change impacts (Adams, personal communication; U.S. Country Studies Program, 1994). On the production side, important data include alternative crops and production techniques (e.g., rainfed farming verses irrigation); nature and extent of resource use (inputs) in different systems of agricultural production; as well as costs of production (e.g., land, labor, inputs). On the consumption side, information needed includes percent of crop consumed locally or nationally, and the percent exported; the role of each crop and commodity (including livestock) in national food consumption; and price movements of commodities. Also crucial to economic analysis is information regarding policy, such as the role of government intervention in setting prices and the contribution of agricultural exports to the foreign exchange earnings of nations.

Simple Economic Approaches

As a starting point, the gathering of available economic and agricultural information provides a framework for assessing the economic vulnerability of the agricultural sector to potential changes in the environment (U.S. Country Studies Program, 1994). Information about production, consumption, and governing policies is needed. The essential information includes the number and nature of alternative crops and of production techniques (such as irrigation), and the nature and extent of resource use in agricultural production systems. Such data may suggest what substitution possibilities exist in the agricultural sector. The greater the number of alternative crops and production techniques in a country, the greater the likelihood of adaptation to altered conditions. Costs of production are needed to develop production functions capable of estimating costs and benefits to farmers and consumers.

Information about government policies is also useful for simple analyses of potential vulnerabilities to climate change (U.S. Country Studies Program, 1994). Government policies typically tend to protect the status quo and, hence, to inhibit rapid adjustment to environmental changes. Removal of governmental intervention may promote greater flexibility and, thus, indirectly induce the agricultural sector to overcome climatic stresses. Understanding the role of present and future government involvement in the agricultural sector is essential to the task of forecasting potential vulnerabilities and to devising better policies.

Microeconomic Models

Farm-level models are designed to simulate the decision-making process of a representative farmer in regard to methods of production and allocation of land, labor, existing infrastructure, and new capital. Such models are based on the goal of maximizing economic returns to inputs. Data on crop productivity and such inputs as fertilizer are embedded in the models, and the consequences of potential changes in yields and yield–fertilizer relationships from climate change may be calculated. Ideally, the outputs from microeconomic farm models include the feasible choices of crop/livestock mix and methods of production, as well as the consequent farm income for given scenarios of climate. Some models include a range of farmer behavior in regard to risk—for example, risk-averse or risk-neutral strategies and their probable results in view of climate vagaries. Kaiser et al. (1993) conducted a combined agronomic and economic farm-level analysis in the American midwest and found yield and revenue responses varied by scenario, crop, and location. Antle (1996) has developed a farm-level decisionmaking model that includes both market and nonmarket impacts associated with environmental change.

Macroeconomic Models

Equilibrium models of the agricultural sector include price–responsive behavior for both consumers and producers. Equations for these relationships are developed based on economic principles that consumers will tend to maximize the utility of their food buying and that producers (farmers) will tend to minimize their costs of production. Such models are usually calibrated for a given year (or set of years) in the recent past.

For climate change purposes, the models allocate domestic and foreign consumption and regional production based on given perturbations in crop production, water supply, and demand for irrigation derived from biophysical techniques (see, e.g., Adams et al., 1990). Population growth and improvements in technology are set exogenously. Equilibrium economic model results include quantities of production, equilibrium prices, and measures of producer and consumer welfare.

General equilibrium economic models are useful because they measure the potential magnitude of climate change impacts on the economic welfare of both producers and consumers of agricultural goods. The predicted changes in production and prices from agricultural sectoral models can then be used in general equilibrium models of the larger economy. Such models do not, however, provide a detailed picture of how the economy will respond over time, and they may overestimate the adjustment of the agricultural economy to climate change.

Economic Regression Models

Recently, regression models have been developed that test for statistical relationships between climate variables and economic indicators. A recent study utilized regression analysis to consider the relationships among climate variables and farm values in a technique known as the "Ricardian" approach (Mendelsohn et al., 1994). An econometric model that regressed land values on climate, soil, and socioeconomic variables of counties in the contiguous United States was developed, and it was tested with a 2.5°C increase in temperature and an 8% increase in precipitation. The results predicted a \$21 to \$34 billion loss in farmland values, depending on the year. The potential behavior of consumers is not included in this approach, which assumes that world food prices and domestic farm output prices (and thus farm revenues dependent on changes in agricultural production inside and outside of the United States) are constant. The advantage of this approach is that it implicitly accounts for farmer adaptation to local climate conditions; it does not, however, yield any insight into the nature and timing of those adaptations.

Integrated Assessment

Integrated studies within the agricultural sector link the biophysical and economic realms and may ideally extend to interactions with other sectors (such as competing demands for water by irrigators and urban users, or shifting patterns of land use between agricultural and other natural ecosystems). This seems to be a more realistic, if more complicated, approach, because individual biophysical and socioeconomic sectors will not be affected by climate change in isolation. Agricultural responses will be sensitive not only to changes in crop yields per se, but also to alterations in water supplies, to competing demands for water from other sectors, and to the possible inundation and salinization of arable land by rising seas.

The assessment of the impact of climate change on a country or region consists of a set of tasks beginning with problem definition and leading to sector analysis that considers adaptation methods and response policies (Figure 7.7) (U.S. Country Studies Program, 1994; IPCC, 1994).

In its fullest sense, integrated assessment attempts to link greenhouse gas emis-

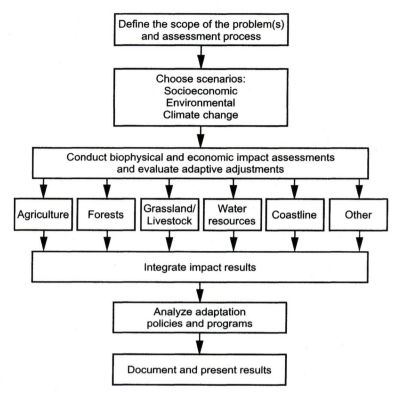

Figure 7.7   Climate change impact assessment process (U.S. Country Studies Program, 1994).

sions caused by human activities with the climatic consequences of the emissions, the impacts of the climate changes on important systems (including agriculture), and the feedback of the impacts back to greenhouse gas emissions. Modeling frameworks have been devised to integrate the causes, impacts, feedbacks, and policy implications of global climate change (Nordhaus, 1992; Peck and Teisberg, 1992; Alcamo et al., 1993; Edmonds et al., 1993; Hulme and Raper, 1993; Manne et al., 1993).

An example of a feedback in such models is the pathway leading from energy consumption to greenhouse gas emissions, to climate warming, to changes in demand for energy (decreases in demand for energy for heating and increases in energy demand for air conditioning), and thence back to changes in energy consumption. The models may be used, for example, to explore the effects of policies that limit greenhouse gas emissions resulting in an ensuing reduction in global warming and the alteration of the impacts of potential climate change.

Examples of Agricultural Integrated Assessment

Parry et al. (1988) reported on integrated agricultural sector studies pertaining to high-latitude regions (Canada, Iceland, Finland, the former Soviet Union, and Japan) that involved teams of meteorologists, agronomists, and economists. The general conclusions of those studies were that warmer temperatures may aid crop production by lengthening the potential growing season at high latitudes, but that higher evapotranspiration and possible drought conditions will tend to counteract the positive effect and may even obviate it in some cases.

Adams et al. (1990) conducted an integrated study for the United States, linking models from atmospheric science, plant science, and agricultural economics. The outcomes depended on the degree of climate change and the compensating effects of carbon dioxide on crop yields. The simulations suggested that irrigated acreage will expand in the Southeast and Great Plains (Figure 7.8) and that regional patterns of agriculture will generally shift northward as temperatures rise (Adams et al., 1990). With the more severe climate change scenario tested, the quantity of farm production available for export was substantially reduced.

The Missouri, Iowa, Nebraska, and Kansas (MINK) study (Rosenberg, 1993) attempted to assess the potential biophysical and economic effects of climate change on agriculture and other sectors. The study used the climate of the 1930s as a historical analog for climate change. The physiological effects of CO_2 enrichment and the ability of farmers to adapt to changed climatic conditions were incorporated. Even with relatively mild warming (1.1°C) and with farmer adaptation and positive CO_2 effects taken into account, regional production declined by 3.3%. Given the IPCC (1990) best estimate of 2.5°C warming for doubled CO_2 conditions (without aerosols), the results of the MINK study imply agricultural production losses of about 10% (Cline, 1992).

Future Scenarios

Climate is not the only factor that will be changing as the 21st century unfolds. Population growth and varying economic and technological conditions are likely to affect world society and the environment no less (and probably even more) than will changes in climate per se. Moreover, socioeconomic and technological conditions will interact with agriculture in ways that may well affect its sensitivity, and that of other sectors, to climate change. However, predicting population growth rates and future economic conditions is as uncertain an exercise as predicting the future climate. Institutions and legal structures may change as well, but these are very hard to predict.

One useful approach is to contrast "optimistic" and "pessimistic" views of the future. In an optimistic scenario, population growth rates are low, world economy grows, and incomes rise at a moderate rate, while both environmental pollution and land degradation are reduced. In a pessimistic scenario, population growth rates are high and economic growth rates and incomes are low, while both environmental pollution and land degradation accelerate. A scenario of no change (i.e., continuation of the present conditions) should also be included. The differential effects of climate change on these alternative scenarios of the future may then be compared and evaluated.

Greenhouse gas emission scenarios are especially important for agriculture because of the need to estimate crop responses to the CO_2 fertilization effect. Projections of sea-level rise are needed as well for assessment of the effects of inundation and salinization in conjunction with climate changes on agricultural production in coastal regions. Environmental factors to be considered include stratospheric and tropospheric

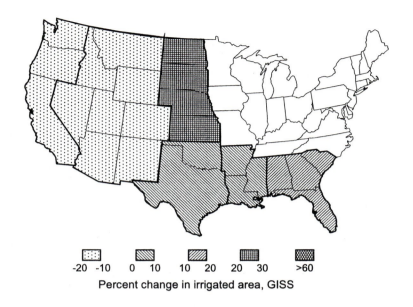

Percent change in irrigated area, GISS

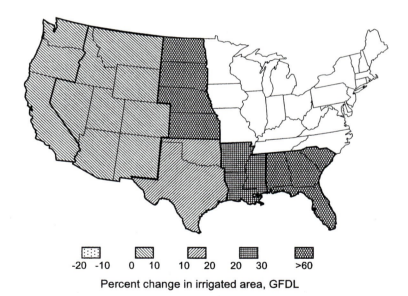

Percent change in irrigated area, GFDL

Figure 7.8 Percentage change in regional irrigation acreage, simulated by an economic model of the U.S. agricultural sector for the GISS and GFDL climate change scenarios with the direct effects of CO_2 on crop yields (Adams et al., 1989).

ozone levels and tropospheric aerosols since these also affect crop productivity. Changes in land use driven by socioeconomic factors should also be taken into account.

CO_2 and Greenhouse Gas Emission Scenarios. Climate modelers require estimates of future levels of atmospheric CO_2 and other trace gases in order to prepare scenarios of future climate. Crop and forest modelers also need such estimates to simulate CO_2 fertilization effects. Global emissions of CO_2 depend primarily on fossil fuel use by three major sectors—electrical power generation, industry, and transportation. A growing world economy includes growth in industrial production, consumption of goods, and travel, altogether tending to contribute to an increase in energy use. Deforestation also contributes to CO_2 emissions and is linked to economic growth, as land is converted from natural ecosystems to agriculture and other uses.

Knowledge of world carbon dioxide emissions from fossil fuels and from deforestation are essential, not only for calculating atmospheric CO_2 levels for climate projections but also for international negotiations that consider limiting CO_2 emissions. Since only a portion of the carbon emitted remains in the atmosphere, models of the entire carbon cycle are necessary to translate carbon emissions into eventual atmospheric levels of CO_2 (Wigley and Raper, 1992). Models that include the effects of CO_2 fertilization, feedback from stratospheric ozone depletion, and the radiative effects of sulfate aerosols have been combined to project the overall radiative forcing of climate, changes in global-mean temperature, and sea level (Wigley and Raper, 1992). Recent projections have tended to lower the rates of warming and sea-level rise, but these rates are still four to five times the rates observed over the past century.

Sea-Level Rise Scenarios. There has been a range of estimates for sea-level rise based on alternative rates of greenhouse gas emissions, climate sensitivities, and uncertainties regarding ocean expansion and glacial melting. Because some coastal areas are subsiding while others are rising, sea-level rise scenarios should be added to current trends. Earlier estimates of sea-level rise were on the order of 1 to 2 meters by the year 2100. Recent estimates are lower, being on the order of a 0.5-m rise by 2100 (Wigley and Raper, 1992; World Coast Conference, 1993).

Socioeconomic Scenarios. In order to place possible changes in climate in the context of potential socioeconomic changes, estimates of population, economic growth, and technological change are needed. These estimates will, in turn, affect future rates of CO_2 and other greenhouse gas emissions. Economic projections beyond the next 10 to 20 years are rather unreliable, especially since changes in population are likely to affect national and per capita income. The IPCC has created scenarios that include estimates of population and economic growth rates (IPCC, 1990, 1992).

The World Bank (1992, 1993) and the United Nations (1996, 1990) have published population estimates by country through 2100 for a range of scenarios, as well as estimates of regional or national changes in income for the coming 10 to 25 years. Pepper et al. (1992) have estimated rates of income growth and population by region up to 2100. Various economic models are used to project such productivity factors as GDP into the future. Population and economic growth will doubtless be accompanied by further urbanization, expansion of agriculture, and mining of natural resources, as well as by accelerating rates of deforestation, habitat fragmentation, desertification, and water and air pollution (Dregne and Chou, 1992; United Nations, 1992; FAO, 1993).

Of Models and Methods

During the last few years, a wide range of methods for the analysis of the impacts of climate change have evolved, from simple regression models to complex integrated systems models. Techniques are becoming ever more complex as additional interacting factors and the propagation of uncertainties are incorporated into the analysis. To comprehensively assess agricultural impacts, the interdependent climatic, biophysical and socioeconomic aspects of agricultural systems should be simulated in a framework appropriate to regional, national, international, and global scales. Spatial analyses and biophysical productivity are relevant, as are studies of potential changes in the socioeconomic welfare of farmers and consumers. Thus, biophysical scientists and social scientists must work together to provide realistic assessments of how climate change might affect agriculture in a broad context. Methodological issues to be resolved include how to generalize from the enormous heterogeneity of agricultural setting and systems, and how to integrate spatial scales and units of analysis from the field to the farm to the region to the nation and on to the global economy. Models must be continually tested, calibrated, validated, and improved for their use to be well-founded. Explicit recognition and inclusion of the transient nature of climate change and of its inevitable uncertainties in the modeling techniques are particularly important.

Another important consideration is to minimize insofar as possible the "black-box" component of modeling. An aspect of modeling too often neglected by those who design models and those who utilize them is the critical analysis of each model's structure, mechanisms, and results that only independent experts can provide. Large-scale modeling efforts are often plagued by "broad-brush" generalizations and simplistically formulated relationships. "Reality checks" by knowledgeable and objective analysts are essential before the results of models can be accepted and applied in practice.

8

Regions at Risk

Regions of the world differ in the biophysical characteristics of their climate, soil, and water resources; in land-use patterns; and in the vulnerability of their agricultural systems to climate change. Any typology of vulnerable regions and impact categories should be based on connections between risks of climate change and existing environmental, economic, and social conditions (Table 8.1) (Schmandt and Clarkson, 1992; IPCC, 1996). Vulnerability differences depend on such current conditions as poverty levels, air and water pollution, availability of agricultural credit, quality and versatility of research, and the rate of population growth.

The Intergovernmental Panel on Climate Change (IPCC, 1996) has defined agricultural vulnerability to climate change as the risk of negative consequences of climate change that are difficult to ameliorate through adaptive measures. Types of vulnerability include risk of large yield reductions that might result from small changes in climate, risk to farmers of profitability loss, risk of economic decline in regions that depend heavily on the farm sector, and risk of hunger for people with limited access to food or means to acquire it. Vulnerability can be defined at different scales, from the household level to the national, regional, and global levels.

Since food is produced and distributed in various ways around the world, regional impacts are likely to differ even among regions undergoing similar climate changes. Variations in current climate, natural resources, production practices, and agricultural policies will engender quite different regional responses.

Analyzing the underlying causes of *vulnerability* to climate is especially important in the semiarid tropics (Ribot et al., 1996). Studies of past and present susceptibility to poverty and hunger are powerful tools for understanding future consequences of climate change since these regions are currently experiencing often severe climate extremes. Ribot et al. (1996) advocate addressing problems of vulnerability in the present by devising ways to achieve equitable and ecologically sound development. An important benefit of such actions is to diminish future vulnerability to climate change.

Table 8.1   Categories of regions vulnerable to climate change (modified from Schmandt and Clarkson, 1992)

Category	Description	Examples
Coastal regions	Vulnerable to sea-level rise. High-risk regions include low-lying and densely populated areas. May also be affected by land subsidence.	U.S. and Mexican Gulf coast, Netherlands, Bangladesh, Guyana, Baltic Sea, Mediterranean Sea, Caribbean Sea.
River deltas	Threatened by sea-level rise, salt water intrusion, reduced fresh water stream-flow, and increased subsidence.	Ganges/Brahmaputra, Nile, Mississippi.
Ocean islands	Vulnerable to sea-level rise. Generally small islands with low elevation.	Maldives, Indonesia, some Pacific islands.
Major food-producing regions	Under current climatic conditions these regions are well-suited to food production. May be vulnerable to more frequent and severe droughts, changes in monsoon patterns or timing of snowmelt, desertification, or other climatic stresses.	U.S. Midwest, Ukraine, Argentina, Australia, India, China.
Marginal agricultural regions	Stressed agricultural systems are highly vulnerable to climatic variations. Typical practices include subsistence agriculture and livestock raising.	Sub-Saharan Africa, northern Mexico, Middle East, northeast Brazil, Australia, China.
Water-stressed regions	Regions of periodic water shortage or where demand for water is increasing and threatening to exceed available supplies.	California, Texas, Mexico and China, the Middle East.

This chapter begins by comparing the effects of climate change on agriculture in temperate and tropical regions. It then reviews, on a continental scale, climate change impact assessments done to date. Finally, the chapter presents case studies of the potential integrated effects of changes in water resources, sea-level rise, and crop production for two countries: Egypt (a country identified as highly vulnerable to climate change) and the United States (a country generally considered less vulnerable).

Temperate and Tropical Agriculture: Differential Responses?

Climate models have predicted larger temperature rises in temperate regions than in tropical regions. These projections have led to the assumption that climate change impacts in the tropics should be less severe than in the temperate zones. This may not be the case, however, due to contrasting biophysical and socioeconomic conditions (Rosenzweig and Liverman, 1992). The projections of changes in the hydrological cycle are uncertain, showing a mixed picture of regional precipitation increases or decreases in parts of both zones. There is some indication, however, that increased drought conditions will be manifested first in tropical regions, and then spread to temperate regions in later decades (Rind et al., 1990).

Climate and Crop Production

The tropics are defined as the geographical band between 23.5°N and 23.5°S latitudes. The temperate regions lie above these parallels and extend to the boreal zones (roughly at 55° and above). Climatologically, the tropics are characterized by perennially high temperatures. Tropical precipitation is primarily convective. In the more humid subregions, annual rainfall is often above 2000 mm and occurs year round. In the drier tropics, however, rainfall tends to be seasonal and its total may not exceed 500 mm. The greater part of the tropical region lies between these precipitation levels, with distinct wet and dry seasons. Agriculture is frequently limited by the seasonal pattern of moisture availability.

In the mid-latitude temperate zone, weather is controlled by alternating or clashing tropical and polar air masses. Precipitation here occurs primarily along fronts of cyclonic storms, although convective storms occur in the summer season. The temperate region also has many different climate subregions with warmer and cooler temperatures, characterized by seasonal patterns of rainfall. Crop growth is often limited during the cooler seasons.

Field experiments have shown that, while the daily rate of growth is typically higher in the tropics, the overall crop-growing period is shorter (Haws et al., 1983). Leaf area expansion and phasic development are faster in the tropics because higher temperatures hasten maturation in annual species, thus shortening the vegetative growth and reproductive stages during which pods, seeds, grains, or bolls can absorb photosynthetic products. This is one reason crop yields are often higher in temperate regions than in the tropics (Table 8.2) (Haws et al., 1983; FAO, 1990). Numerous other factors also contribute to this effect. Soils in the humid tropics tend to be leached of nutrients, and therefore less fertile, because of high temperatures, intense rainfall, and erosion. Soils in the drier tropics, by contrast, are often hampered by accumulations

Table 8.2 Average yields in temperate and tropical regions (Haws et al., 1983)

Crop	Temperate ($kg\ ha^{-1}$)	Tropical ($kg\ ha^{-1}$)
Rice	4109	1958
Wheat	2984	1363
Maize	3993	1351
Sorghum	2270	1249
Soybean	1620	1038
Dry beans	1079	640
Groundnut	1667	1036
Potato	18,056	8704
Sweet potato	13,594	6881
Cassava	11,844	9103
Sugarcane	61,190	53,328

of salt and by lack of water. Although soils of subtropical or temperate regions may be more favorable for agriculture than tropical soils because of higher nutrient levels, there are exceptions in both regions, with highly productive volcanic or fluvial soils found in the tropics, and poorly developed or infertile soils in temperate regions.

The growth of some crops and varieties that require long hours of daylight to reach maturity is limited by the nearly invariable daylength of the tropics. The flux of solar radiation (a factor critical to plant growth) is controlled by the angle of the sun, daylength, and cloudiness. In temperate zones, therefore, it is lower in winter and higher in summer. In the tropics, however, the reception of solar radiation is often reduced by cloudiness during the rainy season. Finally, agricultural production is also limited in many humid tropical regions by the profusion of weeds, pests, and diseases that tend to flourish in consistently warm and moist climates.

Agricultural crops and cropping systems have been developed for, and adapted to, these varied regimes of climate, soil, diseases, and pests (Haws et al., 1983). The main commercial agricultural crops and their adaptations include the following:

1. Cassava and sugarcane grow only in tropical areas and have a crop-growing period of 1 year or longer. Cassava is drought resistant, but sugarcane requires continuous irrigation in dry areas.
2. Sorghum, groundnut, and sweet potato grow in both tropical and subtropical regions in relatively dry seasons.
3. Rice is grown mainly in tropical and subtropical zones in the rainy season or with abundant irrigation.
4. Maize and field beans also grow in both zones, in locations and seasons with enough rain.
5. Wheat, soybean, and potato are mainly crops of the subtropical and temperate zones, but grow in the tropics at high (cooler) elevations.
6. Sugarbeet is grown only in the temperate zone.

A number of "luxury" agricultural crops, especially fruits (such as bananas, mangoes, pineapples), stimulants (coffee, tea), and spices grow only, or best, in the tropics. Tropical regions provide produce for temperate zone markets in the winter season.

In temperate zone agriculture, mechanization, plant breeding, and fertilizer use produced dramatic yield increases for many crops earlier in the 20th century. Similar increases occurred more recently in tropical regions for crops such as wheat, maize, and rice, which benefited from the technological package of improved varieties, fertilizers, irrigation, mechanization, and pesticides, altogether known as the Green Revolution.

In both temperate and tropical regions, irrigation has been developed in areas where dry seasons exist and adequate water can be reserved from other seasons, provided by groundwater or surface sources, or brought in from adjacent regions (Hillel, 1987). Irrigation is an important buffer against climate variability and climate change. Nearly 20% of the world's cropland is currently irrigated, mostly in Asia, producing about 40% of the annual crop production. Improved irrigation practices such as optimal scheduling, high-frequency and low-volume water delivery, adequate drainage and salinity control in agricultural land already irrigated, and extension of appropriate irrigation techniques to rainfed areas will be key elements in expanding food production to meet increased demands in the future (Hillel, 1997). This is especially true for semiarid tropical regions where efforts to assure food security are crucial.

Differences in farming systems, technology, and economics also contribute to the yield differences between temperate and tropical regions. Agriculture in temperate regions is characterized by high levels of inputs (quality seed stock, fertilizers, herbicides, and pesticides), and by a high degree of mechanization, energy use, and capitalization. However, there are wide variations in the use of technology, European agriculture being particularly intensive.

In tropical regions, two broad categories of agricultural systems may be recognized: commercial plantations, with practices similar to those described for temperate regions; and subsistence or semicommercial smallholdings, in which many farmers cannot afford expensive inputs and governments seldom or intermittently subsidize them. In some parts of the tropics, traditional technologies, such as multiple cropping (or intercropping) and terracing may serve to buffer the small-holder farming system against climate variability by conserving soil moisture and fertility and by increasing yields.

Societies in the tropics tend to be more dependent on agriculture than those in the temperate regions, and therefore may be more vulnerable to climatic change. Approximately 75% of the world's people live in the tropics, and two-thirds of them rely on agriculture for their livelihood. With low levels of technology, land degradation, and rapid population growth, societies in some tropical regions may encounter increasing difficulty in providing food security even if climate does not change significantly.

The present social structures of some tropical countries may also exacerbate their vulnerability to climate change. Inequitable land tenure, high numbers of landless rural dwellers, low incomes, and high national debts worsen the negative impacts of climate variability. Growing numbers of people have no extra land, jobs, savings, or government assistance to see them through droughts or other climatic extremes (Jodha, 1989). Moreover, when the economic system is oriented toward export rather than subsistence agriculture, climatic change may threaten a large part of the national economy and thus the ability to import staple foodstuffs. Regions that cannot feed their populations depend on cereal imports. Of the major cereal exporters—the United States, France, Canada, Australia, Argentina, and Thailand (FAO, 1990)—all except Thailand are temperate region countries.

Biophysical Aspects of Climate Change

Numerous investigators have examined the impacts of past climatic variations on agriculture, using case studies, statistical analyses, and simulation models (e.g., Thompson, 1975; Parry, 1978; WMO, 1979; Nix, 1985; Parry et al., 1988a and b). Such studies have clearly demonstrated the sensitivity of both temperate and tropical agricultural systems to climatic variations. In the temperate regions, the impacts of climate variability, particularly drought, on yields of grains (e.g., in North America and the former Soviet Union) have been of particular concern because of their effects on world food security. In tropical and subtropical regions, drought impacts on agriculture and the resulting food shortages have been widely studied, especially when associated with the failure of the monsoon in Asia or of the rains in Sudano-Sahelian Africa (e.g., Nicholson, 1989). In the temperate regions, climatic variations cause economic disruptions. In the tropics, droughts are often associated with famine and widespread social unrest (Pierce, 1990).

Interactions with Thermal Regimes. Because higher temperatures associated with the greenhouse effect will speed crop life cycles (as discussed in Chapter 3), the economic yields of both temperate and tropical crops grown in a warmer environment may not rise substantially above present levels, despite increases in the rate of photosynthesis due to CO_2 enrichment (Rose, 1989). Temperate and tropical regions differ in both current temperature and the temperature rise predicted for climate change; hence, they are also likely to differ with respect to the relative magnitudes of combined CO_2 and temperature effects.

In the midlatitudes, higher temperatures may shift biological process rates toward optima, and beneficial effects are likely to ensue. Regions now limited by cold temperature may become suitable for cultivation (given adequate precipitation regimes) because of earlier snow-melt and longer growing seasons. Lengthened frost-free seasons in temperate regions should allow the growing of crop varieties requiring longer durations, and they offer the possibility of growing successive crops (moisture conditions permitting). Accordingly, agroclimatic zones are expected to shift to higher latitudes and elevations, where longer and warmer growing seasons will allow new or enhanced crop production (soil resources permitting) (Rosenzweig, 1985). Subtropical crops such as citrus should expand their areas of production, while crops such as potatoes that respond to cool temperatures may be limited (Rosenzweig et al., 1996).

Both the mean and extreme temperatures that crops experience during the growing season will change in temperate as well as in tropical areas. In general, higher mean temperatures should reduce cold damage and raise the risk of heat damage. However, studies have shown that winterkill in wheat grown in temperate areas may actually rise due to warmer autumns and consequent suppressed development of cold hardiness (Mearns et al., 1992). In tropical locations where higher temperatures may exceed optimal levels for crops, negative consequences may dominate.

Changes in Hydrological Regimes. In both tropical and temperate regions, the hydrological regimes affecting crop growth are likely to change with global warming. Increased convective rainfall is predicted to occur, particularly in the tropics, caused by stronger convection cells and greater atmospheric moisture. However, the GCM-predicted increases in precipitation may not necessarily result in greater soil moisture. Higher temperatures will induce earlier snowmelt and greater evaporation. Consequently, mid-continental drying in the Northern Hemisphere may take place (Manabe and Wetherald, 1986; Kellogg and Zhao, 1988). Other GCM predictions suggest that the rise in potential evapotranspiration may exceed the rise in rainfall, so that drier regimes may result over extensive areas of the tropics, with subsequent extensions to the midlatitudes (Rind et al., 1990). Global climate change is likely to intensify competing demands for irrigation water in both temperate and tropical regions. Future supplies of needed irrigation water under climate change are uncertain.

Physiological Effects of CO_2. Temperate-zone crops may benefit more than tropical-zone crops from atmospheric CO_2 enrichment. As described in more detail in Chapter 3, in crop species with the C3 pathway (characteristic of nontropical plants such as wheat, soybean, and cotton), CO_2 enrichment has been shown to reduce photorespiration. In contrast, C4 crops, which are particularly characteristic of tropical and warm arid regions (e.g., maize, sorghum, and millet), are generally more efficient

Table 8.3   Crop diseases in temperate and tropical regions (Swaminathan and Sinha, 1986)

Crop	No. of diseases reported	
	Temperate	Tropical
Rice	54	500–600
Maize	85	125
Citrus	50	248
Tomato	32	278
Beans	52	250–280

photosynthetically under current CO_2 levels than are C3 plants (as they fix CO_2 into malate in their mesophyll cells before delivering it to the RuBP enzyme in the bundle-sheath cells). Because of their CO_2-concentrating and photorespiration-avoiding mechanism, C4 plants are known to be less responsive to CO_2 enrichment (Acock and Allen, 1985).

Soils.   In temperate regions where crops are already heavily fertilized, there will probably be no need for any major changes in fertilization practices. Only minor alterations in timing and method of fertilization are expected with changes in temperature and precipitation regimes (Buol et al., 1990). By contrast, in some tropical regions where less fertilizer is applied at present, there will be a need for additional fertilization. Sea-level rise, another predicted effect of global warming, will cause increased flooding, salt-water intrusion, and rising water tables in agricultural lands located near coastlines. This is particularly crucial in tropical countries (such as Bangladesh) with extensive low-lying lands and rural populations concentrated in vulnerable river deltas.

Pests.   Pests are organisms that compete with or attack agricultural plants. They include weeds and certain insects, arthropods, nematodes, bacteria, fungi, and viruses. Because climate variables (especially temperature, wind, and humidity) affect the geographic distribution of pests, climate change is likely to alter their ranges. Insects may extend their ranges where warmer winter temperatures allow their overwintering survival and increase the possible number of generations per season (Stinner et al., 1989). Pests and diseases from low-latitude regions, where they are much more prevalent (Table 8.3) (Swaminathan and Sinha, 1986), may be introduced into higher latitudes. As a possible response to increased pest infestations, there may be a substantial rise in the use of chemical pesticides in both temperate and tropical regions.

Adaptive Responses

In temperate regions, farm-level adaptations may include changes in planting and harvesting dates, tillage and rotation practices, selection of crop varieties or species more appropriate to the modified climate regime, and fertilizer and pesticide applications, as well as in irrigation and drainage systems. Governments can facilitate adaptation to climate change through water development projects; agricultural extension activities; and appropriate incentives, subsidies, and regulations. The provision of crop insurance to farmers affected by changes in the frequency of extreme events may also be necessary.

A wide range of adaptations to possible negative impacts of a warmer, drier climate in the tropics are available (Jodha, 1989; Liverman, 1991). Adaptation options include expansion of irrigation systems; switching to new crops or cultivars (e.g., sowing more drought-tolerant sorghum rather than maize); increased use of fertilizer and pesticides; return to traditional water and soil conserving technologies (e.g., raised fields, agroforestry); diversification of activities (e.g., changing to livestock husbandry or seeking alternative employment); provision of public relief and improved climate information; increase in food imports; and, finally, migration and urbanization.

In Mexico, potential declines in maize yields under global warming could be mitigated if measures are taken to expand and improve irrigation, fertilization, and the use of drought-resistant varieties (Liverman and O'Brien, 1991). However, a major problem in some tropical regions, is the relative inadequacy of resources, institutions, and infrastructure to promote the necessary adaptations. In some cases, traditional technologies and crops in tropical regions may be better able to cope with climate changes. For example, yield stability (i.e., yields with little fluctuation from year to year) may be greater, even though mean yield levels are lower.

Assessment of Major Geographical Regions

Having reviewed the general considerations affecting regions at risk, we now consider how potential climate change might affect agriculture on a continent-by-continent basis. Many national agricultural impact studies have been carried out, for example in the United States (Smith and Tirpak, 1989; Adams et al., 1990), United Kingdom (UK Department of Environment, 1992), Australia (Pearman, 1988), and Japan (Nishioka et al., 1993). Furthermore, the U.S. Country Studies Program, the United Nations Environment Programme (UNEP), and other institutions are encouraging many more nations (especially developing nations) to conduct vulnerability and adaptation assessments as part of the Framework Convention on Climate Change (FCCC) process (U.S. Country Studies Program, 1994; Smith et al., 1996; UNEP, 1996).

Major studies that have characterized regional agricultural effects among various countries include (1) Implications of Climate Change for International Agriculture: Global Food Production, Trade and Vulnerable Regions (Rosenzweig and Iglesias, 1994; Rosenzweig and Parry, 1994) and (2) Modeling the Impact of Climate Change and Rice Production in Asia (Matthews et al., 1994). The United Nations Environmental Programme (UNEP) has also conducted regional studies, including one in Thailand, Indonesia, and Malaysia (Parry et al., 1992). The IPCC Working Group II has comprehensively reviewed regional climate change impact studies in their second assessment (IPCC, 1996).

North America

We consider North America to include Canada, the United States (see Case Study below), and Mexico.

Canada.   Agriculture in Canada is a major industry. The nation contributes significantly to international export markets, being the third largest wheat exporter in the world. The Canadian agricultural sector is diverse, including livestock, dairy, poultry, and potato production in the Atlantic provinces, as well as spring wheat production in the prairie provinces (Manitoba, Saskatchewan, and Alberta).

Climate impact assessments have been carried out for Canadian agriculture, especially the wheat industry in the prairie region (Bootsma et al., 1984; Arthur, 1988; Bootsma and de Jong, 1988; Williams et al., 1988; Smit et al., 1989; Stewart, 1990; Brklacich and Smit, 1991; Singh and Stewart, 1991, Cohen et al., 1992). While some studies show the potential for positive response to climate change resulting from amelioration of cold limitations on the growing season, the prairie region is considered to be vulnerable to climate change due to its drought-prone current climate and soil limitations.

Brklacich et al. (1994) estimated the effects of global climate change on wheat yields in the Canadian Prairie and projected that spring wheat yields would decrease if temperature were to rise and the crop-growing period were to become shorter. However, the positive physiological effects of CO_2 enrichment would likely compensate for those tendencies except under very hot and dry conditions. Adaptation strategies such as irrigation and the substitution of winter for spring wheat could be successful in some areas.

Mexico.   Liverman et al. (1994) and Liverman and O'Brien (1992) estimated the potential impacts of climate change on maize yields in Mexico. Projected climate change caused simulated maize yields to diminish dramatically in two main agricultural regions, but the magnitude of the change varied with climate scenario and the initial set of management variables selected for the simulation. The simulated yield reductions were counteracted to some degree by the inclusion of beneficial physiological CO_2 effects. A sensitivity analysis indicated that global warming would be followed by severe declines in crop yields unless irrigation was provided, fertilizer use was increased, or new varieties that are more heat tolerant were developed and adopted.

South America

Relatively few agricultural climate change impact studies have been done in South America, despite the general importance of agriculture in most of this continent's economies.

Brazil.   According to de Sequeira et al. (1994), wheat and maize production in Brazil will tend to decline under climate change scenarios, but soybean production is likely to remain the same or to increase. Adaptation strategies such as irrigation, changes in planting date, and increased nitrogen fertilization can evidently improve yields, but not enough to compensate entirely for the losses. A hypothetical heat-tolerant cultivar may improve adaptation to warmer climates, but the feasibility of developing such a cultivar remains to be tested through breeding programs. Current climate variability due to El Nino–Southern Oscillation events has a strong effect on many regions of South America, particularly Northeast Brazil. If climate change brings shorter rainy seasons or increased frequency of years when the rainy season fails altogether, this region would suffer.

Argentina and Uruguay.   Sala and Paruelo (1994) examined the potential impacts of global climate change on maize, an important crop in the Pampas of Argentina. Projected changes of climate may reduce maize yields there, but these potential declines may be offset by modifying sowing dates and planting more adapted hybrids. The authors conclude that maize production could expand into areas to the south that are currently limited by frost and that at the northern edge of current maize production, yields may be constrained by high temperature extremes.

Baethgen (1994) and Baethgen and Magrin (1995) estimated the effects of climate change on winter crop production and nitrogen management systems in Argentina and Uruguay. Projected climate change caused simulated barley yields in Uruguay to decrease in all management strategies considered. As in other cases, these potential reductions may be partially countered by the beneficial physiological CO_2 effects on crop growth and water use. The negative effects of the GCM climate change scenarios were worse when no nitrogen fertilizer was applied. The variability of grain yields was found to be larger under the GCM scenarios. A possible adaptation of barley management systems in Uruguay to climate change conditions is an optimization of the soil nitrogen available for the crop, but even considering this adaptation strategy, significant production losses can be expected with high levels of warming.

Chile.   Downing (1992) investigated the heterogeneous region of Norte Chico, Chile, and found mixed results—positive responses for maize and potatoes and negative responses for wheat and grapes.

Western Europe

Considerable work has been accomplished on agricultural climate change impacts in Western Europe (e.g., Parry et al., 1988; Carter et al., 1991; UK Department of the Environment, 1991; Kenny et al., 1993). In general, modeling studies have suggested that simulated grain yields are likely to increase in the north, but to decrease in the Mediterranean area even with agronomic adaptations. The zone of maize production may extend as far north as the UK and central Finland (Kenny and Harrison, 1992a). Vegetable crops may expand in northern and western areas, but decline in southern Europe (Kenny et al., 1993; Olesen et al., 1993). Fruit production may experience reduced winter chilling and loss of production in the south, where grapes in particular are likely to be affected (Kenny and Harrison, 1992b; Kenny et al., 1993). Growing water deficits in southern Europe will intensify the demand for irrigation (Iglesias and Minguez, 1996).

France.   Dellecolle et al. (1994, 1995) reported that under both temperate and Mediterranean climates in France, winter cereal yields can be maintained under future climate conditions, provided that irrigation supply will not be limiting. Under temperate climate, maize yields may well increase, and maize production may shift northward. Under Mediterranean conditions, in contrast, the reduction of phase duration will entail a yield decrease, even under optimal irrigation. Adaptation simulations (change in planting date) produced only slight improvements in yields.

Italy.   Warmer and drier climates in south and central Italy could extend the cultivation range of olive and citrus to the north (Morettini, 1972; Le Houerou, 1992).

Bindi et al. (1993) found that temperature increases and precipitation changes would generally lower yields of winter wheat in northern and central Italy if physiological effects of CO_2 are not taken into account. Rosenzweig et al. (1995) identified the Po Valley, Central Italy (Tuscany and Latium), the Apulean and Sicilian Plains as agricultural regions where integrated climate change impact studies would be especially useful.

Spain and Greece. In Spain, summer irrigated crops such as maize may go out of production due to yield reductions and lack of water availability for irrigation (Iglesias and Minguez, 1995, 1996). The authors suggest that effective irrigation scheduling can minimize water stress during sensitive development phases. For winter dryland crops such as wheat, productivity may increase significantly in some regions and wheat zonation may extend northward.

Kapetanaki and Rosenzweig (1996) have tested potential impacts on maize yields in central and northern Greece and found that while climate change scenarios generally predict decreases in maize yield, analyses of potential adaptations showed that climate change effects may be mitigated by means of earlier sowing dates and the use of varieties with longer growth periods or higher kernel-filling rates.

Former Soviet Union and Eastern Europe

Studies carried out on potential climate change impacts in Russia and other parts of the former Soviet Union include Menzhulin et al. (1995) and Sirotenko et al. (1991). Their results in general indicate a potential for agricultural regions to expand northward and for productivity to improve. Menzhulin et al. (1995) found that winter wheat should be able to replace spring wheat in many locations under warming conditions. Currently, many studies in Eastern Europe are under way through the U.S. Country Studies Program (1994) as reported in Dixon (1997).

Africa

Sub-Saharan Africa is exceedingly diverse in climate, natural resources, and stage of development. Many of the agricultural zones of Africa suffer from periodic drought, among them the subhumid tropical zones, the humid equatorial highlands with bimodal rainfall, and the savanna regions, including the Sahel. Marked intraseasonal and interannual variability of rainfall creates a high-risk environment for agriculture (Schulze et al., 1993). Crop growth in the humid, tropical zones suffers from acid soils (Hillel, 1991) and low solar radiation due to cloudiness.

Agriculture is a dominant activity in many of the countries of Sub-Saharan Africa. Smallholder farmers produce most of the local food supply. Mixed activities include tending woodlots, growing fibers for handicrafts, and raising and grazing livestock, besides the growing of food crops. There are also large-scale plantations and ranches that grow commercial crops such as sugarcane, rice, tea, coffee, and cocoa, and that raise livestock. Countries with strong commercial agricultural sectors include Kenya, Zimbabwe, and South Africa.

Despite the importance of agriculture in Africa, relatively few climate change impact studies have been carried out. Some of the published studies include Zimbabwe

(Muchena, 1994), South Africa (Schulze et al., 1993), Senegal (Downing, 1992), the Gambia (Jallow, 1996), and Kenya (Fischer and van Velthuizen, 1996). Magadza (1994) has examined the impacts of climate change on several sectors, including agriculture and water resources, in southern Africa, and he highlights the sensitivity of water resources, possible loss of important wetland habitats, and projected reduction in biodiversity. Rainfed agricultural systems would be adversely affected with consequent impact on food security. Hulme (1996) has explored potential impacts on natural and managed ecosystems in the Southern Africa Development Region (SADC).

Muchena (1994) found that the probability of obtaining an acceptable yield of maize, the most important food crop in Zimbabwe, decreased under 2°C warming. Simulated maize yields indicated diminished yields, even when the positive physiological effects of higher atmospheric CO_2 were taken into account. Although Muchena's results suggest that farmers in Zimbabwe may be able to offset some of the yield losses by applying fertilizer and irrigation, such solutions may not be very realistic due to the high costs of inputs to farmers with limited financial resources. Some farmers would adapt by switching to crops such as sorghum and millet that are more heat and drought tolerant. In Kenya, by contrast, crop zones may extend to higher altitudes, resulting in a net increase in productivity, although agricultural regions are predicted to shift (Fischer and van Velthuizen, 1996). Vulnerable groups in subhumid and semiarid regions are likely to be negatively affected nonetheless (Downing, 1992).

Fischer and van Velthuizen (1996) have conducted a detailed study of potential impacts of climate change on agricultural potential in Kenya. Utilizing the FAO agroecological zone approach (see Chapter 7 for a description of the methodology), they found that the national-level food productivity potential of Kenya may well increase with higher levels of atmospheric CO_2 and climate change–induced increases in temperature. In particular, in central and western Kenya, temperature increases would result in expansion of cultivation since some higher altitude areas would become suitable for cropping. In the semiarid parts of Kenya, however, if warmer temperatures are not accompanied by higher precipitation, the impact on agricultural productivity could be severe. Thus, the authors conclude that even though the overall effects of climate change in Kenya may be positive, the impacts may intensify regional disparities.

The potential for increases in drought frequency and severity with climate change is of special concern for Sub-Saharan Africa, since droughts in the current climate cause severe disruptions to regional food supplies. Negative impacts would include exacerbated land degradation in the savanna zones, deforestation in more humid regions, and soil erosion. Schulze et al. (1993) emphasize the importance of intraseasonal and interannual variation of rainfall for crop yields in southern Africa. Sivakumar (1993) has documented that recent droughts had shortened the current crop growing seasons by 5 to 20 days.

A recent impacts study for the Southern Africa Development Community (SADC) (Hulme, 1996) emphasized the need to put potential climate change in context with the other problems facing the region. The study concludes that climate change in southern Africa is likely to add further stress to ecosystems and agroecosystems already under severe pressure because of population growth, increasing subsistence needs, droughts, unequal land distribution, and limited coping ability.

Many more studies are needed to characterize the impacts of climate change for

African conditions. Research to be encouraged includes studies of high temperature, water stress, and enhanced CO_2 effects on the subsistence crops of the region such as millet, cassava, and sweet potato (*Ipomoea batatas*). For example, there is some evidence that soil temperature at both the diurnal and seasonal time scales is a key yield determinant of sweet potato, a root crop that is especially important in Africa (Kays, 1985; Phillips, 1995).

Asia

The vast continent of Asia encompasses the populous regions of South and Southeast Asia (Pakistan, India, Bangladesh, Thailand, Myanmar, Vietnam, Cambodia, Malaysia, Indonesia, and the Philippines) and of East Asia (China, Taiwan, North and South Korea, and Japan).

South and Southeast Asia. Agriculture is an important sector in South and Southeast Asia. For example, agriculture in Pakistan accounts for about 30% of the GDP and employs over 50% of the labor force (Qureshi and Iglesias, 1994). Rice is the leading food crop, although other crops, including wheat, soybeans, and maize, are raised in the drier regions. Root tuber, fruit, and vegetable crops are also grown. Plantation crops include tea, cocoa, coffee, and rubber for export. The climate is dominated by seasonal monsoons.

The dominant system for growing rice is in irrigated fields (paddies). Dryland or upland rice is grown in the drier regions. Rice farms are generally small. Improved rice varieties developed primarily at the International Rice Research Institute (IRRI) and greatly increased fertilizer applications have raised rice yields in the region to high levels. Crop production is strongly dependent on irrigation; in fact, irrigation here covers a higher proportion ($\sim$70%) of agricultural land than anywhere else in the world.

Qureshi and Iglesias (1994) used global climate models and dynamic crop growth models to estimate the potential agricultural effects of climate change in Pakistan. Under present climate conditions, wheat is under stress due to high temperatures and arid conditions. Projected climate change is projected to diminish wheat yields dramatically in the major areas of agricultural production, even under fully irrigated conditions. Decreases in simulated grain yields were caused primarily by temperature rises that shortened the life cycle of the crop, particularly the grain-filling period, thus exerting a strong negative effect on yields. These decreases were countered somewhat by the beneficial physiological effects of CO_2 on crop growth. Adaptation strategies such as shifting cultivars and delaying planting mitigated simulated yield losses partially, but not entirely.

Studies of climate change impacts on wheat and sorghum production in India (Rao and Sinha, 1994; Rao et al., 1995) also indicated that yields would generally decrease, although responses varied by crop and season. Wheat yield decreases could have a serious impact on food security in the region, with its growing population and great demand for grain. Most of the wheat production in India comes from the northern plains, where it is almost impossible to increase the present area of wheat under irrigation. Sorghum is apparently less sensitive to temperature rise, and its simulated yields were not affected as much by climate change scenarios as were the yields of wheat.

Bangladesh is vulnerable to many environmental hazards, including frequent floods, droughts, cyclones, and storm surges that damage life, property, and agricultural production. Karim et al. (1994) found that current rice production could be damaged under climate change conditions as projected by GCMs. Climate change also poses the hazard of a rising sea level, leading to flooding of low-lying coastal areas that currently support a dense population and intensive agricultural production. A drop in rice production due to climate effects and sea-level rise, combined with a rapidly growing population, would threaten the food security of Bangladesh (Karim et al., 1994, 1996).

Agriculture is also a key contributor to the economy of Thailand. Agricultural systems (arable crops, rangelands, forestry, and fisheries) employ more than 60% of the labor force during the cropping season and account for some 20% of the national GDP (Tongyai, 1994). Rice is the main food crop and the main agricultural export commodity. Tongyai (1994) tested GCM climate change scenarios on simulated upland and paddy rice and found that yields diminished by 2 to 17% in two of the scenarios (with direct CO_2 effects included); the third GCM scenario projected little change (-1 to $+6\%$).

Agriculture is considered to be the economic lifeline of the archipelago nation of the Philippines (Escano and Buendia, 1994). More than 50% of the working population are engaged in agriculture, and more than 70% of foreign exchange earnings are derived from exports of agricultural products. Rice is the staple food that is grown on 3.4 million ha of land. Agriculture here is very vulnerable to climate hazards, especially the occurrence of tropical cyclones and floods, and to delays in the onset of the rainy season that may cause drought. In January of 1991, the Philippine weather bureau (PAGASA) reported that 10 weeks after the onset of an ENSO-related drought, rice and maize crops suffered estimated damages of $753 million (Escano and Buendia, 1994).

In their climate change study, Escano and Buendia (1994) found that simulated rice yields declined under GCM scenarios, with direct CO_2 effects taken into account, in at least one important agricultural region of the Philippines. Testing of possible adaptation strategies to climate change suggested modifying the existing cropping patterns of rice. Earlier planting dates and changing cultivars may mitigate negative climate change impacts, but they imply changes in the current farming system of the Philippines. For example, farmers in Batac report that planting 1 to 2 months earlier may not be appropriate for rice because strong winds later in the season may coincide with the grain-filling stage and therefore cause lodging of the crop (Escano and Buendia, 1994). Thus, additional climate factors such as wind should be considered in the development of adaptation strategies.

A major study on climate change and rice has estimated the potential impacts of equilibrium doubled-CO_2 climate change scenarios in many countries in Asia (Matthews et al., 1995). This study found that simulated yield effects can vary widely ($+30$ to -38%) across the region and for the different GCM scenarios. Decreased rice yields were projected for low-latitude countries, whereas increased yields were projected for higher latitudes. Such results suggest a possible shift in rice-growing regions away from the equatorial regions to higher latitudes. Panicle sterility can be caused by high temperature in many current rice varieties, and this study showed that this could be a critical factor in rice response to global warming scenarios. Where current conditions are near critical thresholds, a difference in mean temperature of less than 1°

resulted in a simulated positive yield change becoming a large decline (Matthews et al., 1995). There is hope, however, that genetic differences among varieties may be utilized for adaptation to warmer conditions.

The United Nations Environment Programme sponsored a project that considered the potential socioeconomic effects of climate change in Southeast Asia (Parry et al., 1992). Here, also, a range of crop yield changes (including rice, soybean, maize, oil palm, and rubber) (−65 to +10%) were found across Thailand, Indonesia, and Malaysia, culminating in an overall mean loss to farmer income of $10 to $130 per year. Increased cloudiness could reduce radiation and cause yield losses. Sea-level rise was highlighted as a threat to coastal rice-producing areas and to fish, prawn, and shrimp ponds. A sea-level rise of 1 meter would cause a landward retreat of 2.5 km in Malaysia, affecting 4200 ha of agricultural land. However, this is an area less than 1% of the paddy rice and total cereal producing area in Malaysia (Parry et al., 1992). Soil erosion and fertility would also be negatively affected.

Across the entire region of South and Southeast Asia there is concern about how climate change may affect El Niño/Southern Oscillation (ENSO) events, since these play a key role in determining yearly agricultural production. ENSO events, which tend to recur every 2 to 9 years, are related oceanic and atmospheric phenomena, characterized by increases in sea-surface temperatures of the tropical Pacific Ocean, suppression of upwelling nutrient-rich water along the coast of South America, and disruption of the trade winds. The cycle has long been known to constitute a large component of natural interannual climate variability in the tropics and subtropics and, to a lesser extent, in the midlatitude regions. Global "teleconnections" (relationships between weather at two or more distant points) linked to ENSO include lower-than-normal precipitation in western Oceania, India, southeastern Africa, and northeastern South America, and excessive precipitation in western South America and eastern equatorial Africa (Ropelewski and Halpert, 1987).

Changes in ENSO event frequency and severity would be sure to affect the agriculture of the region. Other concerns involve the development of rice cultivars that are more tolerant to heat, particularly at the spikelet stage, and that can mature within a shorter growing season. This is one of the most crucial strategies for adaptation to climate change in the region. Finally, field experimentation on crops such as chickpeas, rapeseed, and mustard is needed to generate the crop phenology data needed to develop crop models for these important local crops.

East Asia. Several climate change impact studies have been conducted in Japan, examining both major production areas and vulnerable regions. Uchijima (1987), Horie (1987), and Uchijima and Seino (1988) have shown that warming could expand the area suitable for agricultural production and increase potential rice yields. Rice is currently cultivated on 2.11 million ha in Japan. Seino (1995) found that adjustment of agricultural practices, such as advancing the planting date and applying supplemental irrigation, could adapt and even augment yields of rice, wheat, and maize at most sites under doubled-CO_2 climate change scenarios. The success of the adaptation measures, however, depends on the magnitude in the change in precipitation and the future improvement of irrigation systems.

China has about 7% of the world's cultivated land, but supports more than one-fifth of the world's population. China is the largest rice producer and consumer in the world. However, increasing population, spread of urbanization, lack of sufficient water resources, and environmental pollution may hinder growth in China's agricultural productivity in the future. Jin et al. (1995) examined potential climate change effects on rice production in Southern China and found that a rise of temperature will extend the northern limits for double-rice and triple-rice cropping systems by 5 to 10° of latitude, depending on the scenario. For paddy rice, adjusting planting dates ameliorated the negative effects of climate change on modeled yields in the northern part of the studied region, but not in the southern part. In climate change scenarios where precipitation decreased, the amount of water needed for full irrigation increased. The study by Matthews et al. (1995) also revealed that simulated rice yields in China tended to vary among regions and with GCM doubled CO_2 scenario (yield changes ranged from 14 to −38%).

Oceania

Oceania includes Australia, New Zealand, Papua New Guinea, and the small island nations of the Pacific. Numerous climate change impact studies have been carried out for Australia and New Zealand, while few studies regarding the Pacific island states have included agricultural effects.

Australia. In Australia, Baer et al. (1994) found that dryland wheat yields may increase in response to greater precipitation, but may be reduced if temperature increases are large. Rice yields are expected to decline slightly due to temperature increases. Successful adaptation strategies may include switching to more tropical varieties of rice and adjusting sowing dates to maximize timely water availability.

The IPCC (1996) has summarized the following potential climate change agricultural effects in Australia: poleward shifts in production, varying impacts on wheat including changes in grain quality, likely inadequate chilling for stone and pome fruits reducing fruit quality, enhanced likelihood of heat stress in livestock (particularly dairy and sheep), greater infestations of tropical and subtropical livestock parasites but possible decreases for other species, livestock improvement due to warmer and shorter winters, increased damage due to floods and soil erosion, more severe drought potential (with wheat and barley more sensitive than oats), changes in severity of outbreaks of downy mildew on grapevines and rust in wheat, and beneficial effects of elevated CO_2 on many agricultural crops. These projections are based on studies by, among others, Hobbs et al. (1988), Wardlaw et al. (1989), Blumenthal et al. (1991), Wang et al. (1992), Hennessy and Clayton-Greene (1995), and Wang and Gifford (1995).

New Zealand. Pastoral systems are very important in both Australia and New Zealand. Climate change studies predict overall gains due to poleward and altitudinal shifts in production areas, more tropical and less temperate pasture regions, and the beneficial effects of CO_2 on growth (Campbell et al., 1995). Subtropical horticultural crops and maize may expand, but temperate crops such as apples and kiwi crops may lose critical vernalization periods (Salinger et al., 1990). Kenny et al. (1995) have developed an integrated model for assessment of climate change on New Zealand ecosystems.

Small Island Nations. Few studies have been conducted on potential agricultural impacts in the small island nations of the Pacific. Singh et al. (1990) have projected that crop yields might suffer from reduced solar radiation due to increased cloudiness, higher mean temperature leading to shorter growth duration and greater incidence of sterility, and both excess and limited water availability depending on changes in climate variability. Flooding could cause maize yield and production losses, while greater cloudiness and sea-water intrusion could damage rice-growing potentials.

Egypt: A Case Study in Vulnerability?

Egyptian agriculture is based entirely on irrigation and, hence, is utterly dependent on a tenuous balance between the supply of water from the Nile, and to a much lesser extent from groundwater, and the demand for water by crops (Hillel, 1994). That balance is mainly dictated by the climate, inasmuch as climate determines both the supply of water by the Nile and the evapotranspirational demand for water imposed on crops by the atmosphere. The water balance is affected secondarily by the pattern of water use (i.e., the specific crops grown and the mode of irrigation), as well as by soil conditions and water quality—both of which appear to be deteriorating.

Any attempt to assess the future of Egyptian agriculture must consider the complex interactions of those factors, as well as the inexorable growth of population (now multiplying at the rate of 2.3% per year) and urban encroachment (currently estimated at 10,000 to 20,000 ha yr^{-1}) (Rosenzweig and Hillel, 1994). The future of Egypt's food supply appears to be problematic, even assuming the continuation of current climate conditions, subject to fluctuations but not to long-term change. The task of ensuring food security will be made all the more difficult if a significant warming trend occurs as a result of the enhanced greenhouse effect.

Egyptian Agriculture

Egypt's favorable conditions of temperature, solar radiation, fertile soils, and abundant water supplies from the Nile River have given rise to a rich agricultural system that has persisted for millenia. Although the total area of Egypt is relatively large (about 1.1 million km^2), agriculture is limited mainly to the Nile Delta and Valley (about 1700 km long and from 3 to 40 km wide). Thus, only about 3% of the country's land is available for agriculture. The Old Lands irrigated directly from the Nile or from groundwater fed by the Nile are found in the Nile Delta and along the banks of the Nile in Middle and Upper Egypt. These soils, predominantly alluvial silt and clay loams, have been farmed continuously throughout the history of Egyptian civilization.

In the last 20 years, the government of Egypt has promoted the expansion of agriculture into New Lands located in desert regions. In these regions, water must be conveyed over some distance from the Nile, or supplied from deep wells. Figure 8.1 shows the locations of the Old and New Lands. The New Lands are distributed west of the Delta (Nubaria), east of the Delta (Salhia and along the western side of the Suez Canal), in the northern Sinai, and in the New Valleys of the Western Desert. The soils here are generally sandy and calcareous, not nearly as naturally fertile as the alluvial soils. However, they are more readily drainable though often more prone to salinity.

Degraded Old Lands that need to be rehabilitated are not included in the New Lands category. The Egyptian government has also established programs to reclaim these long-used areas now salinized or waterlogged through drainage and leaching. These areas are referred to as "New-Old Lands."

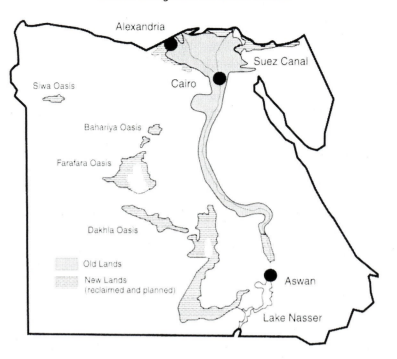

Figure 8.1   Old and New Lands in Egypt (Hillel, 1994).

In the Nile Delta and along its banks, agriculture is characterized by complex yearlong cropping patterns carried out by traditional farmers on small units of land with intricate land-tenure relationships. Two-thirds of the landowners in Egypt own less than 5 feddans (one feddan equals 0.4 ha of 1 acre) (Table 8.4) (CAPMAS, 1993). Agriculture in the Old Lands is so intensively managed that it may be better described by the term "gardening" rather than "farming."

Table 8.4   Ownership of land in Egypt by number of feddan[1] and percentage of landowners (CAPMAS, 1993)

Ownership (feddan)	Landowners (%)
<1	12.6
1–3	34.4
3–5	20.1
5–10	16.0
10–50	16.7
>50	0.2

1 feddan = 0.4 ha

Table 8.5 Cropping systems, seasons, representative crops, and areas in Egypt (CAPMAS, 1994)

System	Cropping season	Major crops[1]	Area[2] (%)
Winter crops	October–December to May	Wheat, barley, beans, clover	43
Summer crops	April–June to October	Cotton, maize, rice, sugar cane	36
Nili (Kharif) crops	July–August to November	Cotton, maize, rice, sugar cane	4
Perennial crops	Planting in March or October	Alfalfa	2
Vegetables	All year	All types	8
Orchards	Transplantings in February	All types	7

1. The table indicates representative crops; 28 different crops can be identified in Egypt.

2. The areas indicate the relative proportion of each cropping system. The available agricultural land is used intensively all year; ~5 million ha of crops are harvested each year from ~2.5 million ha of land. Intercropping (inter-planting of a major crop with a secondary crop) is another form of intensification.

Three cropping periods are utilized per year. Winter crops are sown in October and November; summer crops are sown in April and May; and Nili (or Kharif) crops are sown in July and August. Twenty-eight major seasonal crops may be identified in the current cropping system (Table 8.5), with wheat, maize, clover, cotton, rice, sugarcane, fava bean, and soybean as the most important (El-Shaer et al., 1996). Perennial crops, such as sugarcane and alfalfa, are sown either in the spring (March) or in the autumn (October). Cotton is a relatively long duration summer crop and is planted in March. Vegetables are planted all year long, with spring and autumn plantings added to summer and winter plantings. In the New Lands, major crops are primarily fruit and oil trees and vegetables planted in larger fields. Intensive management of modern irrigation is needed to sustain these crops at highly productive levels.

The Future without Climate Change

In the future even more than in the present, the primary constraints in Egyptian agriculture will be water (availability and quality) and land (soil fertility, salinity, and drainage). Two visions of the future without climate change can be compared: a pessimistic "worst-case" scenario in which environmental degradation proceeds unchecked versus an optimistic "best-case" scenario of wise management of Egypt's rich natural resources.

Pessimistic Case. A continuation of current trends in crop patterns and water use in Egypt will lead to an accelerated loss of agricultural land to waterlogging and salinization, as well as to urbanization. Efficiency values of water applied to the fields (defined as the fraction of the water applied that is beneficially utilized, or transpired, by the crop) are typically well below 50%, and in many cases are below 30% in Egypt.

Such low values imply that more than half (and often two-thirds) of the water applied in the field exceeds the irrigation requirement of the crop. Although much of the excess water applied in upstream areas is reused in downstream areas, such reuse entails loss of energy and quality.

Excess irrigation reduces productivity below potential insofar as it impedes aeration, leaches nutrients, and induces water-table rise and salinization. Moreover, it raises the cost of drainage. Concurrently, irrigation water becomes progressively salinized. Crop yields therefore tend to diminish despite improvements in varieties, fertilization, and pest control.

Especially vulnerable to the progressive degradation of land and water resources are the ill-drained areas of the lower Nile Delta that are already subject to land subsidence, water-table rise, and saline-water intrusion. Combating these processes will require large investments in expensive drainage, along with greater government intervention and regulation. If investments, interventions, and regulations are lacking or are inadequately implemented, these lands seem destined to become unusable for agriculture. The strains to the coastal and delta system may also exacerbate rivalries among agricultural, urban, and industrial sectors.

Optimistic Case.   Through implementation of water-use efficiency and crop management, there is an opportunity and a challenge for Egypt to conserve water and reduce drainage requirements while raising crop yields in both the Old and New Lands. This is the essence of the optimistic vision of Egypt's agricultural future. Although yields in the fertile lands of the Nile Valley and Delta are already high, there is certainly room for improvement in water-use efficiency. The potential increase in productivity inherent in the improvement of irrigation in the Old Lands probably exceeds the potential production increase from the development of New Lands in desert areas outside the Nile Valley and Delta. The latter will, in fact, be enhanced by water conservation in the Old Lands.

The experience of Israel is instructive, particularly in regard to the development of New Lands. In the last 40 years, the average seasonal irrigation applied to Israeli field and orchard crops has been reduced by over 40%, from more than 10,000 to less than 6,000 m^3 ha^{-1}. At the same time, in large measure as a result of the more precise optimization of soil moisture and nutrients (as well as the improvement of crop varieties and microclimate control), average crop yields have approximately doubled. The irrigated crop productivity ratio (defined as the yield (kg) obtained per unit volume of water applied (m^3) has tripled (Figure 8.2).

The actions necessary to achieve the potential improvement in water-use efficiency in Egypt are not easy to undertake and implement. Needed is a strong system of rewards and penalties to create incentives for water conservation and the installation of modern irrigation technology. Water metering and water pricing should be instituted; water should be made available on demand or at high frequency (rather than on a fixed schedule at infrequent intervals); and credit, as well as training, should be offered to farmers willing to modernize their irrigation. In addition, efforts should be made to promote the preferential adoption of high-return, specialized and water-conserving crops instead of the presently grown water-profligate crops such as rice and sugarcane. Given Egypt's already high yields, perhaps water-use efficiency values will not be quadrupled, but they can very probably be doubled.

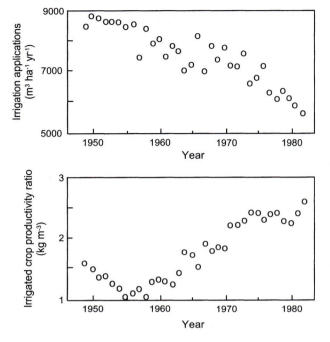

Figure 8.2 Change in annual water application and productivity of irrigated crops in Israel (1948–1982) (Hillel, 1994).

Given a determination to modify irrigated agriculture, to conserve water and maintain water quality, and to substitute high-return crops for the subsistence grain and fodder crops now predominating, a doubling of agricultural productivity in Egypt seems possible in the coming decades.

The Future with Climate Change

Climate change is predicted to bring a significant rise in mean temperature to Egypt. Prediction of hydrological changes is more uncertain, and solar radiation changes are projected to be small for doubled atmospheric CO_2 as simulated by three GCMs (Table 8.6). If no timely measures are taken to adapt Egyptian agriculture to the expectable warming, the effects may be negative and serious. Egypt appears to be particularly vulnerable to climate change because of its dependence on the Nile River as the primary water source, its traditional agricultural base, and its Delta coastline, which is already undergoing both intensifying urban development and land degradation. Contrarywise, if appropriate measures are taken, negative effects on these major resource sectors may be obviated or at least lessened.

Sea-Level Rise. One serious threat is the potential effect of sea-level rise resulting globally from the thermal expansion of seawater and the melting of land-based glaciers (See Chapter 6). Even a slight rise of sea level will exacerbate the already active process of coastal erosion along the shores of the Delta (currently 50 m per year at the head of the Rosetta branch of the Nile at Rashid), a process that has accelerated since

Table 8.6   GCM climate change scenarios for Egypt
(Sakha and Giza)

Climate variable	GISS	GFDL	UKMO
Temperature change (°C)			
Spring	5.1	4.5	4.7
Summer	3.2	4.4	4.1
Autumn	4.4	4.1	4.5
Winter	4.0	3.7	4.5
Annual	4.2	4.2	4.4
Precipitation change (%)			
Spring	−7.1	−19.2	−12.5
Summer	350.0	0.0	−37.0
Autumn	27.3	−20.0	1.2
Winter	5.9	−10.0	−8.9
Annual	55.7	−15.3	−13.8
Solar radiation change (%)			
Spring	−0.3	2.0	6.2
Summer	−4.2	−0.6	6.1
Autumn	−1.2	0.6	1.3
Winter	0.0	0.8	8.7
Annual	1.7	0.6	5.5

GISS = Goddard Institute for Space Studies (Hansen et al., 1983).

GFDL = Geophysical Fluid Dynamics Laboratory (Manabe and Wetherald, 1987).

UKMO = United Kingdom Meteorological Office (Wilson and Mitchell, 1987).

the construction of the Aswan High Dam and its attendant curtailment of sedimentation for replenishing the Delta.

Most vulnerable to sea-level rise are the low-lying lands along the northern strip of the Delta, where the surface elevation is less than 1 m above sea level. Owing to land subsidence (projected to be about 0.1 m in the next half century), as well as expectable sea-level rise (variously estimated to total 0.2 to 0.5 m in the same period), the widening strip of land subject to seawater intrusion may reach 20 km or more. Within this strip, the maintenance of agriculture will become progressively more difficult, and eventually much land will be retired from production. For a 1-m sea-level rise, 12 to 15% of the existing agricultural land in the Delta may be lost (Nicholls and Leatherman, 1994).

Sea-level rise will also accelerate the intrusion of saline water into surface bodies of water (the lagoons and lakes of the Delta), as well as into the underlying coastal aquifer (Sestini, 1992; El-Raey, personal communication). The rise in the base level of drainage will further increase the tendency toward waterlogging and salinization of low-lying lands, with the consequence that significant areas will become unsuitable for agriculture. Drainage to control waterlogging will become increasingly expensive, as it will necessitate pumping rather than reliance on gravity flow. Urban and industrial development, too, will be problematic because of waterlogging, and the ecology and economy of the coastal lagoons will be affected by saline water intrusion.

Crop Yields and Water Use.   Coupled with the deleterious effects described in the pessimistic case without climate change, global warming is likely to reduce agricultural productivity in Egypt significantly. Crop modeling simulations with GCM climate change scenarios at the high end of the IPCC range (~4°C) project that maize and wheat yields may decline in the Delta (Sakha) by as much as 30% and in Middle Egypt (Giza) by more than 50% (Figure 8.3) (Eid, 1994). Incorporation of alternative available crop cultivars and adjusted planting dates into the simulation did not compensate for the projected yield losses under the warmer climate, nor improve crop water-use efficiency. In view of the continuing growth of population, Egypt may suffer a worsening shortage of food and an eventual crisis.

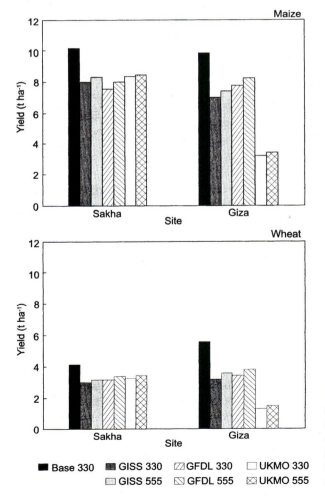

Figure 8.3 Simulated maize and wheat yields under GCM doubled-CO_2 climate change scenarios with (555) and without (330) direct effects of CO_2 on crop growth and yield (Eid, 1994).

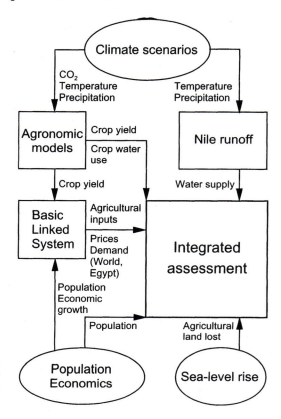

Figure 8.4 Processes examined in an integrated assessment of climate change impacts on Egyptian agriculture (Strezepek et al., 1995).

The impact of climate change on the supply of water (i.e., on the flow of the Nile) is greatly uncertain (Strzepek et al., 1995). We may be certain, however, that a warmer climate will impose a greater evaporational demand and, hence, will increase irrigation water requirements. This effect may be mitigated in part by the higher water-use efficiency of some crops in a CO_2-enriched atmosphere (Rosenzweig and Hillel, 1994). Higher evaporation rates will have the secondary effect of worsening the tendency toward soil salinization by speeding the transport of salts toward the soil surface.

In the traditional regime of infrequent irrigation common in Egypt, sensitive crops are therefore likely to suffer greater moisture stress and salt damage. Hastened maturation in a warmer climate may constrain yields (Eid, 1994), as will greater infestations of pests. Heat-sensitive crops that are already near the limit of their heat tolerance will be especially vulnerable.

Integrated Case Study. A recent study conducted an integrated assessment of the potential impact of changing climate on Egypt's agricultural sector, taking into account changes in sea level, water resources, crop production, and world food trade (Strzepek et al., 1995). The study examined the combined impacts of different biophysical and socioeconomic factors on agriculture and the adaptability of the Egyptian economic system (Figure 8.4). Climate change scenarios, generated by GCMs, were used to simulate changes in Nile River flow and their potential impacts on Egyptian crop

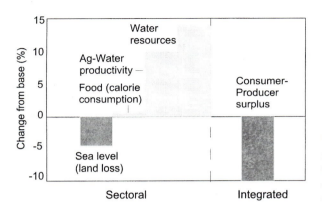

Figure 8.5 Sectoral and integrated results of "lowend" GCM climate change scenario on Egyptian agriculture (Strezepek et al., 1995).

yields and irrigation water use. A world food supply and trade model then simulated the possible economic consequences of the foreseen changes in crop yields relative to yield changes elsewhere. Results from the world food model were combined with a model of the Egyptian economy to assess changes in the agricultural sector, including water supply, land, crops, livestock, and labor. A measure of social welfare, the annual consumer-producer surplus, was used to integrate the potential changes in all factors.

Results from the study suggest that population growth and economic development will be the major drivers of future welfare in Egypt. Climate change, while important, is a less dominant factor. The study demonstrates the importance of placing sectoral impacts in an integrated and international context. Figure 8.5 shows the impacts, relative to a projection with no climate change, of a 2.4°C average global temperature increase, 5% increase in precipitation, and doubled atmospheric CO_2. While land is lost to sea-level rise, impacts on agricultural productivity and water resources are positive. However, the overall indicator of social welfare, namely the consumer-producer surplus, is negative. This counterintuitive result arises from concurrent projections of climate change in other countries: they would reap even greater benefits to agriculture than Egypt, thus causing Egypt to lose its present comparative advantage.

Adaptation to Climate Change. Much can be done in the context of an irrigation-based agricultural economy to mitigate the potentially dire consequences of climate change in Egypt. The first imperative is to improve both the technical water-application efficiency and the agronomic water-use efficiency. This involves nothing less than revamping the entire system of water delivery and control.

Ideally, water should be made available on demand (rather than on a fixed schedule) and be delivered in measured quantities in closed conduits subject to effective monitoring and regulation while avoiding seepage losses. While this will be difficult to achieve in the Old Lands, where traditional systems (open canals, etc.) and traditional concepts are entrenched, it is certainly achievable from the outset in the New Lands.

To facilitate adoption of water conservation, the authorities should provide farmers with explicit guidance regarding optimal crop selection, irrigation, and fertilization, and they should institute strong incentives to avoid excessive water use (including the oft-suggested but seldom-implemented pricing of water in increasing proportion to the amount used). Modern methods of irrigation based on the high-frequency, low-volume application of water and fertilizers directly to the plants need to be adapted to the scale of operation and local practicalities of Egyptian farming. Fortunately, such systems are flexible and lend themselves readily to downsizing so as to accommodate the small-scale nature of most Egyptian farming units. Moreover, such systems can be applied successfully to sandy and even to gravelly desert soils (potential New Lands) that are not considered irrigable by the traditional surface-irrigation methods.

An additional set of measures involves the careful selection and/or breeding of heat-tolerant, salinity-tolerant, water-conserving crops, as well as controlled-environment production methods that minimize water use while maximizing the production of high-value crops (e.g., all-season vegetables and fruits, spices, and medicinals).

Further mitigation measures involve management of the low-lying lands on the northern fringe of the Delta, where the consequences of sea-level rise (submergence and salinization) are certain to wreak their greatest damage. Some of those lands must be retired from agriculture, and the amount of water made available consequently could be diverted to the irrigation of New Lands outside the Nile Valley and the Delta.

The overall effect of the measures listed herewith, in light of the future either with or without climate change, will be to raise the potential and actual productivity of Egyptian agriculture. Thus, climate change may encourage rather than thwart progress toward the goal of providing sufficiently for the Egyptian people. A final caveat, however, is that much depends on whether the rate of population growth in Egypt, which has already begun to decline, continues to do so fast enough to allow agricultural productivity to keep pace with the country's growing needs.

The United States: A Case of Low Vulnerability?

Agriculture in the United States, in contrast to Egyptian agriculture, is regionally diverse and primarily rainfed. Only about 10% of cropland is under irrigation, in contrast to virtually 100% of Egyptian cropland. Also different is the agricultural population, which in the United States is only a small fraction of the national population, and agriculture, per se, accounts for less than 5% of the national GDP. The sector is highly productive, intensively managed, and market based. Farms and the associated processing industries in the United States provide low-cost, high-quality food for domestic consumers and contribute substantially to export earning for the country as a whole. The heavy dependence of world grain reserves on North America (on the order of 80% of the global marketable surplus) has increased the sensitivity of world food supply to the climate of the region. Although the domestic system is primarily market oriented, the government is involved in agriculture through price supports, loan assistance, acreage controls, foreign food aid, domestic food subsidies, and a variety of land and soil conservation programs.

The primary production regions of the United States include (1) major maize, soybean, and livestock production in the Midwest (Illinois, Iowa, Indiana, Missouri, Wisconsin, and Minnesota); (2) large-scale wheat and livestock production in the Great Plains (North Dakota, South Dakota, Colorado, Nebraska, Oklahoma, and

Texas); (3) fruit, vegetable, cotton, and dairy production in large commercial farms in California, Florida, and the Southwest (Arizona and New Mexico); (4) dairy and mixed cropping, including fruits and vegetables, on smaller, owner-operated farms near the Great Lakes (Minnesota, Michigan, Ohio, and New York); (5) soybean, cotton, tobacco, citrus, and vegetables in the South (Mississippi, Alabama, Georgia, North and South Carolina, Tennessee, and Arkansas); (6) wheat and fruit production in the Pacific Northwest (Oregon, Washington, and Idaho); and (7) vegetable, broiler, and mixed production in the Mid-Atlantic states (Delaware, Maryland, and Virginia).

The three most important crops, both in domestic and export markets, are wheat, maize, and soybean. The United States ranks first in world maize and soybean production (accounting for half the world's total) and third in wheat production. Major soybean and maize production occurs in rotation on the best soils in the Midwest. Only 10% of the maize area in the Corn Belt is under irrigation. About 80% of wheat production comes from unirrigated winter wheat; the only major irrigated zone for wheat is in the Pacific Northwest.

Future U.S. Agriculture without Climate Change

U.S. agriculture faces some serious challenges in the coming decades. The most striking of these are moderating domestic demand, loss of comparative advantage vis-à-vis international growers, and rising costs due to environmental protection policies (Easterling, 1996). Domestic demand is projected to grow only 1–1.5% per year in the coming decades due to slowing population growth and to the high level and quality of food already available to consumers (i.e., at current consumption levels demand for food is relatively inelastic) (Easterling, 1996). World agriculture demand is, on the other hand, projected to more than double by 2050 (Crosson, 1989). Thus, the demand for U.S. products for export, especially from developing countries, appears destined to grow, although competition for these markets will intensify. Countries such as Brazil, Argentina, and Thailand, whose labor and other production costs are lower than those of the United States, may well increase their market share.

Environmental protection policies tend to raise the costs of growing food and fiber in the United States compared to other places. Public awareness of agriculture as a less than benign player on the landscape, especially in regard to erosion, contamination of water supplies, chemical residues in food, and pesticide damage to workers (as well as to wildlife) has brought increasingly stringent regulations. Such regulations are likely to continue and to be augmented. Other trends include uncertain rate of productivity growth, structural changes leading to fewer and larger farms, decline of rural communities, and increasingly scarce water supplies (Easterling, 1996). In the western Great Plains and the Arizona Central Valley, water for irrigation has already been preempted by urban users in recent years. In these and other semiarid agricultural regions of the United States, conflicts over water supplies with industrial, commercial, and domestic users may limit agriculture (CAST, 1992).

Future U.S. Agriculture with Climate Change

Climate change will affect U.S. agriculture in many ways, most likely bringing regional shifts in production areas and increased demand for irrigation. Projected U.S.

impacts depend strongly on the severity of climate change scenarios, in regard to both temperature and precipitation, as well as possible changes in climate variability. Studies tend to show worsening effects as temperatures approach the high end of the IPCC range of mean global warming (1.5–4.5°C). (Those estimates have recently been reduced to 1–3.5°C, to better account for the effects of aerosols.)

Methods. A variety of methods have been used to estimate U.S. agricultural climate change impacts. These include dynamic-process crop-growth models for biophysical yield impacts and spatial-equilibrium models of the U.S. agricultural sector. Regression models and econometric approaches have also been used. Spatial analyses have tested how the geographical distribution of crops may change.

Crop Yields. Rosenzweig et al. (1994) considered the potential effects of global climate change on wheat, maize, and soybean production in the United States. Climate scenarios derived from three GCMs near the high end of the IPCC range for global mean temperature rise (1.4–4.5°C) were used in combination with crop-growth models to characterize yield and irrigation water demand changes of the three main crops in major agricultural regions. Under the present management system, projected climate change caused simulated wheat, maize, and soybean yields to decrease at most sites even when the direct effects of CO_2 were included. These decreases were caused primarily by temperature increases, which shortened the duration of the crop life cycles (particularly the grain-filling periods). At some northern sites, yields increased, probably because crop growth is temperature-limited at these higher latitudes. Yield decreases were low to moderate. Adaptation strategies were identified that compensated for the negative effects of climate change at some, but not all, sites. These strategies included changing planting dates and shifting to cultivars more adapted to the projected future climate.

Market Adjustments. Kaiser et al. (1993) found that increases in agricultural commodity prices could offset farm income loss in the midwest for the scenarios tested. Mendelsohn et al. (1994), using a Ricardian approach, found that warming may be generally beneficial for the United States as a whole, although changes in prices were not taken into consideration.

Regional Change. While studies have shown that effects on total U.S. agricultural productivity and economy as a whole are likely to be small to moderate, significant regional change is likely. Warmer temperatures may shift much of the wheat-maize-soybean-producing capacity northward, somewhat reducing U.S. production and increasing production in Canada (Rosenzweig et al., 1994; Crosson, 1989). The northern states could become more productive for annual crops such as maize and soybean because of the lengthening of the frost-free period, while the southern states could become less productive for grain crops, due to heat and moisture stress. The South and Southeast agricultural regions may undergo especially significant change. Here, the yields and production of maize and soybean seem likely to decline, although fruit and vegetable production may be introduced to replace them in part (Rosenzweig et al., 1996). This would constitute a major adjustment to warmer climate conditions, since the capital, infrastructure, and time frame for establishment of fruit-tree plantations, for example, are considerable.

Water Use.   Across all agricultural regions of the United States, the demand for irrigation is likely to increase. Higher temperatures and higher potential evaporation will increase peak crop water demand. Changes in farm profitability, especially if declining yields bring higher farm-gate prices, may further encourage the spread of irrigation systems (a trend already underway). However, the availability of water supplies to satisfy the increased demand from the U.S. farm sector may be limited. There are likely to be competing demands from urban and industrial users in warmer climate conditions. The country's hydrological system is managed within a complicated legal framework that limits flexibility. Moreover, existing irrigation systems may be subjected to climatic conditions for which they were not designed. Even without climate change, U.S. water-allocation institutions need to evolve toward programs that encourage more efficient use of water. Currently there is little incentive to invest in equipment and practices designed to avoid wasting water. The environmental problems (salinity, erosion, and water pollution) derived from excessive irrigation will also need to be addressed more meticulously.

Adaptation to Climate Change in the United States.   U.S. farmers will try to adjust to changing environmental conditions, of course. Agronomic strategies include changes in crop varieties and species, scheduling of operations, and land and water management. Planting and harvesting practices that are adjustable at the farm level include planting dates, choice and diversity of cultivars, depth of planting, timing of harvest, and adjustment of artificial drying. Tests with a variety of farm-level adaptations have shown that yield losses can indeed be largely mitigated (Rosenzweig et al., 1994). Tactics to conserve moisture include conservation tillage, substitution of less water-intensive crops, modification of microclimate, and improved irrigation scheduling. Various options for adapting U.S. agriculture to climate change have been reviewed by Easterling (1996).

However, successful adaptations to climate change may imply significant changes to current agricultural systems, and some of the required changes may be costly. There will be a need for investment in new technologies, infrastructure, and labor. Shifts in international trade may also ensue.

Many U.S. government policies, such as price supports and acreage controls, may hinder resiliency in farmer response to climate changes, through restrictions that limit crop selection. Modifying such policies could increase the adaptive capability of U.S. agriculture to climate change (Easterling, 1996). Government policies regarding crop insurance also need to be reexamined.

Because of its high GDP per capita, the low percentage of the population directly dependent on agriculture, and the small share of agriculture in its national economy, the United States appears to be less vulnerable than many other countries to climate change. In terms of potential risk of hunger or of severe economic distress for the overall economy, this is probably so. The U.S. agricultural system is rich in natural, human, and financial resources, many of which will contribute to adaptive capacity as tangible climate change indeed takes place. However, it is important to remember that climate change may bring significant economic dislocation to the agricultural sector on a regional basis. Finally, if drought frequency were to increase in the Midwest and the Great Plains, as some GCMs predict, the U.S. role as a major provider of food for export may be affected.

Regional Vulnerability

In general, the detailed national and regional studies summarized in this chapter support the hypothesis that tropical regions appear to be more vulnerable to climate change than the temperate regions are. On the biophysical side, temperate-zone C3 crops are likely to be more responsive to increasing levels of CO_2. Although the temperature rise is expected to be smaller in tropical than in temperate regions, tropical crops are closer to their high temperature optima and are more likely to experience stress under rising temperatures. Insects and diseases, already much more prevalent in warmer and more humid regions, may proliferate and become even more widespread and damaging than at present.

Tropical regions also seem to be more vulnerable to climate change because of the often more tenuous economic and social conditions prevailing in many of the countries there. With more people dependent directly on agriculture, widespread poverty, inadequate technologies, and lack of effective political leadership are likely to exacerbate the impacts of climate change in tropical regions.

In the particular case of Egypt, the future is likely to be neither as dire as the pessimistic scenarios predict nor as bright as the optimistic scenarios suggest. Nonetheless, it is clear that this country will be extremely sensitive to the envisaged changes in climate and hydrology. Rapid increases in population and urbanization tend to exacerbate that vulnerability. Given the intertwined nature of the Nile River, the Delta, and its coast and the surrounding deserts, Egypt is addressing the potential impacts of climate change in an integrated way by joining the disciplines of hydrology, agronomy, coastal zone geography, and economics (U.S. Country Studies Program, 1994). Only thus can progress be made in understanding and responding to the critical environmental processes likely to affect the future welfare of the Egyptian people.

In the case of the United States, caution is needed to temper the optimistic view that climate change will be potentially beneficial or at least not damaging to the agricultural sector. Regional responses are likely to differ significantly. If climate variability increases, effects may be more severe. Finally, if warming continues unmitigated through the coming century, even U.S. agriculture will suffer its eventual consequences.

Global Assessments and Future Food Security

The preceding chapter considered regional responses to climate change. The global scale provides a framework in which to embed the regional studies, especially with regard to potential shifts in comparative advantage among regions, and the consequent adjustment of the world food system to climate change–induced alterations in crop production. This chapter considers climate change impacts on a global scale and seeks to answer such questions as: What may happen to the world food supply? How might international trade be affected? Will the risk of famine grow or diminish? We describe one global world food supply and trade study in detail, summarize the agriculture chapter of the contribution of Working Group II (Impacts, Adaptations, and Mitigation) to the Second Assessment of the Intergovernmental Panel on Climate Change (IPCC, 1996), and survey recent work on how food security may be affected by global warming.

Global Assessments

In order to assess climate change impacts on agriculture in a global context, the nature of the world food system needs to be taken into account. Previous chapters have shown that impacts of climate change on agricultural productivity may differ significantly among nations and regions. Here, we review studies in which the determining factors of the regional differences in response include not only the direct impact of climate change on yields but also the global effects on commodity prices and, hence, on the import/export balance of individual countries.

The World Food System

The world food system consists of a complex set of dynamic relationships among producers and consumers, interacting through local, regional, and global markets. Related activities include the production and acquisition of inputs, as well as the transportation, storage, and processing of products. Although there is a trend toward internationalization in the world food system, only about 15% of total world agricultural production currently crosses national borders (Fischer et al., 1990). National governments shape the system by imposing regulations and by investing in agricultural research, infrastructure improvements, and education. The system is directed to meet the demand for food, to increase the efficiency of food production, and to promote trade within and across national borders. Although the system does not guarantee stability, it has effectively generated long-term real reductions in the prices of major food staples (Fischer et al., 1990).

Global Biophysical Assessment

Virtually all studies of climate change impacts concur that there will be significant changes in the patterns of agricultural production. All regions are likely to be affected, but to varying degrees. Studies using a crop zonation approach predict changes in both the regional distribution of crops and in their potential yields. A climate-crop model based on the FAO crop-suitability methodology has been used to test potential change in yield and distribution of major crops under one GCM scenario (Cramer and Solomon, 1993; Leemans and Solomon, 1993). These studies have indicated large differences in regional response. Only high-latitude regions appear to benefit consistently from a climatic change due to projected longer growing periods and increased productivity. Other regions either do not benefit significantly or lose productivity. Potential agricultural declines are due primarily to regional differences in moisture availability.

Combined Biophysical and Economic Analyses

Combined biophysical and economic analyses suggest that the effects of climate change on global agricultural production will depend on the magnitude of the change, the realization of potential physiological effects of CO_2 enrichment on crop growth and water use (such as have been found in experimental settings), and the extent to which appropriate adaptive measures are implemented. The last depends to a large degree on the costs of the adaptive measures. While uncertainties yet exist regarding the direction of change in global agricultural production in various regions, aggregate effects have been generally predicted to be small to moderate (Kane et al., 1992; Reilly and Hohmann, 1993; Reilly et al., 1994; Rosenzweig and Parry, 1994; Darwin et al., 1995).

The small aggregate response (either positive or negative) occurs because reduced production in some areas is likely to be balanced by gains in other areas (Kane et al., 1992). Exporting countries stand to gain if the supplies of agricultural products are restricted and prices rise, whereas low-income countries needing to import food may be hard-pressed to pay for it (Reilly and Hohmann, 1993). There are indications, moreover, that vulnerability to climate change is greater in lower latitude developing countries than in mid- and high-latitude developed countries (Rosenzweig and Parry, 1994). In the former regions, cereal grain crop yields and production have been projected to decline under doubled-CO_2 climate change scenarios (Rosenzweig and Parry, 1994).

Yet, not all impacts in developing countries would be negative. Some areas may receive enhanced water supplies for agriculture and increased crop yields, notwithstanding the overall picture of regional decline. And prospects for adaptation may be available.

A World Food Supply and Trade Study

In 1989, the U.S. Environmental Protection Agency commissioned a 3-year study to assess the potential impacts of climate change on world food supply, trade, and risk of hunger (Fischer et al., 1994; Rosenzweig and Iglesias, 1994; Rosenzweig and Parry, 1994; Rosenzweig et al., 1995b). Agricultural scientists in 18 countries estimated potential changes in national grain crop yields using compatible crop models and consistent climate change scenarios (Rosenzweig and Iglesias, 1994). The crop models were developed by the International Benchmark Sites Network for Agrotechnology Transfer (IBSNAT, 1989), under the auspices of the U.S. Agency for International Development. The crops modeled were wheat, rice, and maize (accounting for approximately 85% of world cereal exports), as well as soybean (accounting for about 67% of world trade in protein cake equivalent) (Table 9.1).

Crop Models. The IBSNAT crop models are based on parametric representations of the major physiological processes responsible for plant growth and development, including evapotranspiration and the partitioning of photosynthates to produce economic yields. The simplified functions permit tracing crop growth as influenced by such factors as genetic potential, climate (daily solar radiation, maximum and minimum temperatures, and precipitation), soils, and management practices. The IBSNAT models include a soil moisture balance submodel so that they can be used to predict crop yields under both rainfed and irrigated conditions. The cereal models also simulate the effects of nitrogen fertilizer on crop growth. The models have been improved to account for the potentially beneficial physiological effects of increased atmospheric CO_2 concentrations on crop growth and water use (Peart et al., 1989).

The crop models were run for current climate conditions, for arbitrarily modified climates (2 and 4°C increases in temperature and +/−20% changes in precipitation), and for the climate conditions predicted by three general circulation models (GCMs) for doubled atmospheric CO_2 levels. The direct effects of carbon dioxide level on crop yields were taken into account. Site-specific estimates of yield changes were aggregated regionally to estimate national crop yield changes, assuming two alternative levels of farmer adaptation. The national yield changes for the modeled crops were then

Table 9.1 Current world crop yield, area, production, and percentage of world production aggregated for countries in the climate change and world food study (Rosenzweig et al., 1995b)

Crop	Yield (t ha^{-1})	Area (ha × 1000)	Production (t × 1000)	Study Countries (%)
Wheat	2.1	230,839	481,811	73
Rice	3.0	143,603	431,585	48
Maize	3.5	127,393	449,364	71
Soybeans	1.8	51,357	91,887	76

extrapolated to provide yield change estimates for other countries and other crops included in the overall food trade analysis.

To conform to a comprehensive framework and attain compatible results, the agricultural scientists participating in the project conducted the following tasks for their respective countries:

1. Definition of geographic boundaries for major production regions; description of the agricultural systems (e.g., rainfed and/or irrigated production, number of crops per year); collection of data on regional and national rainfed and irrigated production of major crops.
2. Creation of observed climate database for representative sites within these regions for the baseline period (1951–1980), or for as many years as data were available; specification of the soil, crop variety, and management inputs necessary to run the crop models at the selected sites.
3. Validation of the crop models with experimental data from field trials, insofar as possible.
4. Running of the crop models with baseline data, values from arbitrary sensitivity tests, and GCM climate change scenarios, with and without taking into account the direct effects of CO_2 on crop growth; simulation of rainfed and/or irrigated conditions as relevant to current or future practices.
5. Identification and evaluation of alteration in farm-level agricultural practices that may lessen adverse consequences of climate change by simulating irrigating production and other adaptive responses (e.g., shifts in planting dates and fertilizer applications, and substitution of crop varieties or species).

Crop model results for wheat, rice, maize, and soybean from the 112 sites in the 18 countries were aggregated by weighting regional yield changes (based on current production) in order to estimate potential changes in total national yields (Tables 9.2 and 9.3). Changes in national yields of other crops and commodity groups and for regions not simulated were estimated by assuming similar response to growing conditions as in the case of the modeled crops. Results from approximately 50 previously published and unpublished regional climate change impact studies and projected temperature, precipitation, and soil moisture changes from GCM climate change scenarios were also used in the estimation process as necessary.

World Food Trade Model. The national grain crop yield change estimates from the climate change scenarios were used as inputs for a world food trade model, designated the Basic Linked System (BLS). This model was developed at the International Institute for Applied Systems Analysis (IIASA) (Fischer et al., 1988). The BLS links 16 national agricultural sector models with a common structure, 4 models with country-specific structures, and 14 regional group models. The 20 models in the first two groups encompass about 80% of the demand, land, and production in the world food system. The remaining 20% are covered by the 14 regional models for groups of countries that have broadly similar attributes based on geographic location, income per capita, and net food trade (e.g., African oil-exporting countries, Latin American high-income exporting countries, and Asian low-income countries).

The BLS is a general equilibrium model system, with representation of all economic sectors, empirically estimated parameters, and no unaccounted supply sources or demand sinks (Fischer et al., 1988). In the BLS, countries are linked through trade, world market prices, and financial flows (Figure 9.1). It is a dynamic iterative model.

Table 9.2 Current production and change in simulated wheat yield under GCM doubled-CO_2 climate change scenarios, with and without accounting for the direct effects of CO_2 (Rosenzweig et al., 1995b)

| Country | Current production | | | | Change in simulated yield[1] | | | | | |
| | | | | | Without CO_2 | | | With CO_2 | | |
	Yield (t ha^{-1})	Area (ha × 1000)	Prod. (t ×1000)	% Total	GISS[2] (%)	GFDL[2] (%)	UKMO[2] (%)	GISS[3] (%)	GFDL[3] (%)	UKMO[3] (%)
Australia	1.38	11,546	15,574	3.2	-18	-16	-14	8	11	9
Brazil	1.31	2,788	3,625	0.8	-51	-38	-53	-33	-17	-34
Canada	1.88	11,365	21,412	4.4	-12	-10	-38	27	27	-7
China	2.53	29,092	73,527	15.3	-5	-12	-17	16	8	0
Egypt	3.79	572	2,166	0.4	-36	-28	-54	-31	-26	-51
France	5.93	4,636	27,485	5.7	-12	-28	-23	4	-15	-9
India	1.74	22,876	39,703	8.2	-32	-38	-56	3	-9	-33
Japan	3.25	237	772	0.2	-18	-21	-40	-1	-5	-27
Pakistan	1.73	7,478	12,918	2.7	-57	-29	-73	-19	31	-55
Uruguay	2.15	91	195	0.0	-41	-48	-50	-23	-31	-35
Former Soviet Union										
Winter	2.46	18,988	46,959	9.7	-3	-17	-22	29	9	0
Spring	1.14	36,647	41,959	8.7	-12	-25	-48	21	3	-25
United States	2.72	26,595	64,390	13.4	-21	-23	-33	-2	-2	-14
World[4]	2.09	231	482	72.7	-16	-22	-33	11	4	-13

1. Results for each country represent the site results weighted according to regional production. The world estimates represent the country results weighted by national production.
2. GCM doubled-CO_2 climate change scenario alone.
3. GCM doubled-CO_2 climate change scenario with direct CO_2 effects.
4. World area and production × 1,000,000.

Table 9.3 Changes in simulated wheat, rice, maize, and soybean yields (Rosenzweig et al., 1995b)

Crop	Change in simulated yields[1]					
	Without CO_2			With CO_2		
	GISS[2] (%)	GFDL[2] (%)	UKMO[2] (%)	GISS[3] (%)	GFDL[3] (%)	UKMO[3] (%)
Wheat	−16	−22	−33	11	4	−13
Rice	−24	−25	−25	−2	−4	−5
Maize	−20	−26	−31	−15	−18	−24
Soybean	−19	−25	−57	16	5	−33

1. Crop yield changes were obtained by weighting site results first by regional production within countries and then by national contribution to total production simulated in the study.
2. GCM climate change scenario alone.
3. GCM climate change scenario with direct effects of CO_2.

A round of exports from all countries is first calculated for an assumed set of world prices, and international market clearance is checked for each commodity. World prices are then revised, using an optimizing algorithm and are again transmitted to the national model. Next, new domestic equilibria are generated and net exports are adjusted. This process is repeated until the world markets for all commodities are cleared. At each stage of the iteration, domestic markets are assumed to be in equilibrium. This process yields international prices as influenced by governmental and intergovernmental agreements. The model is solved in annual increments, simultaneously for all countries. Summary indicators of the BLS include world cereal production and prices, as well as the potential occurrence and extent of hunger that might affect developing countries.

The estimates of climate-induced changes in food production potential were used as inputs to the BLS in order to assess possible impacts of climate change on future levels of food production, food prices, and the number of people at risk of hunger (Figure 9.2). Impacts were assessed for the year 2060, with estimates of population, technology, and economic growth trends projected to that year. Assessments were first made for a reference scenario that assumed no climate change and were subsequently made for three GCM scenarios. The difference between the two assessments is the climate-induced effect. A further set of assessments examined the efficacy of two levels of farmer adaptation in possibly mitigating climate change impacts. Finally, assessments were made of future production as affected by different rates of economic and population growth, and by alterative policies liberalizing the world trade system.

The above projections were compared to a reference scenario, which projected the agricultural system to the year 2060 on the assumption that no climate change and no major changes in the political or economic context of world food trade will occur. The reference scenario assumed medium U.N. population estimates of 10.2 billion people by 2060 (International Bank for Reconstruction Development/World Bank, 1990); 50% trade liberalization (e.g., removal of import restrictions) introduced gradually by 2020; moderate economic growth (ranging from 3.0% per year in 1980–

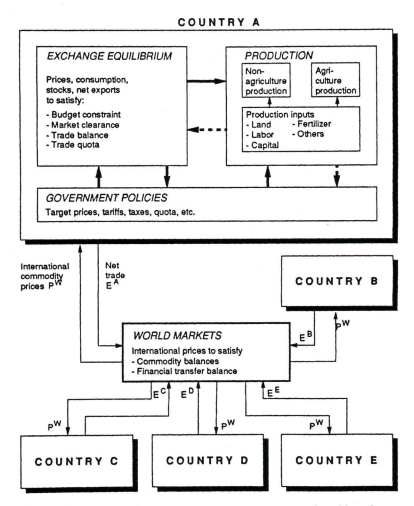

COUNTRY A

EXCHANGE EQUILIBRIUM

Prices, consumption,
stocks, net exports
to satisfy:

- Budget constraint
- Market clearance
- Trade balance
- Trade quota

PRODUCTION

| Non-agriculture production | Agri-culture production |

Production inputs
- Land   - Fertilizer
- Labor   - Others
- Capital

GOVERNMENT POLICIES
Target prices, tariffs, taxes, quota, etc.

International commodity prices P^W

Net trade E^A

COUNTRY B

WORLD MARKETS
International prices to satisfy
- Commodity balances
- Financial transfer balance

E^B P^W

E^C E^D E^E

P^W P^W P^W

COUNTRY C   COUNTRY D   COUNTRY E

Figure 9.1   Relationships between country components and world markets
in the Basic Linked System. Arrows to countries represent international
commodity prices; arrows to world markets represent net trade (Fischer
et al., 1988).

2000 to 1.1% per year in 2040–2060); and technological improvement in crop yields over
time (averaging by 0.7%, 0.9%, and 0.6% per year for world, developing countries, and
developed countries, respectively, throughout for the period 1990–2060) (Table 9.4).

Crop Model Results.   Figure 9.3 shows estimated potential changes in average national grain crop yields for the GISS(Hansen et al., 1983), GFDL (Manabe and
Wetherald, 1986), and UKMO (Wilson and Mitchell, 1987) doubled-CO_2 climate
change scenarios (Table 9.5) with the direct physiological effects of CO_2 on crop
growth and water use. The maps are created from nationally averaged yield changes
for wheat, rice, coarse grains, and protein feed estimated for each country or group of

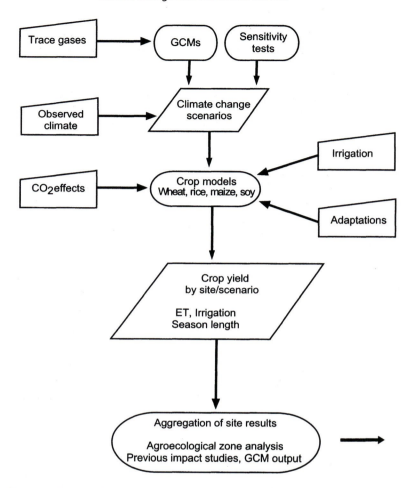

Figure 9.2 Key elements of world food study (Rosenzweig et al., 1995b).

countries. When climate change is considered without the direct effects of CO_2, averaged national crop yields were projected to decline everywhere, although the declines were smaller at midlatitudes and high latitudes. In the simulations with direct CO_2 effects taken into account, yield changes were positive at midlatitudes and high latitudes, and negative at low latitudes for the GISS and GFDL scenarios, which produced yield changes ranging from $+30$ to -30%. The UKMO scenario ($5.2°C$ mean global surface air temperature rise) caused average national crop yields to decline almost everywhere, even with the direct CO_2 effects included.

Several factors contributed to the latitudinal differences in simulated yields. At some sites near the high-latitude boundaries of current agricultural production, simulations of increased temperatures benefited crops otherwise limited by cold temperatures and short growing seasons. In other mid- and high-latitude areas, however, in-

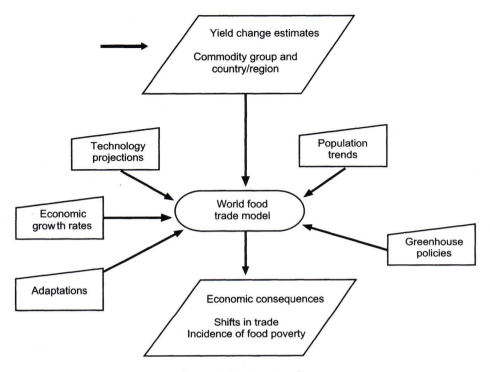

Figure 9.2 (continued).

creased temperatures exerted a negative influence on yields through shortening of crop development stages. In these regions, the inclusion of the direct CO_2 effects was the key factor in yield increases. In contrast, the warming at low latitudes not only accelerated growing periods for crops but also imposed heat and water stress, resulting in steeper yield decreases than at higher latitudes, notwithstanding the beneficial physiological effects of atmospheric CO_2 enrichment.

Crop model adaptation results were grouped into two levels. Level 1 implies little change to existing agricultural practices, other than responses readily achievable

Table 9.4 Projected world growth rates in the Basic Linked System reference scenario (Rosenzweig et al., 1995b)

Growth rate (%)	Years			
	1980–2000	2000–2020	2020–2040	2040–2060
Population	1.7	1.3	0.8	0.5
GDP	2.9	2.0	1.5	1.1
Cereal yield	1.2	0.7	0.5	0.4
Agricultural production	1.8	1.3	1.0	0.7

All growth rates refer to world average annual percentage of growth during the indicated period.

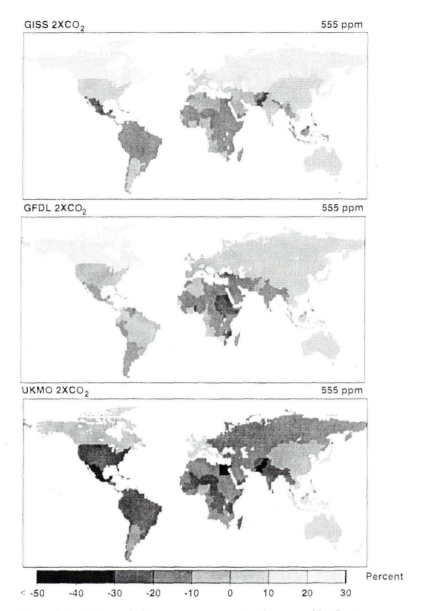

Figure 9.3 Estimated change in average national grain yield (wheat, rice, coarse grains, and protein feed) for the GISS, GFDL, and UKMO climate change scenarios. The direct physiological effects of CO_2 on crop growth are taken into account. Results shown are averages for countries and groups of countries in the Basic Linked System (BLS) world food trade model (Fischer et al., 1988); regional variations within countries are not reflected (Rosenzweig and Parry, 1994).

Table 9.5   GCM doubled-CO_2 climate change scenarios (Rosenzweig et al., 1995b)

				Change in average global	
GCM	Year[1]	Resolution (lat. × long.)	CO_2 (ppmv)	Temperature (°C)	Precipitation (%)
GISS[2]	1982	7.83° × 10°	630	4.2	11
GFDL[3]	1988	4.4° × 7.5°	600	4.0	8
UKMO[4]	1986	5.0° × 7.5°	640	5.2	15

1. When calculated.
2. Goddard Institute for Space Studies (Hansen et al., 1983).
3. Geophysical Fluid Dynamics Laboratory (Manabe and Wetherald, 1986).
4. United Kingdom Meteorological Office (Wilson and Mitchell, 1987).

by individual farmers. Level 1 adaptations included minor shifts in planting date (+/−1 month) that do not imply major changes in crop calendars, additional application of irrigation water to crops already under irrigation, and switches to currently available crop varieties better adapted to the altered climate.

Adaptation level 2 implies much more substantial modifications of agricultural production systems, possibly requiring resources beyond the farmers' means, investment in regional and national agricultural infrastructure, and policy changes. Level 2 adaptations included large shifts in planting date (>1 month), increased fertilizer applications (included here because of their implied costs for farmers in developing countries), installation of irrigation systems, and development of entirely new varieties. (The latter were tested by manipulation of the genetic coefficients in the crop models.) Level 2 represents a much more optimistic view of world agriculture's ability to respond to the expectable change of climatic conditions.

Level 1 adaptation compensates incompletely for the negative effects of all three climate change scenarios, particularly in the developing countries. For the GISS and GFDL scenarios, level 2 adaptation compensates almost fully for the negative impacts of climate change. With the high level of global warming projected by the UKMO climate change scenario, however, neither of the two adaptation levels fully overcomes the projected negative effects of climate change on crop yields in most countries, even when the potentially beneficial CO_2 effects on crop physiology are taken into account (Figure 9.4).

World Food Trade Results.   For the reference scenario assuming that the climate will not change at all in the future, while population growth, economic growth, technological progress, and trade liberalization proceed, world cereal production is estimated to grow from 1,795 million metric tons (mmt) in 1990 to 3,286 mmt in the year 2060. Cereal production in developing countries is predicted to exceed production in developed countries by 2020. Despite slowing gains in yield, food production measured as net calories produced is projected to exceed population growth throughout the simulation period of the reference scenario.

Cereal prices in the reference scenario are predicted to rise gradually to an index of 121 (100 = 1970 value) for the year 2060, thus reversing the falling trend of real ce-

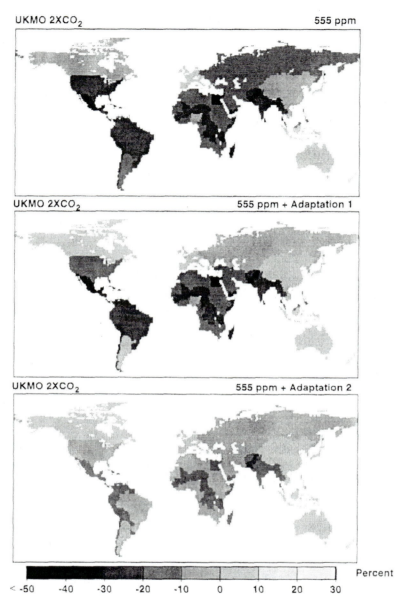

Figure 9.4 Estimated changes in average national grain yield (wheat, rice, coarse grains, and protein feed with direct 555 ppmv CO_2) under two levels of adaptation for the UKMO doubled-CO_2 climate change scenario. Adaptation level 1 signifies minor changes to existing agricultural systems; adaptation level 2 signifies major changes. Results shown are averages for countries and groups of countries in the BLS world food trade model (Fischer et al., 1988); regional variations within countries are not reflected (Rosenzweig and Parry, 1994).

real prices that had prevailed over the last 100 years. The standard reference scenario has two phases of price development. From 1980 to 2020, while trade barriers and protection are still in place but are being reduced, relative prices are seen to increase. This occurs in the medium term because a removal of subsidies leads immediately to lower farm-gate prices and therefore discourages production. While consumers benefit initially from somewhat lower retail prices, the greater demand that follows drives the prices upward. In the longer term, however, price decreases are likely to result from efficiency gains and technical progress.

In the case of climate change without the beneficial CO_2 effects on crop yields, world cereal production is predicted to fall by 11 to 20%. In contrast, inclusion of the direct CO2 effects obviates most of that fall, to between 1 and 8% (Figure 9.5). The overall world production changes, however, mask a disparity in response to climate change between developed and developing countries (Figure 9.6). The largest negative changes are predicted to take place in developing regions, although the extent of decreased production varies greatly from country to country, depending on the specific nature and degree of the local change in clime. In developed countries, by contrast, production is predicted to grow under all but the UKMO scenario. Cereal price increases resulting from climate-induced reductions in yield are estimated to range between 24 and 145% (Figure 9.7).

Globally, both minor and major levels of adaptation can help restore world production levels (especially when CO_2 physiological effects are included), compared to climate change scenarios with no adaptation (see Figure 9.5). With minor (level 1)

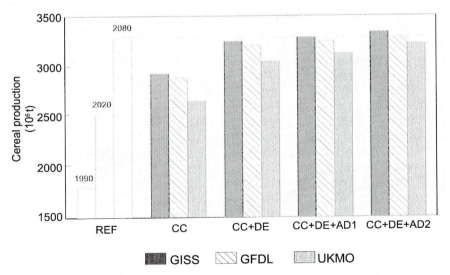

Figure 9.5 World cereal production (million metric tons) projected by the BLS for the reference, GISS, GFDL, and UKMO doubled-CO_2 climate change scenarios, with (CC+DE) and without (CC) direct CO_2 effects on crop yields, and with adaptation levels 1 and 2 (AD1 and AD2). Adaptation level 1 implies minor changes to existing agricultural systems; adaptation level 2 implies major changes (Rosenzweig and Parry, 1994).

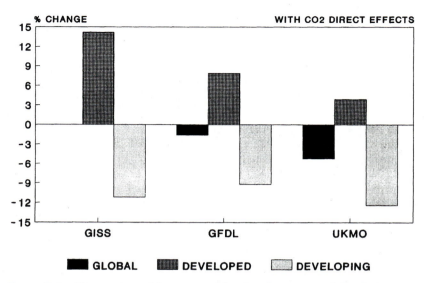

Figure 9.6 Changes in world, aggregated developed country, and developing country cereal production projected by the BLS under GCM climate change scenarios in 2060 for adaptation level 1. Reference scenario for 2060 assumed no climate change (global 3286 mmt, developed 1449 mmt, developing 1836 mmt) (Rosenzweig et al., 1995b).

adaptation, averaged global cereal production is predicted to diminish by up to ~160 mmt (0 to −5%) from the reference scenario projection of 3,286 mmt. With major (level 2) adaptations, however, global cereal production may range from a slight increase of 30 mmt to a slight decrease of ~80 mmt (i.e., from +1% to −2.5%).

Level 1 adaptation largely offsets the negative climate change effects on yields in developed countries, thus improving their comparative advantage in world markets (see Figure 9.6). In these regions, cereal production is calculated to rise by 4 to 14% over the reference scenario. However, developing countries are predicted to benefit little from this level of adaptation, and may experience a negative change of −9% to −12% in cereal production. More intensive (level 2) adaptation may effectively eliminate the overall global reduction of cereal yields foreseen under the GISS and GFDL climate scenarios, while reducing the negative impact under the UKMO scenario to one-third. Whereas under adaptation level 1 cereal price increases range from 10 to almost 100% (see Figure 9.7), under adaptation level 2 the compounding price responses range from a decline of 5% to an increase of 35%.

Alternative assumptions—lowering of trade barriers, low economic growth, and low population growth—were tested both in the absence and in the presence of climate change. Without climate change, the combination of full trade liberalization and low population growth would have beneficial effects on the world food system, whereas the effects of low economic growth would be detrimental. With climate change, the beneficial effects of full trade liberalization and low population growth would be equal to or even exceed the otherwise adverse effects of climate change. Therefore, there may be much to be gained from altering the conditions of trade and

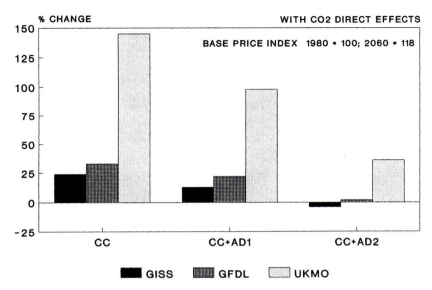

Figure 9.7 Changes in cereal price index in the year 2060 calculated by the Basic Linked System (Fischer et al., 1988) under climate change scenarios for no adaptation and adaptation levels 1 and 2 (AD1 and AD2). Direct effects of CO_2 on crop growth are taken into account. Reference scenario for 2060 assumed no climate change (price index is 18% about 1980 levels) (Rosenzweig et al., 1995b).

development as a strategy for helping to mitigate potentially negative climate change impacts. In the absence of such measures, cereal production would tend to diminish, particularly in the developing world, while prices and population at risk of hunger would increase due to climate change. The alternative assumptions of trade liberalization, economic development, and population growth made little difference with respect to the geopolitical patterns of climate change effects. So did a climate change scenario with a surface air temperature increase of 2.8°C (near the lower end of the IPCC range of 1.5–4.5°C for doubled CO_2) (Table 9.6). The adverse climate impacts would be the least, in any case, if population growth were kept low.

Major IPCC Findings for Agriculture

A different type of "global assessment" of climate change impacts on agriculture is carried on periodically by the Intergovernmental Panel on Climate Change. The IPCC brings international scientists together to review and summarize relevant work on climate change, in order to inform the world policy-making process related to the United Nations Framework Convention on Climate Change (FCCC). These comprehensive reviews were published as a First Assessment in 1990 (IPCC, 1990); the Second Assessment has recently been published in 1996 (IPCC, 1996). Working Group II of the IPCC deals with impacts, adaptation, and mitigation; two chapters of the Working Group II contribution deal with agriculture: one on impacts and adaptation, the other on mitigation. Here, we summarize the impacts and adaptation chapter.

Table 9.6   Change[1] in cereal production, cereal price index, and people at risk of hunger in 2060 for a lowend climate change scenario (Rosenzweig et al., 1995b)

Predicted variable	Reference 2060	LE[a]	LE[b]
World			
Cereal production	3286 mmt	−145 mmt	82 mmt
Cereal price index	121	81	−21
People at risk of hunger	641 mil	265 mil	−84 mil
Developed countries			
Cereal production	1449 mmt	26 mmt	153 mmt
Developing countries			
Cereal production	1836 mmt	−170 mmt	−71 mmt

1. Change relative to Basic Linked System reference scenario in 2060.

LE = "Low end" of the range of temperature changes projected by the IPCC (1996) for doubled CO_2, represented by the climate projected by the GISS GCM for the 2030s (~2.4°C mean surface air temperature rise).

LE[a] = Not accounting for the physiological effects of CO_2 enrichment on crop yields.

LE[b] = Accounting for the physiological effects of CO_2 enrichment on crop yields.

The Working Group II Agriculture chapter of the IPCC Second Assessment (IPCC, 1996) begins by reviewing the biophysical effects of climate change on crops (including alterations in temperature and moisture regimes and the direct effects of increasing CO_2), soils, crop pests, and livestock. Next, since food production is practiced very differently around the world, the chapter then analyzes the potential for differential regional impacts based on variations in current climate and production conditions. Regions analyzed are Africa and the Middle East, South and Southeast Asia, East Asia, Oceania and Small Island Nations of the Pacific, areas of the former Soviet Union, Latin America, Western Europe, and the United States and Canada.

The regional analysis reveals the existence of many different types of vulnerability, defined as the risk of negative consequences of climate change that are difficult to ameliorate through adaptive measures. Types of vulnerability include risk of large yield decreases with small changes in climate, risk to farmers of loss of profitability or viability, risk of regional economic decline due to high level of dependence on the farm sector, and risk of hunger for people with few resources or rights to food.

The chapter goes on to review global climate change assessments in light of the current world agricultural system and its future. While some studies suggest that in the absence of climate change, food supply will continue to expand faster than demand over the coming decades, others are less optimistic, citing limits on further land expansion, resource degradation, and reduced confidence that positive historical trends in yields will continue. Because a significant and unavoidable amount of climate change is projected to occur over the next 50 years due to the current atmospheric accumulation of greenhouse gases, the IPCC Agriculture Chapter ends with consideration of potential adaptations and adjustments to changing climate conditions.

Major Findings at the Biophysical Level

1. For annual crops, warmer temperatures will speed development, shortening the period of growth and lowering yield. Extreme high temperature during flowering can lead to sterility in rice and loss of pollen viability in maize. If extreme climate events (storms with high winds, flooding, heavy rains, severe late or early frosts) increase, severe or total crop losses could become more frequent.
2. On the positive side, experiments have shown that crop yields respond beneficially to elevated CO_2 due to increased photosynthesis and improved water use. Yields of annual C3 crops increase on average about 30% for doubled-CO_2 concentrations. Responses depend, however, on crop species, availability of plant nutrients, temperature and moisture stress, as well as on differences in experimental techniques.
3. The few studies that have investigated the combined effects of climate, CO_2, ozone (O_3), and UV-B radiation on crops have found that responses are usually negative, but that they differ among crop species.
4. Many of the world's soils are potentially vulnerable to soil degradation—such as loss of soil organic matter, leaching of soil nutrients, salinization, and erosion—as a likely consequence of climate change.
5. Changes in the geographic distribution of weeds, insects, and plant diseases and their vigor in current ranges will be likely to affect crops in changed climate conditions. The risk of agricultural losses due to these pests may possibly increase.
6. Warming in warm regions of the world would negatively impact livestock production by reducing animal weight gain, milk production, feed conversion efficiency, and reproduction. Warming during the cold periods in temperate areas would be beneficial to livestock production due to lower feed requirements, increased survival of young, and savings in energy costs.

Major Findings at the Global and Regional Levels

1. Global agricultural production appears to be sustainable in the face of climate change as predicted by GCMs for doubled CO_2 equilibrium scenarios. However, crop yields and productivity changes will vary considerably across regions, bringing repatterning of the geographical distribution of farming activities. Countries in the low latitudes with low incomes show potential for more negative response to climate change than do high-latitude, high-income countries.
2. By most measures, many of the countries in Sub-Saharan Africa appear to be most vulnerable to climate change. The region is home to many low-income populations who depend on isolated farming systems. The climate of the region is already hot, and large areas are arid or semiarid. Average per capita income is among the lowest in the world and has been declining since 1980. Over 60% of the Sub-Saharan population depends directly on farming, and agriculture accounts generally for more than 30% of GDP of the countries in that region.
3. Countries in South and Southeast Asia are also vulnerable to climate change due to heavy dependence on the food-production sector. Agriculture accounts for more than 30% of GDP in most countries in the region, and each hectare of cropland supports about 5.5 people. Increases in tropical storm irregularity and intensity could be particularly destructive here.
4. A third vulnerable regional group is the Pacific Island Nations. Although these countries are small, the potential for loss of coastal land to sea-level rise, saltwater intrusion into water supplies, and increased damage from tropical storms threatens the agriculture of these nations.

5. Agricultural adaptation to climate change is likely. The extent of adaptation depends on the affordability of adaptive measures, access to technology, and biophysical constraints such as water resource availability, soil characteristics, and genetic material for crop breeding. Adapting to climate change could create a serious burden for some developing countries.

6. Many current agricultural policies are likely to discourage effective adaptation and are a source of current land degradation and resource misuse. Government policies, such as price support and pricing programs, marketing orders, loan assistance, acreage controls, export promotion, foreign food aid, domestic food programs, and land and soil conservation programs, may hinder resiliency in farmer response to climate changes.

IPCC Recommendations for Future Research.   The IPCC identifies three high-priority research needs for improving our knowledge of climate change impacts on agriculture. These research areas are improved integrated modeling of biophysical and socioeconomic processes, the use of transient (that is, gradually changing conditions) rather than abrupt equilibrium climate change scenarios, and the inclusion of the effects of changed climatic extremes on crop and livestock production.

Food Security

Most research on agriculture and climate change has focused on potential impacts on regional and global food production, yet few studies have considered how global warming may affect food security. Food security has been defined as "access by all peoples at all times to enough food for an active, healthy life" (World Bank, 1986). The World Food Summit, convened in 1996 by the Food and Agriculture Organization of the United Nations (FAO) in Rome, highlighted the basic right of all people to an adequate diet and the need for concerted action among all countries to achieve this goal in a sustainable manner (FAO, 1996).

How, then, may climate change alter the ability of the world's growing population to gain access to food? While the overall, global impact of climate change on agricultural production may be small, regional vulnerabilities to food deficits may increase, due to growing difficulties in distributing and marketing food to specific regions and groups of people. For subsistence farmers and people who now face a shortage of food, lower yields and yield quality, when and if they occur, may result not only in measurable economic losses but also, indeed, in malnutrition and possibly famine. A compendium of recent work on this topic is found in Downing (1995).

Levels of Food Security

Food security may be defined at several levels of human organization (Chen and Kates, 1994b). At the regional or national level, a calculation of food availability includes production in the agricultural sector, less the amount exported, plus the amount of food imported and food aid received. This is sometimes designated as "national food availability" in terms of kcal day^{-1} capita^{-1} (Bohle et al., 1994).

At the household level, the calculation includes the food that a household raises on its own, the food that a family can buy, plus additional welfare or food assistance and gifts. The mean gross national product (GNP) per household is sometimes used as an approximate measure of household food poverty (Bohle et al., 1994). Climate change impacts related to food security will probably be felt first at the household level, but the risk of climate change will vary greatly among households, based on access to land, water, and government support.

Finally, at the individual level, access to food may vary within a family or household according to age, gender, economic status, and cultural mores. Children's (under 5 years of age) mortality per 1,000 can give an indication of individual food deprivation (Bohle et al., 1994). Nutritional knowledge can improve the quality of the dietary intake of individuals and families.

Measures of Food Security

The term famine generally describes a large and persistent food shortage affecting a significant population. Paradoxically, sufficient food may be available in famine-stricken countries yet fail to reach vulnerable social groups that lack the means to obtain or the entitlement to food (e.g., the right to participate in welfare or assistance programs)(Sen, 1981).

Food security can be threatened by war, as hunger is often used as a weapon. Recent internal and external wars in Bosnia, Ethiopia, and Somalia are harsh examples of this fact.

The number of people suffering from famine at any time is difficult to determine. One generalized measure of famine has been devised based on the total population of countries that report cases of famines to international agencies (Figure 9.8) (Chen and Kates, 1994b). This index suggests that populations suffering famine have been declining in recent years, from almost 800 million people in 1957–1963 to about 100 million people in famine-reporting countries in 1985–1991. The downward trend is

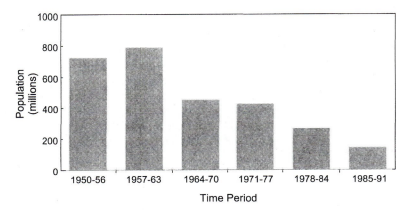

Figure 9.8 The FAMINDEX: average total population of countries where famine was reported in the *New York Times*, 1950–1991 (Chen and Kates, 1994b).

Table 9.7 Recent estimates of global food security (Chen and Kates, 1994b)

Dimension of food security	Population (millions)	Affected (%)	Year
Famine (population at risk)	15–35	0.3–0.7	1992
Undernutrition (chronic and seasonal)			
FAO food poverty (1.2 BMR)	477	9	1990
Updated FAO food poverty (1.54 BMR)	786	20	1988–1990
Child malnutrition (weight below −2 s.d.)	184	34	1990
Micronutrient deficiencies			
Iron deficiency (women, ages 15–49)	370	42	1980s
Iodine deficiency	211	5.6	1980s
Vitamin A deficiency (children <age 5)	14	2.8	1980s
Nutrient-depleting illness			
Diarrhea, measles, malaria (deaths of children <age 5)	6.5	0.8	~1990
Parasites (infected population)[1]			
Giant roundworm	785–1300	15–25	1980s
Hookworm	700–900	13.17	1980s
Whipworm	500–750	10–14	1980s

1. Includes those people expected to have multiple infections.

BMR = basal metabolic rate.

s.d. = standard deviation.

due to shifts in the occurrence of famine from the more populous Asian countries to the less densely populated African countries. In 1992, reported famine occurred only in Somalia and the Sudan.

Besides famine, there are several measures of food security that involve adequacy of nutrition. Recent estimates of famine and these other measures of hunger are given in Table 9.7. It is estimated that more than 700 million people suffer from chronic undernutrition as defined by the FAO (Chen and Kates, 1994a). The FAO criterion for assessing nutrition adequacy is based on basal metabolism rates (calories needed to maintain body weight) for age cohorts in a population group and the ability of that population to purchase food.

Another measure of food security is the status of child nutrition; there are over one hundred million children with below-normal growth (Chen and Kates, 1994b). This may be measured by calculating how many children in a population are two standard deviations away from a reference weight. Beyond undernutrition due to lack of food quantity, there is also the problem of micronutrient deficiency due to inadequate dietary quality. Lack of required iron and iodine is prevalent in many regions of the world.

Finally, there are illnesses that deplete the body's ability to utilize nutrients, such as diarrhea, measles, malaria, and intestinal parasites. The number of people suffering from micronutrient deficiencies and nutrient-depleting illnesses may exceed a billion, far more than the number suffering from famine at any given time. Because climate change may alter both the quality and the quantity of available foodstuffs, these other aspects of food in security may be affected, as well as the incidence of famine per se.

Vulnerability to Famine

Vulnerability to famine is a complex concept that integrates environmental, social, economic, and political aspects. Bohle et al. (1994) suggest that vulnerability has three components: risk of *exposure* to crises, stress and shocks; risk of inadequate *capacity* to cope with crises, stresses, and shocks; and risk of severe *consequences* with associated slow or limited recovery from crises, stresses, and shocks. Thus, groups most vulnerable to climate change in regard to food security may be those who are most exposed to the risk of climate change impacts on crop productivity and changes in commodity prices, with the least capacity to cope with unfavorable changes in agricultural conditions and to access food, and therefore, prone to suffer the severest consequences of famine, undernutrition, and debility.

Groups vulnerable to hunger in the present, and likely to be vulnerable to climate change effects in the future, include rural smallholder farmers, pastoralists, wage laborers, urban poor, refugees, and other destitute groups. Of these, the most vulnerable to the negative effects of climate change are probably rural subsistence farmers, landless wage earners, and urban poor: they have little food security now and therefore are susceptible to even small changes in agroclimatic circumstances or economic status (Bohle et al., 1994). For households that both raise food and work off the farm, climate change may affect the time for crop planting, but this may conflict with critical off-farm employment.

Reducing vulnerability to climate change impacts on agriculture requires lessening the risk of significant climate changes on regional productivity and access to food, enhancing the capacity of vulnerable groups to adapt their farming systems or economic livelihoods to changing agroclimatic and market conditions, improving their ability to recover from temporary food shortages, and minimizing the potential disruptions that may result from either governmental or private donor interventions. Efforts on a broad multidisciplinary front, including the fields of agriculture, health, and the environment, are needed to achieve the goals of lessening vulnerability to hunger and promoting sustainable growth in the agricultural sector, especially in view of the threat posed by climate change (Ruttan et al., 1994).

Country Case Studies

Case studies regarding Zimbabwe, Kenya, Senegal, and Chile have illustrated how climate change can heighten the risk of famine in countries that are already vulnerable (Downing, 1992). Research in Mexico has shown that global warming may present a threat to local and national food security in that country, especially if farmers are unable to adapt to a drier climate and as food imports from other regions become more costly (Liverman and O'Brien, 1991; Appendini and Liverman, 1994). In Southern Africa, sensitivity to climate change is multifaceted, especially as affected by potential declines in water resources, loss of important wetland habitats, reduction in biodiversity, impairment of hydroelectric power generation, spread of disease vectors, deterioration of productivity in the rainfed agricultural sector, and attendant declines in food security (Magadza, 1994).

In Zimbabwe, groups currently vulnerable to hunger include unemployed and partially employed workers in urban areas, as well as altogether communal farmers, farm workers, and landless people in rural areas (Table 9.8) (Christiansen and Stack, 1992; Bohle et al., 1994). Of the population vulnerable to food insecurity (20–40% of total population, depending on annual rainfall), some 6% live in urban areas while the majority live in rural areas. Communal farmers who cultivate small plots in the lowland, semiarid regions of western and southern Zimbabwe are often vulnerable to hunger even in years of average rainfall, due to sandy, infertile soils and low levels of technology. Households headed by women have been identified as particularly vulnerable.

Maize is the main staple crop in Zimbabwe, but only 10–20% of communal farmers consistently produce a surplus (Downing, 1992). Risk of drought is high, and crop failures have been associated with El Niño/Southern Oscillation (ENSO) events (Cane et al., 1994). In 1991–1993, an El Niño event, which was of moderate intensity, corresponded with an extremely severe drought and decrease in maize yields in southeastern Africa. Vulnerability to hunger in rural households tends to double during drought (Downing, 1992).

Climate change may lower maize yields in Zimbabwe due to shortening of the favorable growing period and increasing water stress (Muchena, 1994; Muchena and Iglesias, 1995) (Figures 9.9 and 9.10). Simulations with a crop model that tested the

Table 9.8 Vulnerable livelihood groups in Zimbabwe, 1991 (Bohle et al., 1994)

Group	No. of food-insecure households[1]	Percentage of population
Urban		
Unemployed	72,000	3.7
Informal workers	53,000	2.7
Urban total	125,000	6.4
Rural		
Communal farmers[2]		
Zones I and II	20,000–39,000	1–2
Zone III	22,500–98,000	1–5
Zones IV and V	137,000–450,500	7–23
Landless farm workers and unemployed	210,000	12.5
Rural total		
Average years	389,500	21.5
Poor years	797,500	42.5
Total		
Average years	514,500	27.9
Poor years	922,500	48.9

1. Includes estimates from Christensen and Stack (1992), who report estimates of food insecurity, the confluence of poverty, malnourishment, and variable incomes. This corresponds to the broader definition of vulnerability of Bohle et al. (1994). These provide initial estimates by vulnerable group.
2. The lower number is food security in average years, whereas the higher number suggests additional vulnerability due to crop failure in poor years.

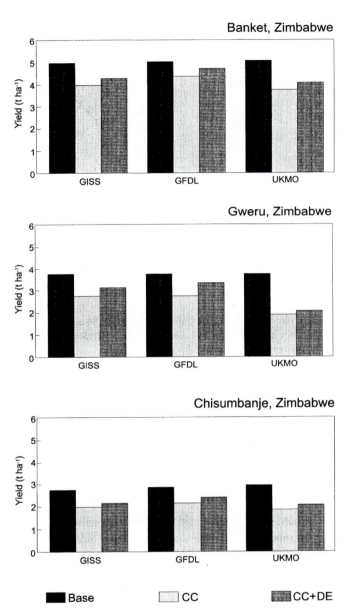

Figure 9.9 CERES-Maize yield at three sites in Zimbabwe for observed baseline climate (~1960–1988) and the GISS, GFDL, and UKMO climate change scenarios. CC indicates climate change simulations without direct CO_2 effects; CC + DE indicates climate change simulations with direct CO_2 effects (Muchena, 1994).

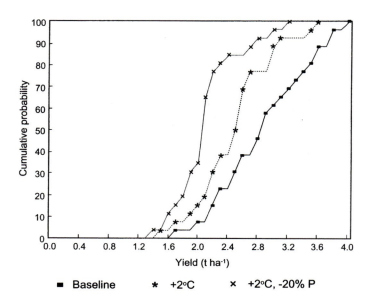

Figure 9.10   Cumulative probability of CERES-Maize yield under baseline (~1960–1988), +2°C, and +2°C and −20% precipitation at Chisumbanje, Zimbabwe (Muchena, 1994).

effect of a 2°C warming for Chisumbanje, a site in the semiarid zone, showed that adequate yields currently expected 70% of the time would occur in fewer than 40% of the years. Furthermore, climate change is likely to cause significant spatial shifts in agricultural capability and land-use zones, with wet zones diminishing and dry zones expanding due primarily to rises in potential evapotranspiration (Downing, 1992).

Risk of Hunger in the World Food Supply and Trade Study

One goal of the climate change and world food study described earlier in this chapter was to evaluate the impact of alternative climate change scenarios on the incidence of hunger in different countries. To accomplish this goal, it was necessary to generate a consistent risk-of-hunger indicator, based both on the need for and the access to food. "Food poverty" is determined by food energy availability relative to nutritional requirements, from the basal metabolism rate of age cohorts within a population. Food availability (or affordability) depends on income and price levels.

In the Basic Linked System, the number of people at risk of hunger was defined as those people in developing countries (excluding China) with an income insufficient to either produce or procure their food requirements. This measure was derived from FAO estimates and methodology for developing market economies (FAO, 1984, 1987). The FAO methodology considers that the distribution of calorie consumption in a typical country is skewed and can be represented by a beta distribution function. The parameters of this function were estimated by the FAO for each country on the basis of country-specific data and cross-country comparisons. Estimates of the minimum per capita energy requirements were obtained from the basal metabolic rate (i.e., during a period spent in a fasting state at complete rest in a warm environment; body weight, age, and sex have an impact on this requirement.) The FAO then provided two alternative estimations of the number of undernourished people, defined as those failing to receive either 1.2 or 1.4 times the basal metabolic rate.

In the standard reference scenario (i.e., the world food system without climate change), the BLS estimates that there were some 500 million undernourished people in the developing world, (excluding China) in 1980. The number of people expected to be at risk of hunger has been estimated at ~640 million (~6% of total population) in 2060. This compares with 530 million in 1990, about 10% of total current population. Without climate change, the incidence of hunger is likely to decrease markedly from an estimated 23% of the population in developing countries (excluding China) in 1980 to about 9% in 2060 (Table 9.9). Yet, despite this expected relative improvement, the absolute number of people at risk of hunger is projected to increase somewhat, from about 500 million in 1980 to almost 600 million in 2000, and some 640 million in 2060. The estimates for a future without climate change show a decline in risk of hunger in Asia in both relative and absolute terms, whereas in Africa there may be a relative decline but an absolute rise in the incidence of hunger. The proportion of African people at risk of hunger may decline from 26% in 1980 to 18% by 2060, even while the total number of people at risk of hunger may increase from about 120 million in 1980 to 415 million in 2060 due to population growth. Even without the detrimental impacts of climate change, Africa can be expected to have more than twice as many people at risk of hunger than the other continents combined.

With climate change, declines in yields in low-latitude regions (where many developing countries are located) are projected to require that net imports of cereals increase under all the scenarios tested. Higher grain prices will affect the number of people at risk of hunger. For the climate change scenarios without adaptation, their estimated number increases ~1% for each 2–2.5% increase in prices. The number of people at risk of hunger grows by 10–60% in the scenarios tested, resulting in an estimated increase of between 60 million and 350 million people in this condition (above the reference scenario projection of 640 million) by 2060.

With less agricultural production in developing countries and higher prices for foodstuffs on international markets, the estimated number of people at risk of hunger

Table 9.9   Number of people at risk of hunger, Basic Linked System reference scenario (Fischer et al., 1994).

Developing countries[1]	No. at risk of hunger (million people)					Fraction of population (%)				
	1980	2000	2020	2040	2060	1980	2000	2020	2040	2060
Total	501	596	716	696	641	23	17	14	11	9
Africa	120	185	292	367	415	26	22	21	19	18
Latin America	36	40	39	33	24	25	17	13	8	4
South and Southeast Asia	321	330	330	232	130	25	17	13	8	4
West Asia	27	41	55	64	72	18	16	14	12	11

1. Estimates do not include China.

Table 9.10   Impact of climate change on people at risk of hunger, year 2060, for three GCM climate change scenarios (Fischer et al., 1994)

Developing countries[1]	Additional million people			% change		
	GISS	GFDL	UKMO	GISS	GFDL	UKMO
Without physiological effects of CO_2	721	801	1446	112	125	225
With physiological effects of CO_2	63	108	369	10	17	58
Adaptation level 1[2]	38	87	300	6	14	47
Adaptation level 2[3]	−12	18	119	−2	3	19

1. Estimates do not include China.
2. Signifies minor changes to existing agricultural systems.
3. Signifies major changes to existing agricultural systems.
GISS = Goddard Institute for Space Studies (Hansen et al., 1983).
GFDL = Geophysical Fluid Dynamics Laboratory (Manabe and Wetherald, 1986).
UKMO = United Kingdom Meteorological Office (Wilson and Mitchell, 1987).

will inevitably increase (Table 9.10). This condition is projected to occur in all scenarios but one. The largest increase occurs with a global mean surface temperature rise of 5°C and with no beneficial physiological effects of CO_2 on crop growth and yield. The smallest change, a decline of 2%, is seen to occur with 4°C warming, full CO_2 physiological effects on yields, and high levels of farmer adaptation.

As a consequence of climate change and adaptation level 1, the number of people at risk of hunger increases by 40–300 million (6–50%) from the reference scenario of 641 million people (Figure 9.11). With a higher level of adaptation by farmers, the number of people at risk of hunger rises by between 12 million for the GISS scenario and 120 million for the UKMO scenario (2 and +20%). These modeling results suggest that, except for the GISS scenario with adaptations implying major changes to current agricultural practices, the simulated farm-level adaptations did not entirely mitigate the negative effects of climate change on the number of people at risk of hunger, even when economic adjustment (that is, the production and price responses of the world food system) are taken into account.

The study also tested the effects of climate change with alternative assumptions including full trade liberalization, lower economic growth rates, and lower population growth rates. Figure 9.12 summarizes the generalized relative effects of different policies of trade liberalization, as well as of economic and population growth rates, on the production of cereals and on the number of people at risk of hunger. In all these cases, cereal production was seen to diminish, particularly in the developing world, while prices and population at risk from hunger increased due to climate change. Lower economic growth resulted in tighter food supplies, and consequently resulted in more prevalent conditions of food poverty. Full trade liberalization in agriculture appears to provide for more efficient resource use and to reduce the number of people likely to be at risk of hunger by about 100 million (from the reference case of about 640 million in 2060). The beneficial effects of trade liberalization and low population growth seem to be of the same order of magnitude as the adverse effects of climate change.

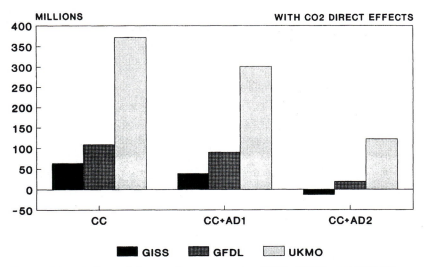

Figure 9.11 Effects of climate change scenarios (GISS, GFDL, and UKMO) and different adaptation levels on risk of hunger. Direct CO_2 effects on crop growth and yield are taken into account (Rosenzweig et al., 1995b).

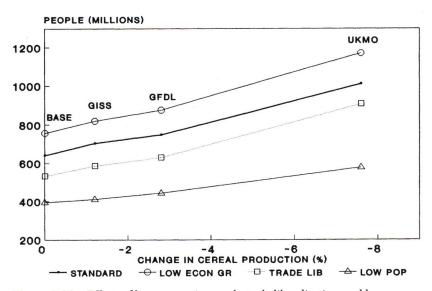

Figure 9.12 Effects of low economic growth, trade liberalization, and low population growth on number of people at risk from hunger calculated by the Basic Linked System (Fischer et al., 1988) under three climate change scenarios. Direct effects of CO_2 are taken into account (Rosenzweig et al., 1995b).

An assumption of low population growth rate minimized the population at risk of hunger in both the presence and absence of climate change in the Basic Linked System simulations.

A Global Framework

It should be emphasized that the results of studies such as those described in this chapter are not true forecasts of the future. They are merely exploratory assessments of the sensitivity of the world food system to a limited number of hypothetical futures, of which there is a much wider array. Very large uncertainties preclude making definite forecasts. Especially significant is the lack of information on possible climate change at the regional level, of population growth rates and trends in demand, of the potential effects of technological changes on agricultural productivity, and of the wide array of possible adaptations.

The implementation of efficient adaptation techniques is far from certain (see Chapter 10). In developing countries, there may be social or technical constraints, and adaptive measure may not necessarily result in sustainable production in the long run. Determining the availability of water supplies for irrigation and the costs of adaptation measures requires further research.

Determining how countries, particularly developing countries, can and will respond to the threat of reduced yields and increased costs of food on the world market is a critical research need arising from global assessments. Will such countries be able to develop alternative economic activities in order to generate payment for importing larger amounts of food? The very uncertainty is likely to reduce food security in many of the developing countries.

Any attempt to project the future requires making critical assumptions about possible technological improvements in crop yields, population growth rates, basic food requirements, income distribution, and the presence or absence of surprises (e.g., wars). However, studies to date have shown that, from a political and social standpoint, climate change threatens to reduce food security in developing countries. The worst situation arises from a scenario of severe climate change, low economic growth, and inadequate farm-level adaptation.

Adaptation, Economics, and Policy

A wide variety of adaptive actions may be taken to lessen or overcome adverse effects of climate change on agriculture. At the farm level, adjustments may include introducing late-maturing crop varieties or species, changing cropping sequences, adjusting timing of planting and other field operations, conserving soil moisture through appropriate tillage methods, and improving irrigation efficiency. Some options such as switching crop varieties may be relatively inexpensive, while other options such as installing new irrigation systems may require major investments (Hillel, 1997). Economic adjustments include shifts in regional production centers and modifications of capital, labor, and land allocations. For example, trade adjustment should help shift commodity production to regions where comparative advantage improves; simultaneously, where comparative advantage declines, labor and capital may shift out of agriculture into more remunerative sectors.

This chapter focuses on the human (societal) aspect of the agricultural system, in terms of adaptive, economic, and political responses to climate change. Translating physical and biophysical consequences into economic terms is essential to providing the information that policy makers need to make appropriate decisions regarding adaptation to climate change. Such information also includes the identification of thresholds and potential "surprises" in climate change impacts.

Adaptation

Adaptation to climate change, though it can be defined in various ways, is usually taken to encompass all types of social and technical accommodation to climate. A more specific definition is "any action that seeks to reduce the negative effects, or to capitalize on the positive effects, of climate change" (Riebsame et al., 1995). Adaptive actions may be either anticipatory or reactive in nature (U.S. Country Studies Program, 1994), and in either case incremental or major in magnitude. Incremental adaptations are minor changes requiring little investment, whereas major adaptations require considerable investment of capital, labor, and time. Innovations to deal with altered climate may be induced as manifestations of changes begin to occur (Ruttan, · 1991).

Anticipation and Reaction

Anticipatory actions in physical or operational aspects of systems are to be made in advance of the impending climate change or of the appearance of its impacts, presumably because resource managers believe that the coming change may indeed threaten the future integrity of the system for which they are responsible, be it an individual farm or an entire agricultural region. An example of an anticipatory adjustment is the development of heat- and drought-tolerant crop varieties with enhanced ability to withstand projected occurrences of heat waves and/or dry spells (Figure 10.1). As indications of the onset of climate change become increasingly evident, there will be greater incentive to undertake anticipatory action.

When and if actual impacts, either positive or negative, do indeed occur, agricultural decision makers will be forced to undertake reactive adaptations. Some adaptations (such as the release of reservoir water, if available, during drought periods for supplemental irrigation) may be relatively quick and easy to accomplish. Others, such as grazing or clipping a crop for fodder instead of awaiting grain ripening, may be quick but painful. Still others, such as planting and establishing orchards or building storage and transportation facilities for commodities appropriate to the new conditions, may take a long time to accomplish and require major investments.

A distinction is sometimes made between *farm-level adaptations*, which can be tested by crop models and result in yield changes, and overall *economic adjustments*, which may be simulated by comprehensive economic models and result in national as well as regional production changes and price responses (Rosenzweig et al., 1995). Farm-level adaptations include shifts in planting dates, the use of more climatically adapted crop varieties, changes in amount and timing of irrigation, and changes in fertilizer application. Possible economic adjustments may include increased investment in agricultural infrastructure, reallocation of existing resources (e.g., land and water) according to economic returns, reclamation of additional arable land, and use of additional inputs as a response to higher commodity prices.

As discussed in Chapter 8, sensitivity and adaptability to climate change differ among regions. The ability of any region's agriculture to adapt to climate change depends on the nature of the change that might be required, on local sensitivity, and on the ability of the agricultural community to marshal the needed resources (Riebsame et al., 1995).

Resilience and Robustness

Two useful concepts pertaining to adaptation that are often used by hazard specialists, ecologists, and systems analysts are resilience and robustness (Riebsame et al., 1995). *Resilience* is the ability of a system to return to a predisturbance state without incurring any lasting, fundamental change. *Robustness* is the ability of a system to continue to function in a wide range of changed conditions.

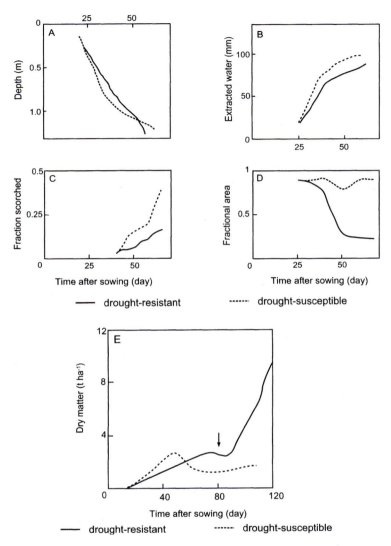

Figure 10.1 Comparison of drought-resistant and drought-susceptible genotypes of sorghum (Squire, 1990).

Resilient resource systems may fail temporarily when perturbed, but then recover to some "normal" range of operation after the perturbation ceases (Riebsame et al., 1995). Robust systems maintain their properties and outputs even under unusual stress, by virtue of strength and control rather than flexibility. System robustness may be increased by "hardening" with increased investment, structural strength, and operational control. There is, generally, a tradeoff between the two characteristics. Overbuilt systems operate normally under a range of different conditions but require large investment; their failures, though rare, are usually more costly and require longer recovery times.

A commonly prescribed adaptation to climate change in agriculture and other sectors is to enhance characteristics that offer flexibility (Riebsame et al., 1995). Flexibility issues are particularly important in regard to the development of water resources for agriculture. Big projects such as dams may actually limit flexibility if they lose effectiveness as regional hydrological water balances undergo major changes. Thus, it may be wise to wait until potential climate change is actualized before certain projects are implemented. Alternatively, as dams and seawalls are built or refurbished, the capacity to withstand the 100-year flood, drought, or storm surge may be enhanced in light of potential changes in climate and climate variability.

Many agricultural planners believe that the appropriate response to the climate change issue, given the present state of knowledge, is further study and monitoring rather than major anticipatory actions. If actions are to be taken at all, they should be only such as can be done with "no regrets"—that is, those that will bring improvements to contemporary conditions whether or not climate changes indeed occur. Recently, projections of the likely magnitude of climate change have come down (from initially high estimates) due to the inclusion of tropospheric aerosols (IPCC, 1996). Consequently, the required adjustments are now seen to be less drastic than before. This situation further encourages a "wait-and-see" attitude, albeit with recognition of the need for continued vigilance in monitoring signs of climate change.

As agricultural planners address the climate change issue in regard to adaptive responses, they should ask the following questions (Riebsame et al., 1995):

- What are the institutional and organizational capacities for adaptation to climate change?
- What secondary problems might be caused by adaptation to climate change?
- What is the range of choices for adapting to climate change?
- How do economics and finance, environmental concerns, and international relations affect the range of choices?

Limits to Adaptation

Including the possibility of adaptation in agricultural impacts studies adds realism, yet many critical uncertainties remain. Although crop models allow testing of some potential improvements in agricultural production, they do not include the full range of possible adaptations (Mendelsohn et al., 1994, 1996), nor other possible technological developments that might be devised in response to negative climate change impacts (Ruttan, 1991). The degree of application and adaptation and the efficacy of various adaptive practices are also uncertain a priori. There may be social or economic reasons why farmers in one region or another are either willing or reluctant to implement certain yield-enhancing measures. For example, increased fertilizer application and improved seed stocks may be too costly or otherwise not compatible with indigenous patterns of production and assumption. Furthermore, such measures may not necessarily result in sustainable production increases (e.g., irrigation may eventually lead to soil salinization).

Recent studies by the U.S. National Research Council and other organizations have emphasized the inherent ability of U.S. farming to adapt to changing conditions (CAST, 1992; National Academy of Sciences, 1992). In the past, technological improvements have indeed been developed and disseminated throughout the agricultural sector. American agriculture has substantial research capabilities and a wide range of adaptive options (Easterling, 1996). Hence, insofar as the United States is concerned, prospects for agricultural adaptation to climate change appear favorable, assuming that water and other essential inputs are and remain available. Considerable investments may be needed, however, to utilize soil and water resources more efficiently in a changed climate. Other countries, particularly in the tropics and semi-tropics, are not so well provisioned with respect to both the research base and the availability of resources (including capital resources).

The potential for adaptation should not lead to complacency (Rosenzweig and Hillel, 1995). Agricultural adaptation to climatic variation is not now and may never be perfect, and changes in how farmers operate or in what they produce may cause significant disruption for people in rural regions. Some adaptive measures may have detrimental impacts of their own. For example, where shifts are made from grain to fruit and vegetable production, farmers may find themselves more exposed to marketing and credit problems brought on by higher capital and operating costs. The considerable social and economic costs that can result from the occurrence of climatic extremes were exemplified by the farmstead damages and crop failures caused by the Mississippi River flood of 1993.

While changes in planting schedules or in crop varieties may be readily applied in an effort to minimize impacts on agricultural incomes, such measures do not ensure equal levels of nutritional quality or food production, nor equal profits for farmers. Expanded irrigation may lead to groundwater depletion, soil salinization, and waterlogging. Increased demand for water by competing sectors may limit the viability of irrigation as a sustainable adaptation to climate change. Expansion of irrigation as a response to climate change will be difficult and costly, even under the best circumstances. Mounting societal pressures to reduce environmental damage from agriculture will likely foster an increase in protective regulatory policies, further complicating adaptations to climate change (Easterling, 1996).

Genetic Resources

A major adaptive response will presumably be an effort to breed heat- and drought-resistant crop varieties by utilizing genetic resources that may be better adapted to warmer and drier conditions. Research is needed to define the current limits to heat and drought resistances and the feasibility of manipulating such attributes through modern genetic techniques. Both crop architecture and physiology may be genetically altered to adapt to warmer environmental conditions. Collections of seeds are maintained in germplasm banks, and their genetic resources may be screened to find sources of resistance to changing diseases and insects, as well as tolerances to heat and water stress and better compatibility with new agricultural technologies. Crop varieties with a higher harvest index (the fraction of total plant matter that is marketable) can help keep irrigated production efficient under conditions of reduced water supplies or greater demands. Genetic manipulation may also help exploit the potentially beneficial effects of CO_2 enhancement on crop growth and water-use efficiency. Biotechnology techniques will likely be brought into play in the effort to raise the rate of germplasm utilization and to transfer genes from distant crop relatives (Jackson and Ford-Lloyd, 1990).

Economics

As discussed in Chapter 7, it is necessary to include economics in the study of climate change impacts on agriculture because biophysical studies per se do not provide data on supply versus demand and on likely prices of agricultural commodities. Analyses that do not include economic adjustments may give quite different results. For example, when yields decline, prices tend to rise, and farmers may benefit from such changes up to a certain point. Even if prices remain constant, information from natural science alone does not yield accurate estimates of output changes when producers can alter production practices and types of output produced (Mendelsohn et al., 1996). The decision-making process, choices, and economic well-being of both producers and consumers can be taken into account in economic studies. Economic assessments may provide information on gains and losses across space and time, as well as on possible benefits and costs to society. Economic information thus plays an important role in developing climate change policies.

As described in the adaptation section in this chapter, farmer and producer adjustments include crop selection, input mixes, alternative technologies, timing of planting, and locational shifts. Consumers of agricultural goods can also adjust by changing the amount and type of commodities bought, both at intermediate levels (e.g., animal feed) and final commodity levels (e.g., food for human consumption). Economic studies often include indicators of how market adjustments such as changes in input and output prices may affect the real net incomes and living standards of producers and consumers, either domestic or international.

If an environmental change affects outputs significantly, price and quality changes can occur, and these, in turn, may lead to further market-induced output changes. Moreover, even if prices remain constant, accurate indications of output changes are needed in cases where individuals can alter production practices and the types of outputs produced. Analyses of the economic consequences of environmental change for agriculture can be valid only if they explicitly recognize the reciprocal relations between physical and biological changes and the potential responses (adaptations) of individuals and institutions.

Economic Assessments

A complete assessment of economic consequences requires three tasks (R. Adams, personal communciation): (1) to measure the differential changes across space and time that climate changes may cause in production and consumption opportunities—such as crop yields, demand for irrigation water, and water supplies; (2) to determine the probable responses of input and output market prices to these changes; and (3) to identify what adaptations (e.g., the input and output changes) can be made by affected producers, consumers, and resource owners in order to minimize their losses or maximize their potential gains from various changes in production and consumption opportunities and in prices. (For example, farmers may substitute inputs and change crops produced, while consumers may change the mix of commodities purchased in response to price signals.) Crop modeling studies provide the information for the first requirement. Evaluation of the latter two requirements represents the economics part of climate change impact studies.

Earlier work in the evaluation of environmental stresses on agriculture has shown that intensifying such stresses increases economic losses (R. Adams, personal communication). Some growers may gain from yield losses because of environmental stress, due to price increases, up to a certain point. Consumer losses are a substantial portion of the total loss from environmental stress. Economic losses in terms of percentage change may be smaller than the underlying biophysical yield changes whenever producers and consumers can adjust their activities. Environmental stresses affect both productivity and demand for inputs, and may have differential effects on the comparative advantage of regions or countries. Consequently, trade flows may be altered, with the result that some economic sectors may gain while others lose.

Findings

Several climate change impact studies have combined biophysical and economic analyses at regional, national, and international scales (Table 10.1). These studies have tested both arbitrary, historical analog, and general circulation model (GCM) based climate change scenarios, some including CO_2 fertilization effects on crop growth and water use. They have also tested such adaptations as varying cropping mixes, as well as shifts in production patterns, planting dates, water supplies, and input use.

Several common themes emerge from these combined biophysical and economic studies of climate change impacts on agriculture. First, adaptations are capable of moderating the potential economic damages and enhancing the potential benefits of climate and CO_2 changes. Projections by Adams et al. (1990, 1995) accounting for economic adjustments suggest that there may be a 50% difference between specific biophysical yield impacts and overall changes in production in the United States. Fischer et al. (1994) showed changes on the order of 10% on a global basis (Tables 10.2 and 10.3).

All studies indicate regional gainers and losers among producers (e.g., Figure 10.2). In general, producers tend to gain while consumers may lose under changed climatic conditions.

The results of the economic studies are sensitive to the levels of temperature change tested, to the GCM scenarios used, and to the assumptions made regarding the beneficial CO_2 effects. Altogether, these studies have shown that economic adjustments can be very important indeed.

However, there are still many uncertainties and limitations in the way economics has been included in climate change impact studies. Interpersonal and intergenerational issues have not thus far been taken into account. As described in Chapter 7, future trends in technology, demand, and population are difficult to project. And, finally, most studies have not included the potential for changes in the variability (in addition to changes in the mean values or ranges) of climate and water supplies.

Challenges

Adams (in U.S. Country Studies Program, 1994) recognized several challenges in the economic assessment of climate change impacts on agriculture. One challenge is to represent the different types of economic processes, markets, and institutions across

Table 10.1 Selected economic assessments of climate change (Adams, personal communication)

Study	Scope of model	Description of economic model	Climate change assumptions	Results
Adams et al., 1990, 1995)	National (U.S.) with 63 production regions	Mathematical programming model of U.S. ag. sector	GCM-based and specific temperature/precipitation changes; varying CO_2 levels from 330 to 550 ppmv	Losses depend on GCM temperature/CO_2 assumptions; possible gainers/losers; role of CO_2 effects critical
Mendelsohn et al., 1994	National, based on country-level data	Econometric model regressing land values on climate, soil and socioeconomic variables	2.5°C increase in temperature, 8% increase in precipitation	$21 to $34 billion loss in farmland values, depending on year
Bowes and Crosson, 1993	Regional (Missouri, Iowa, Nebraska, and Kansas)	Farm-level LP and regional-level input-output models	Temperature/precipitation from 1930s (Dust Bowl)	Results depend on degree of adjustment and CO_2 assumptions
Kaiser et al., 1993	Farm-level; representative farms in Minnesota and Nebraska	Farm-level LP models	2.5° and 4.2°C increases, 10 and 20% precipitation increases, doubled CO_2	Regional gains and losses; +$7 to −$37 billion in aggregate (worldwide)
Rosenzweig and Parry, 1994	Global; 112 sites in 18 countries	Basic Linked System (BLS), consists of country LP models	GCM-based changes (GISS, GFDL and UKMO); major and minor levels of adjustment; doubled CO_2	Varies with GCM; both aggregate gains and losses; developing countries lose due to reductions in cereal production

LP = Linear programming.

Table 10.2 Static climate yield impact in 2060 for three GCM climate change scenarios (Fischer et al., 1994)

World total	GISS Cereals	GISS Other crops	GISS All crops	GFDL Cereals	GFDL Other crops	GFDL All crops	UKMO Cereals	UKMO Other crops	UKMO All crops
Without physiological effects of CO_2	−22.1	−21.8	−22.0	−25.4	−24.3	−25.0	−33.6	−33.4	−33.5
With physiological effects of CO_2	−5.1	3.1	−0.1	−9.0	0.5	−2.8	−18.2	−9.0	−12.2
Adaptation level 1[1]	−1.7	3.1	0.9	−5.5	0.6	−1.7	−12.9	−8.3	−10.1
Adaptation level 2[2]	1.4	4.8	3.2	−1.1	2.5	1.0	−6.1	−3.2	−4.4

1. Signifies minor changes to existing agricultural systems.
2. Signifies major changes to existing agricultural systems.
GISS = Goddard Institute for Space Studies.
GFDL = Geophysical Fluid Dynamics Laboratory.
UKMO = United Kingdom Meteorological Office.

countries in a uniform manner. Individual regions or countries may have quite different characteristics, and these need to be represented adequately in terms of common economic relationships. Another challenge relates to the paucity of biophysical and economic data across different climatic, geographical, and temporal domains, and how such lacunae may be filled in methodologically consistent ways.

The challenge in measuring the global consequences of climate change on agriculture is to include diverse economic structures into a framework that provides a common measure of social welfare (U.S. Country Studies Program, 1994). Both market-oriented and planned economies have a similar goal—namely, the efficient production of a socially optimal level of agricultural output. Market-oriented economies work toward this goal through the market signals created by the interacting forces of supply and demand. Command economies, in contrast, rely on indirect approaches based on planning goals (e.g., perceived consumption needs) and purposeful allocations of resources to meet those goals.

Table 10.3 Impact of economic adjustment for three GCM climate change scenarios (Fischer et al., 1994)

World total	Cereals production (% change) GISS	Cereals production (% change) GFDL	Cereals production (% change) UKMO	GDP agriculture (% change) GISS	GDP agriculture (% change) GFDL	GDP agriculture (% change) UKMO
Without physiological effects of CO_2	−10.9	−12.1	−19.6	−10.2	−11.7	−16.4
With physiological effects of CO_2	−1.2	−2.8	−7.6	−0.4	−1.8	−5.4
Adaptation level 1	0.0	−1.6	−5.2	0.2	−1.2	−4.4
Adaptation level 2	1.1	−0.1	−2.4	1.0	0.0	−2.0

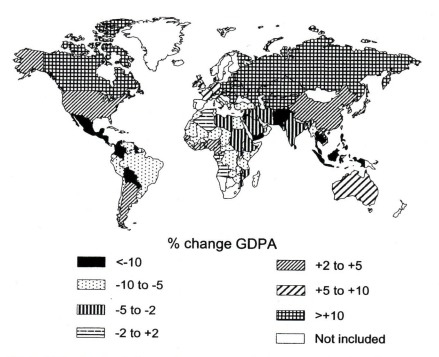

% change GDPA

■	<-10	▨	+2 to +5
▦	-10 to -5	▨	+5 to +10
▥	-5 to -2	▦	>+10
▤	-2 to +2	▢	Not included

Figure 10.2   Impact of climate change on gross domestic agricultural production (GDPA) with economic adjustment in 2060, for the GISS climate change scenario (Fischer et al., 1994).

A fundamentally different decision-making problem confronts the subsistence farmer, who is both a producer and consumer of output. Numerous behavioral models have been devised to capture this dual role. In general, such growers are risk averse. They produce first for family or local consumption by using traditional techniques. In this setting, environmental stresses, such as pollution or climate change, affect the individual in ways distinct from both the market-oriented and planned economy cases. As climate change may reduce the growers' quantity (yield) and quality of commodities, the effects on the individual welfare of subsistence farmers may be quite severe.

First, any productivity loss to the subsistence farmer reduces marketable surplus and possibly even the subsistence component of production (U.S. Country Studies Program, 1994). Second, changes in quantity and quality of production may affect labor productivity if the growers' health is influenced, given that the growers (and their immediate families) provide the major input—labor—into the agricultural production process. Finally, climate-induced yield reductions may intensify risk-averse behavior, and would in any case deprive the grower of the means needed to invest in improved technology. These effects would therefore tend to reduce the likelihood of the grower adopting new, yield-increasing methods. The unique dimensions of subsistence agricultural economies suggest that measuring the economic effects of climate change will require a production function defined at the household level. Such functions are likely to vary across climate zones.

Thresholds and Surprises

The search for thresholds involves analysis of climate impacts on a system or activity, as well as the identification of possible discontinuities in response (Parry et al., 1996). Determination of critical levels of climate change for agriculture encompasses both the biophysical and the socioeconomic realms. As described in Chapter 3, in the biophysical realm, critical temperatures (minimum, optimum, and maximum) have been defined for many individual crop processes. Crop models have been developed to map out the potential effects of various temperature regimes on crop growth and yield. In the socioeconomic realm, a much more difficult challenge is to define the interplay of supply, demand, and prices and to take into account the characteristic adaptability of agriculture as a managed human system. Here, determining critical levels of warming involves defining relative impacts on producers and consumers from diverse geographic and social groups, an especially challenging task.

Arbitrary sensitivity tests and dynamic process crop models have been used to test critical levels of temperature change on crop yields. A modeling study utilizing International Benchmark Sites for Agrotechnology Transfer crop models (IBSNAT, 1989) described in Chapter 9 considered the potential effects of a 2 and a 4°C temperature rise on wheat, rice, maize, and soybeans at study sites from a range of latitudes (Figure 10.3) (Rosenzweig et al., 1993). To aggregate the results, changes in simulated crop yields were weighted by current national production. Without the direct physiological effects of CO_2 enrichment, aggregate crop yields showed a progressively negative response to rising temperatures, with the percentage reduction in yields ap-

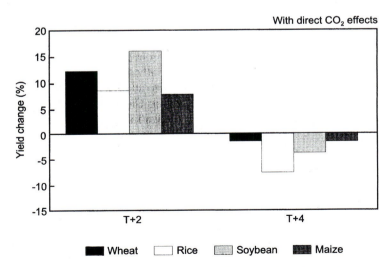

Figure 10.3   Aggregated crop model yield changes for +2°C and +4°C temperature increase. Country results are weighted by contribution of national production to world production. Direct effects of CO_2 on crop growth and water use are taken into account (Rosenzweig et al., 1995).

proximately doubling from the +2 to +4°C cases. On the other hand, when the positive effects of CO_2 were included in the simulation, it indicated an enhancement of globally aggregated yields under a 2°C temperature rise but a reduction under a 4°C rise. The globally aggregated crop yields may mask localized effects, however, as some locations in semiarid and subtropical regions indicated yield declines even with a 2°C rise. These results suggest the possible occurrence of a temperature threshold affecting global grain yields (given current cultivars and management techniques) near the high end of the range of warming predicted currently for 2100 (with aerosols) by the Intergovernmental Panel on Climate Change, namely 1.0 to 3.5°C (IPCC, 1996).

A Caveat Related to Thresholds

The concept of thresholds is sometimes carried into the policy arena through the suggestion that arbitrary levels for atmospheric trace gas concentrations, emission rates, or temperatures might serve as upper limits of acceptability in response to the overall climate change issue. In this case, the term "threshold" may be misleading since artificially contrived levels are specified rather than biophysical thresholds. Proponents of this approach contend that such levels, if generally agreed upon, can serve as quantitative criteria or guideposts for directing national and international efforts to contain potentially harmful consequences of the greenhouse effect. The concept is, in fact, a double-edged sword, since it can be used to either justify or to delay societal action on global warming. The shadow side of such a policy is the implication, however unintended, that amounts under the specified levels are harmless and that the consequences of the enhanced greenhouse effect do not become manifest or significant until these levels are exceeded. Misconstrued, this concept can give license to continue "business as usual," with no need for societal action until the aribtrary level is about to be exceeded.

A more prudent principle may be the quite plausible assumption that global warming and its manifestations will in some manner proportionate to the increase of trace gas concentrations and that the eventual consequences of any significant human alteration of the earth's energy balance is potentially serious. This principle, were it accepted, would encourage responsible agencies to adopt a policy aimed at reversing current trends rather than implicitly sanctioning the continued enhancement of greenhouse gas emissions until such time as the warming effect becomes clearly evident.

Risk, Uncertainty, and Surprise

An attempt was made to define a typology of risk, uncertainty, and surprise regarding climate change at the Aspen Global Change Institute (AGCI, 1994). When there is exposure to the possibility of loss, injury, or other adverse circumstance, and the events, processes, or outcomes involved and the probabilities of their occurrence are known, a *risk* may be defined. When the probabilities of adverse events are not known, there is *uncertainty*. *Surprise* occurs when events in reality depart qualitatively from expectations. Surprises related to global climate change may be either physical or societal in nature. Table 10.4 lists some potential surprises of each type that have been identified (AGCI, 1994).

Table 10.4 Possible global change surprises (from Aspen Global Change Institute, 1994).

Anthropogenic Forcing Functions
 several catastrophic nuclear plant accidents lead to ban on nuclear power before cheap noncarbon
  backstop technology is available
 land-cover change stabilizes (minimal deforestation)
 synergism of habitat fragmentation, chemical assault, introduction of exotic spp., and anthropogenic
  climate change negatively affect biodiversity
 spatially varying (regional scale) competing forces create unforseen regional climate anomalies (land-
  use changes, aerosols, or tropospheric ozone)

Nonanthropogenic Forcing Functions
 gradual reduction in "conveyor belt" oceanic overturning leading to cooling at high latitudes despite
  general (but slower) global warming
 heat stored in the ocean at intermediate depths released to the atmosphere, leading to rapid warming
 Antarctic volcanoes induce ice-stream flow causing glacial surge and rapid sea level rise
 surge of Greenland ice sheet toward the northeast
 change in volcanism induced by change in climate

Environmental Consequences
 regional environmental degradation has global impacts on economic and political systems
 differential movement of species ranges in response to global environmental change causes irreversible
  or very long-term ecological damage (extinctions or cascading effects)
 enhanced hydrological cycle leads to unanticipated extreme floods or droughts
 changes in hurricane intensity with warming

Human Response to the Advent or Prospect of Global Change
 geoengineering adopted
 climate convention increases funding for low cost noncarbon backstop technologies
 creation of wildlife reserves and migration corridors lowers impact on biodiversity
 CO_2 build-up stalls for 5 years, derailing the current convention process

The incorporation of uncertainty into climate change impact studies is an important new development (Dowlatabadi and Morgan, 1993). Earlier studies have often used "best estimate" scenarios, which represent the midpoint of the predicted range of change (e.g., of temperature). Including a spectrum of scenarios encompassing upper and lower bounds of the predicted effects permits the propagation of uncertainty throughout the operation of a model system. Further, probability distributions of different events may be defined, with contrasts between low probability catastrophic events (surprises) and higher probability gradual changes in climate trends.

Complex systems and chaos theory provide conceptual and analytical tools for anticipating and preparing for surprises. One "surprise" may lead to another in a cascade, since subsystems are connected. Identifying potential surprises and communicating them to the public and to policy makers should promote preparedness and resilience to surprise.

Anticipation of surprises in the analysis of global climate change may be encouraged by efforts to integrate across disciplines, to support a multiplicity of research approaches, and to focus on outlier outcomes and unconventional views (AGCI, 1994). Beyond the anticipation of surprise, it seems worthwhile to promote the ability of social structures to adapt to, and recover from, unexpected or uncertain perturbations. Societal preparedness might include diversification of productive and technological systems, establishment of disaster-coping and entitlement systems, and development of adaptive management systems capable of learning from surprises (AGCI, 1994).

Policy

Policies relating to climate change and agriculture may aim at mitigation and/or adaptation. Mitigation is understood to mean efforts to reduce greenhouse gas emissions and thus to minimize the extent and effect of climate change; adaptation is usually defined as human responses to climate changes before or after such changes have actually begun. Since it is unlikely that climate change can be completely averted, policies may be set in place that recognize the potential damages that may arise from climate change and that commit current resources to moderate future impacts, alleviate damages, and foster efficient reactive adaptations when climate change occurs. This section focuses on governmental policies involving agricultural adaptation to climate change.

Adaptation Policy Options

Government policies may tend to thwart rapid adjustment to environmental changes. In such cases, removal of governmental intervention often promotes greater flexibility and thus reduces vulnerability to climatic stress. Therefore, an understanding of the actual and potential role of government policy in the agricultural sector can be useful in forecasting the impacts of climate change. Internationally, lowering barriers to trade in agricultural commodities, as has been pursued under the General Agreement on Tariffs and Trade (GATT), should help the world food system adjust to climate changes more rapidly (Reilly et al., 1994).

In the United States, some present agricultural institutions and policies tend to discourage farm management adaptation strategies, such as altering the mix of crops that are grown and investing in water-conserving technologies (Lewandrowski and Brazee, 1993). At the policy level, obstacles to change are created by supporting prices of crops that are not well suited to a changing climate, by providing disaster payments when crops fail, and by restricting competition through import quotas. Programs could be modified, however, to promote flexibility in the choice of crops, to remove institutional barriers to the development of water markets, and to improve the basis for crop disaster payments.

Wise policy must recognize that regional agricultural development is a dynamic process involving both physical and social systems. Studies of climate–agriculture relationships should be based not just on climate averages and simple trends but on approaches that take into account short-term fluctuations, medium-term shocks or extreme events, and long-term trends. An example of an important phenomenon that calls for greater understanding is the so-called El Niño/Southern Oscillation (ENSO) event that tends to recur every 2 to 9 years, and which already perturbs current climate and is likely to interact with global climate change. Needed are evaluations of past responses by farmers and by regional, national, and international bodies to experiences of climate extremes, which in some cases are related to ENSO events.

The task of assessing adaptation policies includes analyzing the effectiveness or shortcomings of current policies in coping with climate change, calculating the costs and benefits of alternative policies designed to anticipate or mitigate climate change impacts, and identifying policies with highest priority for immediate implementation. Policy makers should be invited to participate in this assessment (U.S. Country Studies Program, 1994). A policy maker is defined as anyone with significant influence over the management of resources vulnerable to climate change, including national, regional, and local officials as well as officials from donor nations or multilateral institutions that influence resource management (U.S. Country Studies Program, 1994). High priority in adaptation policy should be placed on avoiding irreversible or catastrophic impacts (such as eradication of species), preventing climatic extremes from damaging long-term projects (such as dams or seawalls), and halting trends that are likely to thwart future adaptation to climate change.

In view of the present uncertainties over the pace and magnitude of climate change, the most promising policy options for agricultural adaptation are ones for which benefits accrue even if no climate change takes place. Such policy options include the following:

1. *New crop varieties and species.* Breeding objectives should include heat-tolerant and low-water-use crops, as well as crops with high value per unit volume of water used. Salt-tolerant crops should be introduced in regions with brackish water supplies or vulnerable to soil salinization (such as might be caused by water-table or sea-level rise). More tolerant crops may enable farmers to diversify and to produce profitably even under adverse conditions.

2. *Maintenance of seed banks.* Collections of seeds are maintained in germplasm banks around the world. These genetic resources must be maintained in order to allow future screening for sources of resistance to changing diseases and insects, as well as tolerances to heat and water stress and better compatibility with new agricultural technologies.

3. *Liberalization of trade.* Removing barriers to international trade in agricultural commodities, as under the General Agreement on Tariffs and Trade (GATT), should help the world food system to adjust to climate changes more efficiently and rapidly.

4. *Flexibility of commodity support programs.* Commodity support programs often discourage farmers from changing cropping systems, and this may hinder adaptation to climate change. Efforts to stabilize food supplies and maintain farm incomes should avoid disincentives for farmers to switch and rotate crops. Such a policy will induce greater efficiency in farming practices and promote flexibility in the face of future climate change.

5. *Agricultural drought management.* Drought is an intrinsic feature of the climate prevalent in many agricultural regions. Mistakenly, it is commonly viewed as a natural disaster rather than as an inevitable occurrence (though of irregular frequency, duration, and severity). Drought management can be improved by providing information about climatic conditions and patterns, sound preparatory practices and options for the eventuality of drought, and appropriate insurance programs. Farm disaster relief and other government subsidies, however, may distort markets and encourage the continuance and even expansion of farming in marginal lands.

6. *Promoting efficiency of irrigation and water use.* Presently wasteful surface irrigation systems may be converted to more efficient sprinkle, drip, and micro-spray techniques. Drainage water and wastewater may be reused for irrigation.

Evaporation and seepage losses can be reduced by encouraging use of nighttime irrigation, lining of canals, closed conduits, delivery of water in measured quantities, and charging for water in proportion to the volume used. Finally, and perhaps most important, water conservation should be promoted by means of public education and consciousness raising.

7. *Dissemination of conservation management practices.* Conservation tillage, furrow diking, terracing, contouring, and planting windbreaks protect fields from water and wind erosion, and retain soil moisture by reducing evaporation and increasing infiltration. Improving rainfed (dryland) farming can reduce dependence on irrigation, save water, and allow greater resiliency in adapting to climate change.

(Office of Technology Assessment, 1993)

The Human Side

Combined biophysical and economic analysis suggests that, in general, agricultural adaptation and market adjustments can moderate the negative impacts of climate change. However, adaptation cannot be taken for granted: improvements in agriculture have always depended on the investment made in agricultural research and infrastructure. Research can identify the specific ways that farmers now adapt to present variations in climate, whether by applying more fertilizer, more mechanization, or more labor. Information of this nature is needed to assess potentialities for coping with the threat of a more severe climate in the future. Success in adapting to possible climate change will depend on a better definition of what changes will occur where, and on prudent investments, made in a timely fashion. A variety of policy adaptations are available that may enhance the agricultural system's flexibility in preparing for climate change, while improving the efficiency and sustainability of food production.

The Global Harvest: A Summation

The trend toward rising concentrations of greenhouse gases, if allowed to continue, seems bound to result in significant warming during the coming decades. The complacent view that "science" will surely solve the problem is a dangerous delusion. Equally dangerous is the fatalistic acceptance that nothing can be done. No technological panacea (like the recently proposed fertilization of the ocean with iron) is likely to remedy global warming in a single stroke. Rather, the feasible approach is to apply a variety of potential means in combination, through trial and error and retrial.

To begin, we need substantive action on the primary root cause: the high rate of carbon dioxide emissions. These result from the combustion of fossil fuels, the eradication of forests—an action that both releases CO_2 from existing biomass and reduces its absorption into new biomass—and the cultivation ("reclamation") of virgin land that results in decomposition of the initially present soil organic matter.

The enhanced greenhouse effect presents a serious threat to natural ecosystems. Here, the adjustment processes of changing species composition and zonation will be thwarted by the fragmentation and island-like isolation of the remaining "pristine" domains, due to the human usurpation of the greater fraction of the planet's terrestrial surface. Many species, perhaps even entire biotic communities, may not be able to survive a change in climate if they cannot migrate. Consequently, the role played by natural ecosystems in environmental processes (e.g., O_2-CO_2 exchange, runoff, evaporation, groundwater recharge, nutrient recycling, and more) may be jeopardized.

Agricultural systems may be more adjustable, being subject to our control—but only if we do the right thing in fitting management practices to altered circumstances. Shifting agricultural zones onto so-called virgin land may in turn impinge on the remaining natural habitats.

The role of carbon dioxide in agriculture is complex in that it can be positive in some respects and negative in other respects. CO_2 concentration affects crop production directly by influencing the physiological processes of photosynthesis and transpiration; therefore, it has the potential to stimulate plant growth. The magnitude of that stimulation will vary greatly among species of differing photosynthetic pathway, and it will depend on growth stage and on water and nutrient status. However, the resulting climatic effects (including warmer temperatures, changed hydrological regimes, and altered frequencies and intensities of extreme climatic events) may inhibit crop production. Agricultural pests, overall, are likely to thrive under conditions of increasing atmospheric CO_2 concentrations and rapid climate change. All these changes, in concert, could have major impacts on the prospects for food security—in some cases positive and in other cases negative.

Some experts believe that the gradual change of climate will exert its influence so slowly that its effects will hardly be noticeable, given the capacity of modern society to make technological and economic adjustments. However, many studies (reported throughout this book) integrating climate, crop, and market dynamics suggest strongly that the anticipated changes are likely to have large and far-reaching consequences, especially in less-developed regions.

Linking the main interactions at each scale of activity and at successively expanding scales is key to the assessment of how local effects are likely to coalesce in the global context. Crop plants conjoin to form crop stands in farmers' fields, and the latter combine successively to constitute or affect farms, farming regions and agroecological zones, national agricultural sectors, and the international food trade system. Climatic, agronomic, economic, and political factors thus create complex agricultural econo-ecosystems. Comprehensive studies of the integrated responses of such systems to changes in climate and land-use patterns are only now evolving.

There is grave danger in concluding that "in general" climate change does not pose a threat to world or national agriculture. Climate change will gradually (and at some point may even abruptly) affect the range of options available for agriculture in any given region. Farmers' strategies grow out of experience. Under progressively changing climate conditions, the past will be a less reliable predictor of the future, and the accumulated experience obtained to date will be less useful as a tool for coping with what might be a very different future. In some regions, agriculture may be changed radically and may even disappear due to rising sea-level and saltwater intrusion, or to unfavorable climate regimes (including extreme variability or aridity). In other regions, the opposite may be the case, but the gains of some regions may not fully compensate for the losses of other regions.

As NEW AREAS become suitable for crop production while old agricultural areas become less so, the geographic shift of agriculture may encroach on virgin lands and natural ecosystems. Even apart from agriculture, climate change is likely to modify the zonation and bio-productivity of forests, grasslands, savannas, wetlands, tundras, and other biomes. A warmer regime might disrupt the prior adaptation of native plants and animals to their existing habitats. The flooding and waterlogging of some areas and the aridification of others could weaken currently vigorous biotic communities. For example, the thawing of permafrost could dry out tundras, just as the invasion of seawater can doom freshwater wetlands (estuaries, deltas, marshes, lagoons) near coastlines.

The rate of climate change may be too rapid to allow some natural communities to adjust, and where evolving climate becomes increasingly unfavorable there could be a large-scale die-back of forests. Associated species that depend on these forest ecosystems may then be threatened with extinction. In this manner, climate change constitutes a threat to biodiversity in general and to the survival of vulnerable or endangered species in particular. Conversely, some types of forests and other biomes may expand and become more vigorous as a consequence of the warming trend and enhanced photosynthesis, demonstrating beneficial effects as well.

UNCERTAINTIES ABOUND in the climate change issue. Some stem from the intrinsically disordered behavior of the physical climate system. Others derive from our lack of complete understanding of that system, especially in regard to ocean, cloud, and ice responses. Still other uncertainties derive from the fast pace and unknowable directions of future social, political, and technical changes. The world is certain to be quite different in the coming century, and unpredictable developments may change how agriculture responds to climate change. Questions about population growth rates (e.g., How many people will the agricultural system need to feed?), consumption patterns (e.g., What type of diet will people prefer?), and technological change (e.g., Can agricultural productivity continue to improve?) are especially apropos in this regard and, hence, can only be projected within a wide range of uncertainty.

Global change is a deceptively simple expression for what is actually an exceedingly complex array of dynamic processes, with specific combinations or interactions in each region. Climate change, sea-level rise, and increases in carbon dioxide, ultraviolet radiation, and tropospheric ozone are but a few of the potentially fateful factors involved. While many studies have investigated these factors singly, there is much to be gained from studying them interactively, but that is an extremely difficult task.

Climate change is predicted to take place over a period of decades to perhaps centuries. Society may have some time to study the potential impacts and develop effective mitigation and adaptation actions. Most proposals to mitigate the CO_2 buildup involve some combination of conserving energy, substituting alternative energy sources (e.g., solar, wind, and hydropower) for fossil fuels, and reducing tropical deforestation. Others propose to establish large areas of tree plantations in order to fix excess CO_2 through photosynthesis, thus helping postpone the CO_2 buildup and, hence, delay global warming. Adaptive strategies include developing drought-resistant hybrids, soil-moisture conserving cultivation practices, and coastal protection.

The time available for evaluation of such strategies and decision making is not unlimited and should not be wasted. If this time is not used appropriately, sooner than we like, it may be too late to avoid upheavals and, for some nations, suffering. Climate change could lead either to cooperation or to conflict over the world's major resources. Integrated planning is perhaps the highest global challenge, now motivated by the potential for environmental and social transformation caused by climate change.

The dilemma is that remaining passive while awaiting a complete understanding of climatic processes and absolute proof of global warming may prevent timely and effective action to protect our agroecosystems from serious deterioration. A certain amount of warming appears practically inevitable, since a stabilization of the earth's atmospheric concentration of CO_2, the primary greenhouse gas, would require a substantial reduction in current emissions. The dilemma is compounded, however, because not only are we faced with the possibility that we may fail to act when we should act but also that we may act unnecessarily or ineffectually. Because many of the proposed mitigation and adaptation actions are costly (e.g., developing alternative sources of energy), prudence is needed. This situation leads some to argue for "no regrets" actions—undertaking only those strategies that make sense for reasons other than just climate change mitigation or adaptation alone.

The immediate issue before the international community is whether, when, and how to mitigate the projected rise by limiting emissions of the greenhouse gases. This was one of the key issues of the United Nations Conference on the Environment and Development held in Rio de Janeiro in the summer of 1992. At that conference, the United Nations Framework Convention on Climate Change (1992) was organized. The agreement's objective is to bring about "stabilization of greenhouse gas concentrations in the atmosphere at a level that would prevent dangerous anthropogenic interference with the climate system." Explicit commitments to targets and timetables of greenhouse gas limitations have yet to be specified, but they are now under discussion.

While environmental policy for agriculture has traditionally been tied to water quality and soil conservation, these policies may be expanded to limit emissions of greenhouse gases—especially carbon dioxide, methane, and nitrous oxide—from agricultural activities. Further, policies aimed at encouraging carbon sequestration through agroforestry may become important for the industry.

The climate change issue brings the recognition that current and future levels of energy use (especially the burning of fossil fuels) and changing land use (especially the clearing of forests for cultivation) can have profound effects on the global environment. This presents a challenge to human society to manage its future development so as to protect and maintain the conditions of life on earth, and thus the welfare of all nations and future generations.

NOTES

Chapter 1   The Greenhouse Effect and Global Warming

1. Stefan-Boltzmann law states that the rate at which an object radiates heat is proportional to the fourth power of its absolute temperature: $T^4 = (I)(123)(10)^8$, where T is the absolute temperature and I is the intensity of radiation in units of calories per square centimeter per minute.

Wien's law states that as the temperature of an object increases, the wavelength of the most intense radiation that it emits decreases: $L_i = 2900/T$, where L_i is the most intense wavelength in microns and T is the absolute temperature.

2. The greenhouse analogy is not a perfect representation of the atmosphere, however. In the greenhouse, the glass forms a physical barrier to prevent the heated air from rising, and heat is also lost via the process of conduction. In the atmosphere, the so-called radiatively active gases are not concentrated in a thin layer but are diffused throughout the atmosphere. Furthermore, these gases differ in their relative absorptive properties toward infrared radiation. Finally, air currents through the atmosphere cause advective and convective heat fluxes not present in the enclosure of a greenhouse. These differences notwithstanding, the term "greenhouse effect" is commonly applied to the atmosphere.

3. Methane hydrate is a solid form of natural gas composed of crystallized water molecules surrounding molecules of methane gas, present mainly in arctic permafrost and continental shelf regions.

4. A recent New York Times article reported that there is a flourishing black market in CFCs produced in various countries in direct contravention of the Montreal Protocol (Halpert, pp. 1, 31).

5. As stratospheric ozone is scavenged by molecules of CFCs and other halogenated compounds, the earth's surface temperature is affected in two opposing ways: (1) More solar radiation reaches the earth's surface, thus warming it; (2) since, on the other hand, the stratosphere absorbs less radiation, it is cooler than it would be otherwise, and it radiates less longwave radiation to the surface. The first process depends on the total amount of ozone in the atmosphere, and the second depends on its vertical distribution, since ozone is a strong greenhouse gas primarily in the upper troposphere. These two opposing effects are similar in magnitude; hence,

calculating the thermal effect of ozone changes is a difficult task. With the greenhouse effect of the CFCs taken into account, it now appears that the cooling effect of stratospheric ozone depletion may be approximately balanced by the greenhouse forcing of the CFCs causing the ozone losses (IPCC, 1992).

6. Clouds affect the energy balance in at least two opposing ways. By reflecting part of incoming sunlight, clouds reduce the energy flux reaching the ground surface, thereby having a cooling effect. By partially blocking the outgoing heat emitted by the earth, clouds also have a warming effect. Low clouds tend to have a cooling effect, while high clouds tend to have a warming effect.

7. The radiative effect of the aerosols is about $2 \ W \ m^{-2}$, roughly comparable in magnitude to the enhanced greenhouse emissions but with a cooling, rather than a warming, effect. An apt description of the two opposing effects is given by Kerr, 1995, p. 802: "Human beings may be turning up earth's thermostat as they pump carbon dioxide and other greenhouse gases into the atmosphere, but they are drawing the shades against the sun's warmth at the same time."

Chapter 2   Agricultural Emissions of Greenhouse Gases

1. The typical organic matter content of soils is between 1% and 5% by mass. For purposes of illustration, let us assume that a virgin soil containing 2% carbon in its top 0.4 m layer is converted to farming and then loses half its carbon content within, say, 10 years. The fraction lost— that is, 1%—constitutes $0.01 \times 0.4 \ m \times 10,000 \ m^2 \ ha^{-1} \times 1.25 \ t \ m^{-3} = 50 \ t \ ha^{-1}$, or $5 \ t \ ha^{-1}$ yr^{-1}. If the process takes place over 100 years, then the average rate is $0.5 \ t \ ha^{-1} \ yr^{-1}$. For 1.6×10^9 ha (world cropland area), the latter rate is globally equivalent to $0.8 \times 10^9 \ t \ yr^{-1}$. This is a rough estimate to show the approximate magnitude of the process.

2. "Epiaquic" describes a soil saturated with water in one or more layers within 200 cm of the mineral soil surface and with one or more unsaturated layers below the saturated layer. The zone of saturation—the water table—is perched on top of a relatively impermeable layer. "Anthric" saturation is associated with controlled flooding (for such crops as wetland rice and cranberries), which causes reduction processes in the saturated, puddled surface soil and oxidation of reduced and mobilized iron and manganese in the unsaturated subsoil (Soil Survey Staff, 1992).

Chapter 3   Carbon Dioxide, Climate Change, and Crop Yields

1. A third group of plants exhibits a modified C4 photosynthetic pathway known as crassulacean acid metabolism (CAM) (Tolbert and Zelitch, 1983). Some succulent plants such as cacti are characterized by this mechanism, which allows them to conserve water. Such plants fix CO_2 at night and then process it during the day using the C3 pathway. Limited data suggest that higher CO_2 levels enhance photosynthesis in CAM plants. Pineapple is an example of a CAM crop plant.

2. Photorespiration is the rapid oxidation of sugars recently formed by photosynthesis in the light, a process that inhibits photosynthesis. Photorespiration occurs because the chief photosynthetic enzyme, ribulose 1,5-bisphosphate (RuBP or Rubisco), functions both as an oxygenase or a carboxylase. Under current ambient CO_2 concentrations the photorespiratory oxygenase reaction is favored; under elevated CO_2 concentrations more carboxylation occurs, resulting in higher rates of photosynthesis (Tolbert and Zelitch, 1983).

3. Mass of dry matter produced generally refers to assimilation minus respiration.

4. Perhaps the simplest formulation for growing degree days is:

$$GDD/GS = \sum N_d \,(GS) \left(\frac{T_{\max} + T_{\min}}{2} - T_{\text{base}} \right)$$

5. Hardening is defined as acclimation exhibiting increased tolerance of stress such as heat or cold. Hardening may be caused by gradual exposure to stress. Changes may be phenotypic or occur at the protoplasmic level (Hale and Orcutt, 1987).

6. Little wheat is grown where annual precipitation is greater than 1200 mm. Soft wheat types, which are used for pastry flour, are grown in areas with moderate precipitation (760 to 1200 mm yr^{-1}). Hard wheat types, used for bread, are grown in drier areas. As hydrologic regimes are altered with climate change, the geographic distribution of wheat types, as well as of other crops, is likely to change accordingly (Rosenzweig, 1985).

7. In cold temperate regions, winter wheat is normally planted in the middle of fall. By late autumn, the wheat attains a height of 5–10 cm and the vegetation covers about 10% of the field. In unusually severe winters, a high percentage of winter wheat is likely to be killed, requiring reseeding to spring wheat in the spring. Extreme cold is one of the primary causes of winterkill of wheat. Other factors that affect the susceptibility of wheat to winterkill are drought, wind erosion, snow cover, flooding, hail, disease, and insect pests (Caprio, 1984).

Chapter 6   Water Resources and Sea-Level Rise

1. Distinctions are useful in consideration of alternative types of efficiency in irrigation management. Three efficiencies may be defined: system water application efficiency, field water application efficiency, and agronomic water-use efficiency. All three types of efficiency offer possibilities for improvement, both with and without climate change. (See Rosenzweig and Hillel, 1994.)

System water application efficiency is the fraction of the volume of water taken from the source (generally, a river or groundwater aquifer) that is used consumptively by crops in an entire irrigation district or region. For example, in the Nile Valley irrigation system in Egypt, system water application efficiency may be as high as 70%. Although much water is lost in conveyance (generally in open, unlined canals), the system efficiency is enhanced by the repeated use in successive downstream sites of water drained from upstream sites. However, this seemingly high value of efficiency has its drawbacks. With each successive use, the water reused undergoes degradation in energy and quality—that is, it loses elevation and becomes progressively salinized.

Field water application efficiency is the fraction of the water applied to a given field that is consumed by the crop (i.e., by evapotranspiration). In practice, field application efficiency values cannot attain 100%, nor should that be the aim, since a certain fraction of the water applied must be allowed to seep away and leach the salts that would otherwise accumulate in the root zone. However, with careful management, field water-application efficiency values of 90% are possible, and of 80% are practicable. At present, typical values of field water-application efficiency in Egypt, for example, are considerably below 50% and in many irrigated areas are even below 30%.

An entirely different concept is the *agronomic irrigation water-use efficiency*, defined as the economic yield obtained per unit volume of irrigation water applied. As such, agronomic irrigation water-use efficiency is a truer measure of the productivity of irrigation. It is not expressed in percentage terms but in weight of produce per unit volume of water. Because of excessive irrigation, poor drainage, salinization, and nonoptimal management (e.g., insufficient or inappropriate fertilization; poor pest control; poor choice of crop; or poor germination), agronomic irrigation water-use efficiency may be much below the potential productivity attainable under optimal conditions.

Chapter 7   Analysis of Climate Change Impacts

1. Richardson's (1981) stochastic weather generator simulates daily times series of maximum and minimum temperature, incident solar radiation, and precipitation. Daily precipita-

tion occurrence is represented by a two-state first-order Markov chain model. It accounts for the stochastic dependence of the series of wet and dry days. Parameters estimated are two transition probabilities: P_{11} and P_{01}, the probability of a wet day following a wet day, and the probability of a wet day following a dry day. Rainfall amounts (x) are simulated for rain days using the gamma distribution. Maximum and minimum temperature and solar radiation are modeled as a multivariate first-order autoregressive process. The seasonal cycle for the means and standard deviations is determined by two-harmonic Fourier series. Since maximum temperature and solar radiation are conditioned on the occurrence of precipitation, separate models (including different harmonics, means, and variances) are used for values occurring on rain days and dry days.

2. *Exogenous variables* are input variables, independent of the internal state of the system, such as climate (daily minimum and maximum temperature, precipitation, and solar radiation), soil (hydraulic properties of soil layers), and management (planting date, plant population, and cultivar type). Exogenous variables thus represent external factors that are imposed on the system and that induce changes within it. These factors are often called "forcing functions" or "driving functions." As such, they can be either uncontrollable (incoming solar radiation available for crop growth) or controllable (irrigation or fertilization).

Endogenous variables are dependent, output variables ("responses") of the model, generated by the effect of the exogenous variables on the system's state variables. Examples are the time trend of evaporation from a field, as well as crop yield. *State variables* are those which characterize the state of the system and directly determine the processes that bring about changes in the endogenous variables. Examples of state variables are the soil water content and standing crop biomass. State variables are the intermediaries through which exogenous variables eventually influence the endogenous variables.

Rate variables control the rates at which dynamic responses are generated by various changes in the system's state. Rate variables are often called parameters. They represent the coefficients of the governing equations or laws describing the functional dependencies of the endogenous variables upon their controlling state variables.

See Hillel (1977) for further explanation.

BIBLIOGRAPHY

Chapter 1 The Greenhouse Effect and Global Warming

Alcamo, J., G. J. Van den Born, A. F. Bouwman, B. de Haan, K. Klein Goldewijk, O. Klepper, J. Krabec, R. Leemans, J. G. J. Olivier, A. M. C. Toet, H. J. M. de Vries, and H. J. van der Woerd. 1994. Modeling the global society-biosphere-climate system. Part 2: Computed scenarios. *Water, Air, and Soil Pollution* 76:37–78.

Arrhenius, S. 1896. On the influence of carbonic acid in the air upon the temperature of the ground. *Philosophical Magazine* 41:237–276.

Barnola, J. M., D. Raynaud, Y. S. Korotkevich, and C. Lorius. 1987. Vostok ice core provides 160,000-year record of atmospheric CO_2. *Nature* 329:408–414.

Berger, A. L. 1981. Spectrum of climatic variations and possible causes. In A. Berger (ed.). *Climatic Variations and Variability: Facts and Theories*. D. Reidel. Dordrecht. pp. 411–432.

Blake, D. R. and F. S. Rowland. 1988. Continuing worldwide increase in tropospheric methane, 1978–1987. *Science* 239:1129–1131.

Boer, G. J. 1993. Climate change and the regulation of the surface moisture and energy budgets. *Climate Dynamics* 8:225–239.

Bradley, R. S., H. F. Diaz, J. K. Eischeid, P. D. Jones, P. M. Kelly, and C. M. Goodess. 1987. Precipitation fluctuations over Northern Hemisphere land areas since the mid-19th century. *Science* 237:171–175.

Chappellaz, J., J. M. Barnola, D. Raynaud, Y. S. Korotkevich, and C. Lorius. 1990. Ice-core record of atmospheric methane over the past 160,000 years. *Nature* 345:127–131.

Charlson, R. J., S. E. Schwartz, J. M. Hales, R. D. Cess, J. A. Coakley, Jr., J. E. Hansen, and D. J. Hofmann. 1992. Climate forcing by anthropogenic aerosols. *Science* 255:423–430.

Del Genio, A. D., A. A. Lacis, and R. A. Ruedy. 1991. Simulations of the effect of a warmer climate on atmospheric humidity. *Nature* 351:382–385.

Delmas, R. J., J.-M. Ascencio, and M. Legrand. 1980. Polar ice evidence that atmospheric CO_2 20,000yr BP was 50% of present. *Nature* 284:155–157.

Diaz, H. F., R. S. Bradley, and J. K. Eischeid. 1989. Precipitation fluctuations over global land areas since the late 1800's. *Journal of Geophysical Research* 94:1195–1210.

Dickinson, R. E. 1986. Impact of human activities on climate—A framework. In W. C. Clark and R. E. Munn (eds.). *Sustainable Development of the Biosphere.* Cambridge University Press. Cambridge. pp. 252–291.

Dixon, R. K., S. Brown, R. A. Houghton, A. M. Solomon, M. C. Trexler, and J. Wisniewski. 1994. Carbon pools and flux of global forest ecosystems. *Science* 263:185–190.

Emanuel, K. A. 1987. The dependence of hurricane intensity on climate. *Nature* 326:483–485.

Fourier, J. B. J. 1824. Remarques générales sur les températures du globe terrestre et des espaces planétaires. *Annales de Chemie et de Physique* 27:136–167.

Gedzelman, S. D. 1980. *The Science and Wonders of the Atmosphere.* Wiley. New York. 535 pp.

Gornitz, V., C. Rosenzweig, and D. Hillel. 1994. Is sea level rising or falling? *Nature* 371:481.

Grotch, S. L. 1988. *Regional Intercomparisons of General Circulation Model Predictions and Historical Climate Data.* DOE/NBB-0084. U.S. Department of Energy. Washington, DC. 291 pp.

Halpert, Julie Edelson. 1995. Freon smugglers find big market. *New York Times,* April 30.

Hansen, J., G. Russell, D. Rind, P. Stone, A. Lacis, S. Lebedeff, R. Ruedy, and L. Travis. 1983. Efficient three-dimensional global models for climate studies: Models I and II. *Monthly Weather Review* 111(4):609–662.

Hansen, J. E. 1988. The greenhouse effect: Impacts on current global temperature and regional heat waves. Testimony to U.S. Senate, Committee on Energy and Natural Resources. June 23, 1988.

Hansen, J. , I. Fung, A. Lacis, D. Rind, S. Lebedeff, R. Ruedy, and G. Russell. 1988. Global climate changes as forecast by Goddard Institute for Space Studies three-dimensional model. *Journal of Geophysical Research* 93:9341–9364.

Hansen, J. E. and A. A. Lacis. 1990. Sun and dust versus greenhouse gases: An assessment of their relative roles in global climate change. *Nature* 346:713–719.

Hansen, J., A. Lacis, D. Rind, G. Russell, P. Stone, I. Fung, R. Ruedy, and J. Lerner. 1984. Climate sensititivy: Analysis of feedback mechanisms. In *Climate Processes and Climate Sensitivity.* J. E. Hansen and T. Takahashi (eds.). American Geophysical Union. Washington, DC. pp. 130–163.

Hansen, J., A. Lacis, R. Ruedy, and M. Sato. 1992. Potential climate impact of Mount Pinatubo eruption. *Geophysical Research Letters* 19(2):215–218.

Hansen, J. , A. Lacis, R. Ruedy, M. Sato, and H. Wilson. 1993. How sensitive is the world's climate? *Research & Exploration* 9(2):142–158.

Hillel, D. and C. Rosenzweig. 1989. *The Greenhouse Effect and Its Implications Regarding Global Agriculture.* Research Bulletin Number 724. Massachusetts Agricultural Experiment Station. University of Massachusetts at Amherst. 36 pp.

Houghton, R. A. and D. L. Skole. 1990. Carbon. In B. L. Turner, II, W. C. Clark, R. W. Kates, J. F. Richards, J. T. Mathews, and W. B. Meyer (eds.). *The Earth as Transformed by Human Action.* Cambridge University Press. Cambridge. pp. 393–408.

IPCC, 1990. *Climate Change: The IPCC Scientific Assessment.* J. T. Houghton, G. J. Jenkins, and J. J. Ephraums (eds.). Intergovernmental Panel on Climate Change. Cambridge University Press. Cambridge. 365 pp.

IPCC, 1992. *Climate Change 1992: The Supplementary Report to the IPCC Scientific Assessment.* J. T. Houghton, B. A. Callander and S. K. Varney (eds.). Intergovernmental Panel on Climate Change. Cambridge University Press. Cambridge. 200 pp.

IPCC, 1995. *Climate Change 1994: Radiative Forcing of Climate Change and an Evaluation of the IPCC IS92 Emission Scenarios.* J. T. Houghton, L. G. Meira Filho, J. Bruce, Hoesung Lee, B. A. Callander, E. Haites, N. Harris, and K. Maskell (eds.). Intergovernmental Panel on Climate Change. Cambridge University Press. Cambridge. 339 pp.

IPCC. 1996. *Climate Change 1995: The Science of Climate Change.* J. T. Houghton, L. G. Meira Filho, B. A. Callander, N. Harris, A. Kattenberg, and K. Maskell (eds.). Intergovernmental Panel on Climate Change. Cambridge University Press. Cambridge. 572 pp.

Kalkstein, L. S. (ed.). *Global Comparisons of Selected GCM Control Runs and Observed*

Climate Data. U.S. Environmental Protection Agency. Office of Policy, Planning and Evaluation. 21P-2002. Washington, DC. pp. 251.

Karl, T. R., R. W. Knight, D. R. Easterling, and R. G. Quayle. 1995. Trends in U.S. climate during the twentieth century. *Consequences* 1(1):2–12.

Katz, R. W. and B. G. Brown. 1992. Extreme events in a changing climate: Variability is more important than averages. *Climatic Change* 21:289–302.

Keeling, C. D., R. B. Bacastow, A. F. Carter, S. C. Piper, T. P. Whorf, M. Heimann, W. G. Mook, and H. Roeloffzen. 1989. A three-dimensional model of atmospheric CO_2 transport based on observed winds: 1. Analysis of observational data. In: D. H. Peterson (ed.). *Aspects of Climate Variability in the Pacific and the Western Americas*. Geophysical Monograph 55. American Geophysical Union, Washington, DC. pp. 165–236.

Kellogg, W. W. and Z.-C. Zhao. 1988. Sensitivity of soil moisture to doubling of carbon dioxide in climate model experiments. Part I: North America. *Journal of Climate* 1:348–366.

Kerr, R. A. 1995. Study unveils climate cooling caused by pollutant haze. *Science* 268:802.

Lashof, D. A. and D. A. Tirpak (eds.). 1990. *Policy Options for Stabilizing Global Climate*. Report to Congress. 21P-2003.1. U.S. Environmental Protection Agency. Office of Policy, Planning and Evaluation. Washington, DC.

Lindzen, R. S. 1990. Some coolness concerning global warming. *Bulletin of the American Meteorological Society* 71:288–299.

Manabe, S. and R. T. Wetherald. 1987. Large-scale changes of soil wetness induced by an increase in atmospheric carbon dioxide. *Journal of the Atmospheric Sciences* 44:1211–1235.

McFarlane, N. A., G. J. Boer, J. P. Blanchet, and M. Lazare. 1992. The Canadian Climate Centre Second General Circulation Model and its equilibrium climate. *Journal of Climate* 5:1013–1044.

Mearns, L. O., R. W. Katz, and S. H. Schneider. 1984. Extreme high temperature events: Changes in their probabilities with changes in mean temperature. *Journal of Climate and Applied Meteorology* 23:1601–1613.

Mearns, L. O., S. H. Schneider, S. L. Thompson, and L. R. McDaniel. 1990. Analysis of climate variability in general circulation models: Comparison with observations and changes in variability in $2XCO_2$ experiments. *Journal of Geophysical Research* 95:20,469–20,490.

Meehl, G. A. 1990. Seasonal cycle forcing of El Niño-Southern Oscillation in a global, coupled ocean-atmosphere GCM. *Journal of Climate* 3:72–98.

Mitchell, J. F. B. and D. A. Warrilow. 1987. Summer dryness in northern mid-latitudes due to increased CO_2. *Nature* 330:238–240.

Mitchell, J. F. B., C. A. Wilson, and W. M. Cunnington. 1987. On CO_2 climate sensitivity and model dependence of results. *Quarterly Journal of the Royal Meteorological Society* 113:293–322.

Mitchell, J. F. B. 1989. The "greenhouse" effect and climate change. *Reviews of Geophysics* 27(1):115–139.

Mitchell, J. F. B., T. J. Johns, J. M. Gregory, and S. B. F. Tett. 1995. Climate response to increasing levels of greenhouse gases and sulphate aerosols. *Nature* 376:501–504.

Montreal Protocol on Substances that Deplete the Ozone Layer. Adopted and opened for signature Sept. 16, 1987. Reprinted in 26 *International Legal Materials* 1541 (1987) (entered into force Jan. 1, 1989).

Neftel, A., H. Oeschger, J. Schwander, B. Stauffer, and R. Zumbrunn. 1982. Ice core sample measurements give atmospheric CO_2 content during the past 40,000 yr. *Nature* 295:220–223.

Palmer, W. C. 1965. *Meteorological Drought*. Research Paper 45. U.S. Department of Commerce, Weather Bureau. Washington, DC. 58 pp.

Parker, E. E., P. E. Jones, C. K. Folland, and A. J. Bevan. 1994. Interdecadal changes of surface temperature since the late nineteenth century. *Journal of Geophysical Research* 99:14373–14399.

Ramanathan, V., R. J. Cicerone, H. B. Singh, and J. T. Kiehl. 1985. Trace gas trends and their potential role in climate change. *Journal of Geophysical Research* 90:5547–5566.

Ramanathan, V., B. R. Barkstrom, and E. F. Harrison. 1989. Climate and the earth's radiation budget. *Physics Today* 42(May):22–32.

Raval, A. and V. Ramanathan. 1989. Observational determination of the greenhouse effect. *Nature* 342:758–761.

Rind, D., R. Goldberg, J. Hansen, C. Rosenzweig, and R. Ruedy. 1990. Potential evapotranspiration and the likelihood of future drought. *Journal of Geophysical Research* 95:9983–10,004.

Rind D., E.-W. Chiou, W. Chu, J. Larsen, S. Oltmans, J. Lerner, M. P. McCormick, and L. McMaster. 1991. Positive water vapour feedback in climate models confirmed by satellite data. *Nature* 349:500–503.

Robinson, P. J. 1991. Comparisons of Rand climatology and GCM outputs for Australia and tropical Asia. In L. S. Kalkstein (ed.). *Global Comparisons of Selected GCM Control Runs and Observed Climate Data*. U.S. Environmental Protection Agency, Office of Policy, Planning and Evaluation. 21P-2002. Washington, DC. pp. 139–197.

Rosenzweig, C. and D. Hillel. 1993a. Agriculture in a greenhouse world. *Research and Exploration* 9(2):208–221.

Rosenzweig, C. and D. Hillel. 1993b. The dust bowl of the 1930s: Analog of greenhouse effect in the Great Plains? *Journal of Environmental Quality* 22:9–22.

Sahrawat, K. L. and D. R. Keeney. 1986. Nitrous oxide emission from soils. *Advances in Soil Science* 4:103–148.

Smith, T. M. and H. H. Shugart. 1993. The transient response of terrestrial carbon storage to a perturbed climate. *Nature* 361:523–526.

Spencer, R. W. and J. R. Christy. 1992. Precision and radiosonde validation of satellite gridpoint temperature anomalies. Part II: A tropospheric retrieval and trends during 1979–1990. *Journal of Climate* 5:858–866.

Tans, P. P., I. Y. Fung, and T. Takahashi. 1990. Observational constraints on the global atmospheric CO_2 budget. *Science* 247:1431–1438.

Trenberth, K. E. (ed.). 1992. *Climate System Modeling*. Cambridge University Press. Cambridge. 788 pp.

Tyndall, J. 1861. On the absorption and radiation of heat by gases and vapours, and on the physical connexion of radiation, absorption and conduction—The Bakerian Lecture. *Philosophical Magazine* and *Journal of Science, Series 4*, 22(146):169–194 and 22(147):273–285.

Tyndall, J. 1863. On radiation through the Earth's atmosphere. *Philosophical Magazine* and *Journal of Science, Series 4*, 25(167):200–206.

Vinnikov, K. Ya., P. Ya. Groisman, and K. M. Lugina. 1990. Empirical data on contemporary global climate changes (temperature and precipitation). *Journal of Climate* 3:662–677.

Watson, R. T. 1986. Atmospheric ozone. In J. G. Titus (ed.). *Effects of Changes in Stratospheric Ozone and Global Climate*. Vol. 1: Overview. U.S. Environmental Protection Agency and United Nations Environment Programme. Washington, DC. pp. 69–82.

Watterson, I. G., M. R. Dix, H. B. Gordon, and J. L. McGregor. 1995. The CSIRO 9-level Atmospheric General Circulation Model and its equilibrium present and doubled CO_2 climates. *Australian Meterological Magazine* 44:111–125.

Wetherald, R. T. and S. Manabe. 1988. Cloud feedback processes in a general circulation model. *Journal of Atmospheric Science* 45:1397–1415.

Whetton, P. H., A. M. Fowler, M. R. Haylock, and A. B. Pittock. 1993. Implications of climate change due to the enhanced greenhouse effect on floods and droughts in Australia. *Climatic Change* 25:289–317.

Wilson, C. A. and J. F. B. Mitchell. 1987. A doubled CO_2 climate sensitivity experiment with a global climate model including a simple ocean. *Journal of Geophysical Research* 92:13315–13343.

Chapter 2   Agricultural Emissions of Greenhouse Gases

Anastasi, C., M. Dowding, and V. J. Simpson. 1992. Future CH_4 emissions from rice production. *Journal of Geophysical Research* 97:7521–7525.

Anderson, I. C., J. S. Levine, M. A. Poth, and P. J. Riggan. 1988. Enhanced biogenic emissions of nitric oxide and nitrous oxide following surface biomass burning. *Journal of Geophysical Research* 93:3893–3898.

Andreae, M. O., E. V. Browell, M. Garstang, G. L. Gregory, R. C. Harriss, G. F. Hill, D. J. Jacob, M. C. Pereira, G. W. Sachse, A. W. Setzer, P. L. Silva Dias, R. W. Talbot, A. L. Torres, and S. C. Wofsy. 1988. Biomass-burning emissions and associated haze layers over Amazonia. *Journal of Geophysical Research* 93:1509–1527.

Bergen, W. G. and D. B. Bates. 1984. Ionophores: Their effect on production efficiency and mode of action. *Journal of Animal Science* 58:1465–1483.

Blaxter, K. L. and J. L. Clapperton. 1965. Prediction of the amount of methane produced by ruminants. *British Journal of Nutrition* 19:511–522.

Bouwman, A. F. (ed.). 1990. *Soils and the Greenhouse Effect.* Wiley. Chichester. 575 pp.

Bouwman, A. F. 1991. Agronomic aspects of wetland rice cultivation and associated methane emissions. *Biogeochemistry* 15:65–88.

Brady, N. C. 1990. *The Nature and Properties of Soils.* 10th ed. Macmillan. New York. 621 pp.

Bremner, J. M. and A. M. Blackmer. 1978. Nitrous oxide: Emission from soils during nitrification of fertilizer nitrogen. *Science* 199:295–296.

Burke, L. M. and D. A. Lashof. 1990. Greenhouse gas emissions related to agriculture and land-use practices. In B. A. Kimball, N. J. Rosenberg, and L. H. Allen, Jr. (eds.). *Impact of Carbon Dioxide, Trace Gases, and Climate Change on Global Agriculture.* American Society of Agronomy Special Publication No. 53. Madison, WI. pp. 27–43.

Byrnes, B. H., C. B. Christianson, L. S. Holt, and E. R. Austin. 1990. Nitrous oxide emissions from the nitrification of nitrogen fertilizers. In A. F. Bouwman (ed.). 1990. *Soils and the Greenhouse Effect.* Wiley. Chichester, UK. pp. 489–495.

Cicerone, R. J. and J. D. Shetter. 1981. Sources of atmospheric methane: Measurements in rice paddies and a discussion. *Journal of Geophysical Research* 86:7203–7209.

Cicerone, R. J., J. D. Shetter, and C. C. Delwiche. 1983. Seasonal variation of methane flux from a California rice paddy. *Journal of Geophysical Research* 88:11,022–11,024.

Cicerone, R. J. and R. S. Oremland. 1988. Biogeochemical aspects of atmospheric methane. *Global Biogeochemical Cycles* 2:299–327.

Council for Agricultural Science and Technology. 1992. *Preparing U.S. Agriculture for Global Climate Change.* Task Force Report No. 119. Ames, IA. 93 pp.

Coyne, M. S., R. A. Gilfillen, and R. L. Blevins. 1994. Nitrous oxide flux from poultry-manured erosion plots and grass filters after simulated rain. *Journal of Environmental Quality* 23:831–834.

Crutzen, P. J., L. E. Heidt, J. P. Krasnec, W. H. Pollock, and W. Seiler. 1979. Biomass burning as a source of atmospheric gases CO, H_2, N_2O, NO, CH_3Cl, and COS. *Nature* 282:253–256.

Crutzen, P. J., I. Aselmann, and W. Seiler. 1986. Methane production by domestic animals, wild ruminants, other herbivorous fauna, and humans. *Tellus* 38B:271–284.

Dixon, R. K. and K. Andrasko. 1992. Integrated systems: Assessment of promising alternative land-use practices to enhance carbon conservation and sequestration. Proceedings of a Workshop on Assessing Technologies and Management Systems for Agriculture and Forestry in Relation to Global Climate Change. Canberra, Australia, January 20–23, 1992. Intergovernmental Panel on Climate Change Response Strategies Working Group. World Meteorological Organization and United Nations Environment Programme.

Dixon, R. K., K. J. Andrasko, F. G. Sussman, M. A. Lavinson, M. C. Trexler, and T. S. Vinson. 1993. Forest sector carbon offset projects: Near-term opportunities to mitigate greenhouse gas emissions. *Water, Air and Soil Pollution* 70:561–577.

Doorenbos, J. and A. H. Kassam. 1979. *Yield Response to Water*. Food and Agriculture Organization of the United Nations. Rome. 193 pp.

Duxbury, J. M., L. A. Harper, and A. R. Mosier. 1993. Contributions of agroecosystems to global climate change. In L. A. Harper, A. R. Mosier, J. M. Duxbury, and D. Rolston (eds.). *Agricultural Ecosystem Effects on Trace Gases and Global Climate Change*. American Society of Agronomy Special Publication No. 55. Madison, WI. pp. 1–18.

Eichner, M. J. 1990. Nitrous oxide emissions from fertilized soils: Summary of available data. *Journal of Environmental Quality* 19:272–280.

Ehrlich, A. 1990. Agricultural Contributions to Global Warming. In J. Leggett (ed.). *Global Warming: The Greenpeace Report*. Oxford University Press. Oxford. pp. 400–420.

FAO. 1993. *AGROSTAT-PC*. Computerized Information Series: Statistics. Food and Agriculture Organization of the United Nations. Rome.

Hao, W. M., D. Scharffe, P. J. Crutzen, and E. Sanhueza. 1988. Production of N_2O, CH_4, and CO_2 from soils in the tropical savanna during the dry season. *Journal of Atmospheric Chemistry* 7:93–105.

Harriss, R. C., D. I. Sebacher, K. B. Bartlett, D. S. Bartlett, and P. M. Crill. 1988. Sources of atmospheric methane in the south Florida environment. *Global Biogeochemical Cycles* 2:231–243.

Holzapfel-Pschorn, A. and W. Seiler. 1986. Methane emission during a cultivation period from an Italian rice paddy. *Journal of Geophysical Research* 91:11,803–11,814.

Houghton, R. A., J. E. Hobbie, J. M. Melillo, B. Moore, B. J. Peterson, G. R. Shaver, and G. M. Woodwell. 1983. Changes in the carbon content of terrestrial biota and soils between 1860 and 1980: A net release of CO_2 to the atmosphere. *Ecological Monographs* 53(3):235–262.

Houghton, R. A., R. D. Boone, J. R. Fruci, J. E. Hobbie, J. M. Melillo, C. A. Palm, B. J. Peterson, G. R. Shaver, G. M. Woodwell, B. Moore, D. L. Skole, and N. Myers. 1987. The flux of carbon from terrestrial ecosystems to the atmosphere in 1980 due to changes in land use: Geographic distribution of the global flux. *Tellus* 39B:122–139.

IPCC. 1990. *Climate Change: The IPCC Scientific Assessment*. J. T. Houghton, G. J. Jenkins, and J. J. Ephraums (eds.). Intergovernmental Panel on Climate Change. Cambridge University Press. Cambridge. 365 pp.

IPCC. 1992. *Climate Change 1992: The Supplementary Report to the IPCC Scientific Assessment*. J. T. Houghton, B. A. Callander, and S. K. Varney (eds.). Intergovernmental Panel on Climate Change. Cambridge University Press. Cambridge. 200 pp.

IPCC. 1995. *Greenhouse Gas Inventory Reporting Instructions*. Vol. 2: *Greenhouse Gas Inventory Workbook*. United Nations Environment Programme, Organisation for Economic Cooperation and Development, International Energy Agency, and Intergovernmental Panel on Climate Change. IPCC WG1 Technical Support Unit. Bracknell, UK.

IPCC. 1995. *Climate Change 1994: Radiative Forcing of Climate Change and an Evaluation of the IPCC IS92 Emission Scenarios*. J. T. Houghton, L. G. Meira Filho, J. Bruce, Hoesung Lee, B. A. Callander, E. Haites, N. Harris, and K. Maskell (eds.). Intergovernmental Panel on Climate Change. Cambridge University Press. Cambridge. 339 pp.

IPCC. 1996. *Climate Change 1995: Impacts, Adaptations and Mitigation of Climate Change: Scientific-Technical Analyses*. R. T. Watson, M. C. Zinyowera, and R. H. Moss (eds.). Intergovernmental Panel on Climate Change. Cambridge University Press. Cambridge. 879 pp.

Jackson, R. B., IV. 1992. On estimating agriculture's net contribution to atmospheric carbon. *Water, Air, and Soil Pollution* 64:121–137.

Jackson, R. B., IV. 1993. Greenhouse gases and agriculture. In R. A. Geyer (ed.). *A Global Warming Forum: Scientific, Economic, and Legal Overview*. CRC Press. Boca Raton, FL. pp. 417–444.

Keller, M., W. A. Kaplan, and S. C. Wofsy. 1986. Emissions of N_2O, CH_4 and CO_2 from tropical forest soils. *Journal of Geophysical Research* 91:11791–11082.

Keller, M., W. A. Kaplan, S. C. Wofsy, and J. M. Da Costa. 1988. Emissions of N_2O from tropical forest soils: Response to fertilization with NH_4^+, NO_3^-, and PO_4^{3-}. *Journal of Geophysical Research* 93:1600–1604.

Khalil, M. A. K., R. A. Rasmussen, M.-X. Wang, and L. Ren. 1991. Methane emissions from rice fields in China. *Environment Science and Technology* 25:979–981.

Knowles, R. 1993. Methane: Processes of production and consumption. In L. A. Harper, A. R. Mosier, J. M. Duxbury, and D. Rolston (eds.). *Agricultural Ecosystem Effects on Trace Gases and Global Climate Change.* American Society of Agronomy Special Publication No. 55. Madison, WI. pp. 145–156.

Lashof, D. A. and D. Tirpak (eds.). 1990. *Policy Options for Stabilizing Global Climate.* Report to Congress. US Environmental Protection Agency. Office of Policy, Planning, and Evaluation. 21P-2003.1. Washington, DC.

Lerner, J., E. Matthews, and I. Fung. 1988. Methane emission from animals: A global high-resolution database. *Global Biogeochemical Cycles* 2:139–156.

Lindau, C. W., W. H. Patrick, Jr., and R. D. DeLaune. 1993. Factors affecting methane production in flooded rice soils. In L. A. Harper, A. R. Mosier, J. M. Duxbury, and D. Rolston (eds.). *Agricultural Ecosystem Effects on Trace Gases and Global Climate Change.* American Society of Agronomy Special Publication No. 55. Madison, WI. pp. 157–165.

Luizão, F., P. Matson, G. Livingston, R. Luizão, and P. Vitousek. 1989. Nitrous oxide flux following tropical land clearing. *Global Biogeochemical Cycles* 3:281–285.

Matthews, E., I. Fung, and J. Lerner. 1991. Methane emission from rice cultivation: Geographic and seasonal distribution of cultivated areas and emissions. *Global Biogeochemical Cycles* 5:3–24.

Matson, P. A. and P. M. Vitousek. 1987. Cross-system comparisons of soil nitrogen transformations and nitrous oxide flux in tropical forest ecosystems. *Global Biogeochemical Cycles* 1:163–170.

Mosier, A., D. Schimel, D. Valentine, K. Bronson, and W. Parton. 1991. Methane and nitrous oxide fluxes in native, fertilized and cultivated grasslands. *Nature* 350:330–332.

Neue, H. U., P. Becker-Heidmann, and H. W. Scharpenseel. 1990. Organic matter dynamics, soil properties, and cultural practices in rice lands and their relationship to methane production. In A. F. Bouwman (ed.). 1990. *Soils and the Greenhouse Effect.* Wiley. Chichester, UK. pp. 457–466.

Patrick, W. H., Jr. and R. D. DeLaune. 1977. Chemical and biological redox systems affecting nutrient availability in the coastal wetlands. *Geoscience and Man* 18:131–137.

Prinn, R., D. Cunnold, R. Rasmussen, P. Simmonds, F. Alyea, A. Crawford, P. Fraser, and R. Rosen. 1990. Atmospheric emissions and trends of nitrous oxide deduced from 10 years of ALE-GAGE data. *Journal of Geophysical Research* 95:18369–18385.

Sahrawat, K. L. and D. R. Keeney. 1986. Nitrous oxide emission from soils. *Advances in Soil Science* 4:103–148.

Sampson, R. N. and T. E. Hamilton. 1992. Forestry opportunities in the United States to mitigate the effects of global warming. In R. N. Sampson and D. Hair (eds.). *Forests and Global Change. Volume One: Opportunities for Increasing Forest Cover.* American Forests. Washington, DC. pp. 231–245.

Schlesinger, W. H. 1986. Changes in soil carbon storage and associated properties with disturbance and recovery. In J. R. Trabalka, and D. E. Reichle (eds.). *The Changing Carbon Cycle: A Global Analysis.* Springer-Verlag. New York. pp. 194–220.

Schlesinger, W. H. 1991. *Biogeochemistry: An Analysis of Global Change.* Academic Press. San Diego. 443 pp.

Schütz, H., W. Seiler, and H. Rennenberg. 1990. Soil and land use related sources and sinks of methane (CH_4) in the context of the global methane budget. In A. F. Bouwman (ed.). 1990. *Soils and the Greenhouse Effect.* Wiley. Chichester, UK. pp. 269–285.

Sedjo, R. A. 1989. Forests: A tool to moderate global warming? *Environment* 31(1):14–20.

Seiler, W. and R. Conrad. 1987. Contribution of tropical ecosystems to the global budgets of trace gases, especially CH_4, H_2, CO, and N_2O. In R. E. Dickinson (ed.). *The Geophysiology of Amazonia*. Wiley. New York. pp. 133–160.

Singh, S. and R. Prasad. 1985. Studies on the nitrification inhibitor, dicyandiamide (DCD), for increasing the efficiency of nitrogen applied to rice. *Journal of Agricultural Science, Cambridge* 104:425–428.

Soil Survey Staff. 1992. *Keys to Soil Taxonomy*. 5th ed. Agency for International Development, U.S. Department of Agriculture, Soil Conservation Service, and Soil Management Support Services. SMSS Technical Monograph No. 19. Pocahontas Press. Blackburg, VA. 541 pp.

Sprent, J. I. 1987. *The Ecology of the Nitrogen Cycle*. Cambridge University Press. Cambridge. 151 pp.

Takai, Y. and E. Wada. 1990. Methane formation in waterlogged paddy soils and its controlling factors. In H. W. Scharpenseel, M. Schomaker, and A. Ayoub (eds.). *Soils on a Warmer Earth*. Elsevier. Amsterdam. pp. 101–107.

Trexler, M. C., C. A. Haugen, and L. A. Loewen. 1992. Global warming mitigation through forestry options in the tropics. In R. N. Sampson and D. Hair (ed.). *Forests and Global Change. Volume One: Opportunities for Increasing Forest Cover*. American Forests. Washington, DC. pp. 73–96.

Watanabe, I. 1984. Anaerobic decomposition of organic matter in flooded rice soils. In *Organic Matter and Rice*. International Rice Research Institute. Manila, Philippines. pp. 237–258.

Wilson, A. T. 1978. Pioneer agriculture explosion and CO_2 levels in the atmosphere. *Nature* 273:40–41.

Yagi, K. and K. Minami. 1991. Emission and production of methane in the paddy fields of Japan. *Japan Agricultural Research Quarterly* 25:165–171.

Chapter 3   Carbon Dioxide, Climate Change, and Crop Yields

Acock, B. and L. H. Allen, Jr. 1985. Crop responses to elevated CO_2 concentrations. In B. R. Strain and J. D. Cure (eds.). *Direct Effects of Increasing CO_2 on Vegetation*. DOE/ER-0238. U.S. Department of Energy. Washington, DC. pp. 53–97.

Acock, B. 1991. Modeling canopy photosynthetic response to carbon dioxide, light interception, temperature, and leaf traits. In K. J. Boote, and R. S. Loomis (eds.). *Modeling Crop Photosynthesis—From Biochemistry to Canopy*. Crop Science Society of America Special Publication No. 19. Madison, WI. pp. 41–55.

Acock, B. and M. C. Acock. 1993. Modeling approaches for predicting crop ecosystem responses to climate change. In D. R. Buxton, R. Shibles, R. A. Forsberg, B. L. Blad, K. H. Asay, G. M. Paulsen, and R. F. Wilson (eds.). *International Crop Science I*. Crop Science Society of America. Madison, WI. pp. 299–306.

Akita, S. and D. N. Moss. 1973. Photosynthetic responses to CO_2 and light by maize and wheat leaves adjusted for constant stomatal apertures. *Crop Science* 13:234–237.

Allen, L. H., Jr., P. Jones, and J. W. Jones. 1985. Rising atmospheric CO_2 and evapotranspiration. In *Advances in Evapotranspiration*. Proceedings of the National Conference on Advances in Evapotranspiration. December 16–17, 1985. American Society of Agricultural Engineers. St. Joseph, MI. pp. 13–27.

Allen, L. H., Jr. 1990. Plant responses to rising carbon dioxide and potential interactions with air pollutants. *Journal of Environmental Quality* 19:15–34.

Allen, S. G., S. B. Idso, B. A. Kimball, J. T. Baker, L. H. Allen, Jr., J. R. Mauney, J. W. Radin, and M. G. Anderson. 1990. *Effects of Air Temperature on Atmospheric CO_2—Plant Growth Relationships*. DOE/ER-0450T. U.S. Department of Energy. Washington, DC. 61 pp.

Amthor, J. S. 1989. *Respiration and Crop Productivity*. Springer-Verlag. New York. 215 pp.

Arp, W. J. 1991. Effects of source-sink relations on photosynthetic acclimation to elevated CO_2. *Plant, Cell and Environment* 14:869–875.

Baker, J. T., L. H. Allen, Jr., and K. J. Boote. 1992. Response of rice to carbon dioxide and temperature. *Agricultural and Forest Meteorology* 60:153–166.

Bauer, A. 1972. *Effect of Water Supply and Seasonal Distribution on Spring Wheat Yields.* Bulletin 490. North Dakota Agricultural Experiment Station. Fargo, ND.

Bazzaz, F. A. 1990. The response of natural ecosystems to the rising global CO_2 levels. *Annual Review of Ecology and Systematics* 21:167–196.

Bazzaz, F. A. and E. D. Fajer. 1992. Plant life in a CO_2-rich world. *Scientific American* January pp. 68–74.

Butterfield, R. E. and J. I. L. Morison. 1992. Modeling the impact of climatic warming on winter cereal development. *Agriculture and Forest Meteorology* 62:241–261.

Caprio, J. M. 1984. *Study to Improve Winterkill Parameters for a Winter Wheat Model.* Montana State University. Boseman, MT. NASA. Houston, TX. 120 pp.

Chaudhuri, U. N., R. B. Burnett, E. T. Kanemasu, and M. B. Kirkham. 1986. *Effect of Elevated Levels of CO_2 on Winter Wheat Under Two Moisture Regimes.* Response of Vegetation to Carbon Dioxide. 029. U.S. Department of Energy. Washington, DC. 71 pp.

Chaudhuri, U. N., E. T. Kanemasu, and M. B. Kirkham. 1989. *Effect of Elevated Levels of CO_2 on Winter Wheat Under Two Moisture Regimes.* Response of Vegetation to Carbon Dioxide. 050. U.S. Department of Energy. Washington, DC. 49 pp.

Conway, G. R. and J. N. Pretty. 1991. *Unwelcome Harvest: Agriculture and Pollution.* Earthscan. London.

Cure, J. D. and B. Acock. 1986. Crops responses to carbon dioxide doubling: A literature survey. *Agriculture and Forest Meteorology* 38:127–145.

Cure, J. D., D. W. Israel, and T. W. Rufty, Jr. 1988. Nitrogen stress effects on growth and seed yield of nonnodulated soybean exposed to elevated carbon dioxide. *Crop Science* 28:671–677.

Drake, B. G., H. H. Rogers, and L. H. Allen, Jr. 1985. Methods of exposing plants to elevated CO_2 concentrations. In B. R. Strain and J. D. Cure (eds.). *Direct Effects of Increasing CO_2 on Vegetation.* DOE/ER-0238. U.S. Department of Energy. Washington, DC.

Drake, B. G. and P. W. Leadley. 1991. Canopy photosynthesis of crops and native plant communities exposed to long-term elevated CO_2. *Plant, Cell and Environment* 14:853–860.

Ellis, R. H., P. Hadley, E. H. Roberts, and R. J. Summerfield. 1990. Quantitative relations between temperature and crop development and growth. In M. T. Jackson, B. V. Ford-Lloyd, and M. L. Parry. (eds). *Climatic Change and Plant Genetic Resources.* Belhaven Press. London. pp. 85–115.

Enoch, H. Z. and B. Kimball. 1986. CO_2 *Enrichment of Greenhouse Crops.* Vols. I and II. CRC Press. Boca Raton, FL.

Fitter, A. H. and R. K. M. Hay. 1987. *Environmental Physiology of Plants.* Second Edition. Academic Press. London. 423 pp.

Gates, D. M. 1980. *Biophysical Ecology.* Springer-Verlag. New York. 611 pp.

Gifford, R. M. 1979. Growth and yield of carbon dioxide-enriched wheat under water-limited conditions. *Australian Journal of Plant Physiology* 6:367–378.

Gifford, R. M. and J. I. L. Morison. 1993. Crop responses to the global increase in atmospheric CO_2 concentration. In *International Crop Science I.* Crop Science Society of America. Madison, WI. pp. 325–331.

Gleick, P. H. 1987. Regional hydrologic consequences of increases in atmospheric CO_2 and other trace gases. *Climatic Change* 10:137–161.

Hale, M. G. and D. M. Orcutt. 1987. *The Physiology of Plants Under Stress.* Wiley. New York. 206 pp.

Hall, A. E. and L. H. Allen, Jr. 1993. Designing cultivars for the climatic conditions of the next century. In *International Crop Science I.* Crop Science Society of America. Madison, WI. pp. 291–297.

Hansen, J., M. Sato, and R. Ruedy. 1995. Long-term changes of diurnal temperature cycle: Implications about mechanisms of global climate change. *Atmospheric Research* 37:175–209.

Hendrey, G. R. 1993. *Free-air CO_2 Enrichment for Plant Research in the Field.* C. K. Smoley. Boca Raton, FL. 308 pp.

Hodges, T. (ed.). 1991. *Predicting Crop Phenology.* CRC Press. Boca Raton, FL. 233 pp.

Idso, S. B., B. A. Kimball, and J. R. Mauney. 1987. Atmospheric carbon dioxide enrichment effects on cotton midday foliage temperature: Implications for plant water use and crop yield. *Agronomy Journal* 79:667–672.

Jarvis, P. G. and K. G. McNaughton. 1986. Stomatal control of transpiration: Scaling up from leaf to region. *Advances in Ecological Research* 15:1–49.

Karl, T. R., G. Kukla, V. N. Razuvayev, M. J. Changery, R. G. Quayle, R. R. Heim, Jr., D. R. Easterling, and C. B. Fu. 1991. Global warming: Evidence for asymmetric diurnal temperature change. *Geophysical Research Letters* 18:2253–2256.

Katz, R. W. and B. G. Brown. 1992. Extreme events in a changing climate: Variability is more important than averages. *Climatic Change* 21:289–302.

Kimball, B. A. 1983. Carbon dioxide and agricultural yield: An assemblage and analysis of 430 prior observations. *Agronomy Journal* 75:779–788.

Kimball, B. A. and S. B. Idso. 1983. Increasing atmospheric CO_2: Effects on crop yield, water use, and climate. *Agricultural Water Management* 7:55–72.

Kramer, P. J. 1983. *Water Relations of Plants.* Academic Press. San Diego.

Lawlor, D. W. and R. A. C. Mitchell. 1991. The effects of increasing CO_2 on crop photosynthesis and productivity: A review of field studies. *Plant, Cell and Environment* 14:807–818.

Lemon, E. R. 1983. *CO_2 and Plants: The Response of Plants to Rising Levels of Atmospheric Carbon Dioxide.* Westview Press. Boulder, CO. 280 pp.

Lincoln, D. E., N. Sionit, and B. R. Strain. 1984. Growth and feeding response of *Pseudoplusia includens* (Lepidoptera: Noctuidae) to host plants grown in controlled carbon dioxide atmospheres. *Environmental Entomology* 13:1527–1530.

Mearns, L. O., R. W. Katz, and S. H. Schneider. 1984. Extreme high temperature events: Changes in their probabilities with changes in mean temperature. *Journal of Climate and Applied Meteorology* 23:1601–1613.

Mearns, L. O., C. Rosenzweig, and R. Goldberg. 1992. Effect of changes in interannual climatic variability on CERES-Wheat yields: Sensitivity and $2\times CO_2$ general circulation model studies. *Agricultural and Forest Meteorology* 62:159–189.

Mearns, L. O., C. Rosenzweig, and R. Goldberg. 1996. The effect of changes in daily and interannual climatic variability on CERES-Wheat: A sensitivity study. *Climatic Change* 32:257–292.

Monteith, J. L. 1981. Climatic variation and the growth of crops. *Quarterly Journal of the Royal Meteorological Society* 107:749–774.

Morison, J. I. L. 1985. Sensitivity of stomata and water use efficiency to high CO_2. *Plant, Cell and Environment* 8:467–474.

Norman, J. M. and T. J. Arkebauer. 1991. Predicting canopy photosynthesis and light-use efficiency from leaf characteristics. In K. J. Boote and R. S. Loomis (eds.). *Modeling Crop Photosynthesis — From Biochemistry to Canopy.* Crop Science Society of America. CSSA Special Publication No. 19. Madison, WI. pp. 75–94.

Patterson, D. T. and E. P. Flint. 1990. Implications of increasing CO_2 and climate change for plant communities and competition in natural and managed ecosystems. In B. A. Kimball, N. J. Rosenberg, and L. H. Allen, Jr. (eds.). *Impact of CO_2, Trace Gases, and Climate*

Change on Global Agriculture. ASA Special Publication No. 53. American Society of Agronomy. Madison, WI. pp. 83–110.

Paulsen, G. M. 1994. High temperature responses of crop plants. In K. J. Boote, J. M. Bennett, T. R. Sinclair, and G. M Paulsen (eds.). *Physiology and Determination of Crop Yield.* American Society of Agronomy. Madison, WI. pp. 365–389.

Poorter, H. 1993. Interspecific variation in the growth response of plants to an elevated ambient CO_2 concentration. *Vegetatio* 104/105:77–97.

Prior, S. A., H. H. Rogers, G. B. Runion, B. A. Kimball, J. R. Mauney, K. F. Lewin, J. Nagy, and G. R. Hendrey. 1995. Free-air carbon dioxide enrichment of cotton: Root morphological characteristics. *Journal of Environmental Quality* 24:678–683.

Ramirez, J., C. M. Sakamoto, and R. E. Jensen. 1975. *Impacts of Climate Change on the Biosphere.* Climate Impact Assessment Program (CIAP), Monograph 5, Part 2, Climatic Effects. U.S. Department of Transportation. Washington, DC.

Reddy, K. R., H. F. Hodges, and J. M. McKinion. 1995. Carbon dioxide and temperature effects on pima cotton growth. *Agriculture Ecosystems and Environment* 54:17–29.

Rind, D., R. Goldberg, J. Hansen, C. Rosenzweig, and R. Ruedy. 1990. Potential evapotranspiration and the likelihood of future drought. *Journal of Geophysical Research* 95:9983–10,004.

Rogers, H. H., G. B. Runion, and S. V. Krupa. 1994. Plant responses to atmospheric CO_2 enrichment with emphasis on roots and the rhizosphere. *Environmental Pollution* 83:155–189.

Rosenzweig, C. 1985. Potential CO_2-induced climate effects on North American wheat-producing regions. *Climatic Change* 7:367–389.

Rosenzweig, C. 1990. Crop response to climate change in the Southern Great Plains: A simulation study. *Professional Geographer* 42:20–37.

Rosenzweig, C. and D. Hillel. 1993. Agriculture in a greenhouse world. *Research and Exploration* 9:208–221.

Rosenzweig, C. and F. N. Tubiello. 1996. Effects of changes in minimum and maximum temperature on wheat yields in the central U.S.: A simulation study. *Agricultural and Forest Meteorology* 80:215–230.

Schonfeld, R., C. Johnson, and D. M. Ferris. 1989. Development of winter wheat under increased atmospheric CO_2 and water limitation at tillering. *Crop Science* 29:1083–1086.

Shibles, R. M., I. C. Anderson, and A. H. Gibson. 1975. Soybean. In L. T. Evans (ed.). *Crop Physiology.* Cambridge University Press. London. pp. 151–189.

Sionit, N., D. A. Mortensen, B. R. Strain, and H. Hellmers. 1981. Growth response of wheat to CO_2 enrichment and different levels of mineral nutrition. *Agronomy Journal* 73:1023–1027.

Sionit, N. 1983. Response of soybean to two levels of mineral nutrition in CO_2-enriched atmosphere. *Crop Science* 23:329–333.

Stitt, M. 1991. Rising CO_2 levels and their potential significance for carbon flow in photosynthetic cells. *Plant, Cell and Environment* 14:741–762.

Stooksbury, D. E. and P. J. Michaels. 1994. Climate change and large area corn yield in the Southeastern United States. *Agronomy Journal* 86:564–569.

Tolbert, N. E. and I. Zelitch. 1983. Carbon metabolism. In E. R. Lemon (ed.). *CO_2 and Plants: The Response of Plants to Rising Levels of Atmospheric Carbon Dioxide.* Westview Press. Boulder, CO. pp. 21–64.

Waggoner, P. E. 1986. How changed weather might change American agriculture. In J. G. Titus (ed.). *Effects of Changes in Stratospheric Ozone and Global Climate.* Vol. 3. U.S. Environmental Protection Agency and United Nations Environment Programme. Washington, DC. pp. 59–71.

Warrick, R. A. 1984. The possible impacts on wheat production of a recurrence of the 1930s drought in the U.S. Great Plains. *Climatic Change* 6:5–26.

Chapter 4 Effects on Weeds, Insects, and Diseases

Bazzaz, F. A. and K. Garbutt. 1988. The response of annuals in competitive neighborhoods: Effects of elevated CO_2. *Ecology* 69:937–946.

Boag, B., J. W. Crawford, and R. Neilson. 1991. The effect of potential climatic changes on the geographical distribution of the plant-parasitic nematodes *Xiphinema* and *Longidorus* in Europe. *Nematologica* 37:312–323.

Cammel, M. E. and J. D. Knight. 1991. Effects of climate change on the population dynamics of crop pests. *Advanced Ecological Research* 22:117–162.

Carter, D. R. and K. M. Peterson. 1983. Effects of a CO_2-enriched atmosphere on the growth and competitive interaction of a C3 and C4 grass. *Oecologia* 58:188–193.

Dennis, R. L. H. and T. G. Shreeve. 1991. Climatic change and the British butterfly fauna: Opportunities and constraints. *Biological Conservation* 55:1–16.

Elias, S. A. 1991. Insects and climate change: Fossil evidence from the Rocky Mountains. *BioScience* 41:552–559.

Fielding, D. J. and M. A. Brusven. 1990. Historical analysis of grasshopper (Orthoptera: Adrididae) population responses to climate in southern Idaho, 1950–1980. *Environmental Entomology* 19:1786–1791.

Flint, E. P. and D. T. Patterson. 1983. Interference and temperature effects on growth in soybean *(Glycine max)* and associated C3 and C4 weeds. *Weed Science* 31:193–199.

Flint, E. P., D. T. Patterson, D. A. Mortensen, G. H. Riechers, and J. L. Beyers. 1984. Temperature effects on growth and leaf production in three weed species. *Weed Science* 32:655–663.

Furtick, W. R. 1978. Weeds and world food production. In D. Pimentel (ed.). *World Food, Pest Losses, and the Environment.* AAAS Selected Symposium 13. Westview Press. Boulder, CO. pp. 51–62.

Holm, L. G., D. L. Plucknett, J. V. Pancho, and J. P. Herberger. 1977. *The World's Worst Weeds: Distribution and Biology.* East-West Center. University Press of Hawaii. Honolulu.

IPCC. 1996. Agriculture in a Changing Climate: Impacts and Adaptation. In *Climate Change 1995. Impacts, Adaptations and Mitigation of Climate Change: Scientific-Technical Analyses.* R. T. Watson, M. C. Zinyowera, and R. H. Moss (eds.). Contribution of Working Group II to the Second Assessment Report of the Intergovernmental Panel on Climate Change. Cambridge University Press. Cambridge. pp. 427–467.

Kimball, B. A. 1985. Adaptation of vegetation and management practices to a higher carbon dioxide world. In B. R. Strain and J. D. Cure (eds.). *Direct Effects of Increasing Carbon Dioxide on Vegetation.* U.S. Department of Energy. DOE/ER-0238. Washington, DC. pp. 185–204.

Lincoln, D. E., N. Sionit, and B. R. Strain. 1984. Growth and feeding response of *Pseudoplusia includens* (Lepidoptera: Noctuidae) to host plants grown in controlled carbon dioxide atmospheres. *Environmental Entomology* 13:1527–1530.

Marks, S. and K. Clay. 1990. Effects of CO_2 enrichment, nutrient addition and fungal endophyte-infection on the growth of two grasses. *Oecologia* 84:207–214.

Mattson, W. J. and R. J. Haack. 1987. The role of drought in outbreaks of plant-eating insects. *BioScience* 37(2):110–118.

Morison, J. I. L. and R. A. Spence. 1989. Warm spring conditions and aphid infestations. *Weather* 44(9):374–380.

Overdieck, D., D. Bossemeyer, and H. Lieth. 1984. Long-term effects of an increased CO_2 concentration level on terrestrial plants in model-ecosystem: I. Phytomass production and competition of *Trifolium repens* L. and *Lolium perenne* L. *Progress in Biometeorology* 3:344–352.

Overdieck, D. and F. Reining. 1986. Effect of atmospheric CO_2 enrichment on perennial ryegrass (*Lolium perenne* L.) and white clover (*Trifolium repens* L.) competing in managed model-ecosystems. *Acta Œcologica/Œcologia Plantarum* 7:357–366.

Parker, C. and J. D. Fryer. 1975. Weed control problems causing major reductions in world food supplies. FAO Plant Protection Bulletin 23:83–95.

Patterson, D. T. 1993. Implications of global climate change for impact of weeds, insects, and plant diseases. In *International Crop Science I*. Crop Science Society of America. Madison, WI. pp. 273–280.

Patterson, D. T. and E. P. Flint. 1990. Implications of increasing carbon dioxide and climate change for plant communities and competition in natural and managed ecosystems. In B. A. Kimball, N. J. Rosenberg, and L. H. Allen, Jr. (eds.). *Impact of Carbon Dioxide, Trace Gases, and Climate Change on Global Agriculture*. American Society of Agronomy. ASA Special Publication No. 53. Madison, WI. pp. 83–110.

Pedgley, D. E. 1989. Weather and the current desert locust plague. *Weather* 44(4):168–171.

Porter, J. H., M. L. Parry, and T. R. Carter. 1991. The potential effects of climatic change on agricultural insect pests. *Agricultural and Forest Meteorology* 57:221–240.

Prescott-Allen, R. and C. Prescott-Allen. 1990. How many plants feed the world? *Conservation Biology* 4:365–374.

Salt, D. T., B. L. Brooks, and J. B. Whittaker. 1995. Elevated carbon dioxide affects leaf-miner performance and plant growth in docks (*Rumex* spp.). *Global Change Biology* 1:153–156.

Sasek, T. W. and B. R. Strain. 1990. Implications of atmospheric CO_2 enrichment and climate change for the geographical distribution of two introduced vines in the U.S.A. *Climatic Change* 16:31–51.

Sparks, A. N. (ed.). 1986. *Long-Range Migration of Moths of Agronomic Importance to the United States and Canada: Specific Examples of Occurrence and Synoptic Weather Patterns Conducive to Migration*. U.S. Department of Agriculture, Agricultural Research Service ARS-43. National Technical Information Service. Springfield, VA. 104 pp.

Stinner, B. R., R. A. J. Taylor, R. B. Hammond, F. F. Purrington, D. A. McCartney, N. Rodenhouse, and G. W. Barrett. 1989. Potential effects of climate change on plant–pest interactions. In J. B. Smith and D. A. Tirpak (eds.). *The Potential Effects of Global Climate Change on the United States*. EPA-230-05-89-053. Appendix C Agriculture Vol. 2. Washington, DC. pp. 8-1–8-35.

Stinner, R. E., K. Wilson, C. Barfield, J. Regniere, A. Riordan, and J. Davis. 1982. Insect movement in the atmosphere. In J. L. Hatfield and I. J. Thomason. (eds.). *Biometeorology in Integrated Pest Management*. Academic Press. New York. pp. 193–209.

Sutherst, R. W., G. F. Maywald, and D. B. Skarrate. 1995. Predicting insect distributions in a changed climate. In R. Harrington and N. E. Stork (eds.). *Insects in a Changing Environment*. Academic Press. London. pp. 59–91.

USDA. 1972. *Extent and Cost of Weed Control with Herbicides and an Evaluation of Important Weeds, 1968*. Economic Research Service, Extension Service, and Agricultural Research Service. ARS-H-1. U.S. Department of Agriculture.

Worner, S. P. 1988. Ecoclimatic assessment of potential establishment of exotic pests. *Journal of Economic Entomology* 81:973–983.

Chapter 5   The Role of Soil Resources

Arnold, R. W. 1990. Processes that affect soil morphology. In H. W. Scharpenseel, M. Schomaker, and A. Ayoub (eds.). *Soils on a Warmer Earth*. Elsevier. Amsterdam. pp. 31–38.

Aveyard, J. M. 1988. *Climatic Changes—Implications for Land Degradation in New South Wales*. Department of Conservation and Land Management, New South Wales. Chatswood, NSW.

Bumb, B. L. and C. A. Baanante. 1996. *The Role of Fertilizer in Sustaining Food Security and Protecting the Environment to 2020*. Food, Agriculture, and the Environment Discussion Paper 17. International Food Policy Resesarch Institute. Washington, DC. 54 pp.

Buol, S. W., P. A. Sanchez, S. B. Weed, and J. M. Kimble. 1990. Predicted impact of climatic warming on soil properties and use. In B. A. Kimball, N. J. Rosenberg, and L. H. Allen, Jr.

(eds.). *Impact of Carbon Dioxide, Trace Gases, and Climate Change on Global Agriculture.* American Society of Agronomy Special Publication No. 53. Madison, WI. pp. 71–82.

Cure, J. D. and B. Acock. 1986. Crop responses to carbon dioxide doubling: A literature survey. *Agricultural and Forest Meteorology* 38:127–145.

Dickinson, R. K. and R. J. Cicerone. 1986. Future global warming from atmospheric trace gases. *Nature* 319:109–115.

FAO. 1978. *Report on the Agro-Ecological Zones Project.* Vol. I: *Methodology and Results for Africa.* Food and Agriculture Organization of the United Nations. Rome.

FAO. 1990. *FAO-Unesco Soil Map of the World. Revised Legend.* Food and Agriculture Organization of the United Nations. Rome. 119 pp.

Friedland, A. J. and A. H. Johnson. 1985. Lead distribution and fluxes in a high-elevation forest in northern Vermont. *Journal of Environmental Quality* 14:332–336.

Hartig, E. K., O. Grozev, and C. Rosenzweig. 1997. Climate change, agriculture and wetlands in Eastern Europe. *Climatic Change* 36:107–121.

Hillel, D. 1980. *Fundamentals of Soil Physics.* Academic Press. New York.

Houghton, R. A. and D. L. Skole. 1990. Carbon. In B. L. Turner, II, W. C. Clark, R. W. Kates, J. F. Richards, J. T. Mathews, and W. B. Meyer (eds.). *The Earth as Transformed by Human Action.* Cambridge University Press. Cambridge. pp. 393–408.

Hudson, N. 1995. *Soil Conservation.* 3rd Ed. Iowa State University Press. Ames.

IGBP. 1989. *Effects of Atmospheric and Climate Change on Terrestrial Ecosystems.* Report of a Workshop Organized by the IGBP Coordinating Panel on Effects of Climate Change on Terrestrial Ecosystems at CSIRO, Division of Wildlife and Ecology, Canberra, Australia 29 February–2 March, 1988. Compiled by B. H. Walker and R. D. Graetz. IGBP Report No. 5. Lidingo Tryckeri KB. Lidingo, Sweden.

Jenkinson, D. S., D. E. Adams, and A. Wild. 1991. Model estimates of CO_2 emissions from soil in response to global warming. *Nature* 351:304–306.

Jenny, H. F. 1941. *Factors of Soil Formation.* McGraw Hill. New York.

Kimball, B. A. 1983. Carbon dioxide and agricultural yield: An assemblage and analysis of 430 prior observations. *Agronomy Journal* 75:779–788.

Kimball, B. 1985. Adaptation of vegetation and management practices to a higher carbon dioxide world. In B. R. Strain and J. D. Cure (eds.). *Direct Effects of Increasing Carbon Dioxide on Vegetation.* U.S. Department of Energy. DOE/ER-0238. Washington, DC. pp. 185–204.

Lal, R. 1995. Global soil erosion by water and carbon dynamics. In R. Lal, J. Kimble, E. Levine, and B. A. Stewart (eds.). *Soils and Global Change.* Advances in Soil Science. CRC Lewis Publishers. Boca Raton, FL. pp. 131–142.

Lal, R., J. Kimble, E. Levine, and C. Whitman. 1995. World soils and greenhouse effect: An overview. In R. Lal, J. Kimble, E. Levine, and B. A. Stewart (eds.). *Soils and Global Change.* Advances in Soil Science. CRC Lewis Publishers. Boca Raton, FL. pp. 1–7.

Lamborg, M. R., R. W. F. Hardy, and E. A. Paul. 1983. Microbial Effects. In E. R. Lemon (ed.). *CO_2 and Plants: The Response of Plants to Rising Levels of Carbon Dioxide.* Westview Press. Boulder, CO. pp. 131–176.

Lee, J. L., D. L. Phillips, and R. F. Dodson. 1996. Sensitivity of the U.S. corn belt to climate change and elevated CO_2: II. Soil erosion and organic carbon. *Agricultural Systems* 52:503–521.

Li, C., S. Frolking, and T. A. Frolking. 1992a. A model of nitrous oxide evolution from soil driven by rainfall events. 1. Model structure and sensitivity. *Journal of Geophysical Research* 97:9759–9776.

Li, C., S. Frolking, and T. A. Frolking. 1992b. A model of nitrous oxide evolution from soil driven by rainfall events. 2: Model applications. *Journal of Geophysical Research* 97:9777–9783.

Li, C., S. Frolking, and R. Harriss. 1994. Modeling carbon biogeochemistry in agricultural soils. *Global Biogeochemical Cycles* 8(3):237–254.

Phillips, D. L., D. White, and C. B. Johnson. 1991. *Climate Change and Soil Erosion in the United States*. U.S. Environmental Protection Agency. Corvallis, OR.

Post, W. M., W. R. Emanuel, P. J. Zinke, and A. G. Stangenberger. 1982. Soil carbon pools and world life zones. *Nature* 298:156–159.

Raich, J. W. and W. H. Schlesinger. 1992. The global carbon dioxide flux in soil respiration and its relationship to vegetation and climate. *Tellus* 44B:81–99.

Russell, E. W. 1973. *Soil Conditions and Plant Growth*. 10th ed. Longman. New York.

Rykbost, K. A., L. Boersma, H. J. Mack, and W. E. Schmisseur. 1975. Yield response to soil warming: Agronomic crops. *Agronomy Journal* 67:733–738.

Schlesinger, W. H. 1986. Changes in soil carbon storage and associated properties with disturbance and recovery. In J. R. Trabalka and D. E. Reichle (eds.). *The Changing Carbon Cycle: A Global Analysis*. Springer-Verlag. New York. pp. 194–220.

Schlesinger, W. H., J. F. Reynolds, G. L. Cunningham, L. F. Huenneke, W. M. Jarrell, R. A. Virginia, and W. B. Whitford. 1990. Biological feedbacks in global desertification. *Science* 247:1043–1048.

Soil Survey Staff. 1992. *Keys to Soil Taxonomy*. 5th Ed. SMSS Technical Monograph No. 19. Pocahontas Press. Blacksburg, Virginia. 556 pp.

U.S. Department of State (USDS). 1992. *National Action Plan for Global Climate Change*. Bureau of Oceans, and International Environmental and Science Affairs. Publ. 10026. Washington, DC. 129 pp.

Varallyay, G. Y. 1990. Influence of climatic change on soil moisture regime, texture, structure and erosion. In H. W. Scharpenseel, M. Schomaker, and A. Ayoub (eds.). *Soils on a Warmer Earth*. Elsevier. Amsterdam. pp. 39–49.

Williams, L. E., T. M. DeJong, and D. A. Phillips. 1981. Carbon and nitrogen limitations on soybean seedling development. *Plant Physiology* 68:1206–1209.

Wischmeier, W. H. 1976. Use and Misuse of the Universal Soil Loss Equation. *Journal of Soil and Water Conservation* 31:5–9.

Workshop held in Woods Hole, Massachusetts. 1991. *Soil-Warming Experiments in Global Change Research*. September 27–28, 1991. National Science Foundation (Grant BSR 91–01031).

Chapter 6   Water Resources and Sea-Level Rise

Allen, R. G., F. N. Gichuki, and C. Rosenzweig. 1991. CO_2-induced climatic changes and irrigation-water requirements. *Journal of Water Resources Planning and Management* 117(2):157–178.

Asian Development Bank. 1994. *Climate Change in Asia*. Executive Summary. Manila, The Philippines. 121 pp.

Cohen, S. J. 1991. Possible impacts of climatic warming scenarios on water resources in the Saskatchewan River sub-basin, Canada. *Climatic Change* 19:291–317.

Cooter, E. J. and W. S. Cooter. 1990. Impacts of greenhouse warming on water temperature and water quality in the southern United States. *Climate Research* 1:1–12.

Dugan, J. T., T. McGrath, and R. B. Zelt. 1994. *Water-Level Changes in the High Plains Aquifer — Predevelopment to 1992*. U.S. Geological Survey. Water-Resources Investigations Report 94-4027. Lincoln, NE.

Frederick, K. D. 1993. Climate change impacts on water resources and possible responses in the MINK region. *Climatic Change* 24:83–115.

Frederick, K. D. and J. C. Hanson. 1982. *Water for Western Agriculture*. Resources for the Future. Washington, DC. 241 pp.

Glantz, M. H. and J. H. Ausubel. 1984. The Ogallala Aquifer and carbon dioxide: Comparison and convergence. *Environmental Conservation* 11(2):123–131.

Gleick, P. H. 1987. Regional hydrologic consequences of increases in atmospheric CO_2 and other trace gases. *Climate Change* 10:137–161.

Gleick, P. H. 1990. Vulnerability of water systems. In P. E. Waggoner (ed.). *Climate Change and U.S. Water Resources*. Wiley. New York. pp. 223–240.

Gornitz, V. M., R. C. Daniels, T. W. White, and K. R. Birdwell. 1994. The development of a coastal risk assessment database: Vulnerability to sea-level rise in the U.S. Southeast. *Journal of Coastal Research* Special Issue No. 12, Coastal Hazards, pp. 327–338.

Gornitz, V., T. M. White, and R. M. Cushman. 1991. Vulnerability of the U.S. to future sea-level rise. *Proceedings of 7th Symposium on Coastal and Ocean Management*. ASCE. pp. 2354–2368.

Gornitz, V., C. Rosenzweig, and D. Hillel. 1994. Is sea level rising or falling? *Nature* 371:481.

Gornitz, V., C. Rosenzweig, and D. Hillel. 1997. Effects of anthropogenic intervention in the land hydrologic cycle on sea level rise. *Global and Planetary Change* 14:147–161.

Goudriaan, J. and M. H. Unsworth. 1990. Implications of increasing carbon dioxide and climate change for agricultural productivity and water resources. In *Impact of Carbon Dioxide, Trace Gases, and Climate Change on Global Agriculture*. ASA Special Publication No. 53. American Society of Agronomy, Crop Science Society of America, and Soil Science Society of America. Madison, WI. pp. 111–130.

Hale, M. G. and D. M. Orcutt. 1987. *The Physiology of Plants Under Stress*. Wiley. New York. 206 pp.

High Plains Associates. 1982. *Six-State High Plains Ogallala Aquifer Regional Resources Study: Summary*. Camp Dresser and McKee. Austin, TX.

Hillel, D. 1980. *Applications of Soil Physics*. Academic Press. New York. 385 pp.

Hillel, D. 1991. *Out of the Earth: Civilization and the Life of the Soil*. Free Press. New York.

Hillel, D. 1994. *Rivers of Eden: The Quest for Water and the Struggle for Peace in the Middle East*. Oxford University Press. New York.

Huq, S., S. I. Ali, and A. A. Rahman. 1995. Sea-level rise and Bangladesh: A preliminary analysis. *Journal of Coastal Research*. Special Issue No. 12, Coastal Hazards, pp. 44–53.

IPCC. 1990a. *Climate Change: The IPCC Scientific Assessment*. J. T. Houghton, G. J. Jenkins, and J. J. Ephraums (eds.). World Meteorological Organization and United Nations Environment Programme. Intergovernmental Panel on Climate Change. Cambridge University Press. Cambridge.

IPCC. 1990b. *Climate Change: The IPCC Impacts Assessment*. W. J. McG. Tegart, G. W. Sheldon, and D. C. Griffiths (eds.). Intergovernmental Panel on Climate Change. Canberra.

IPCC. 1992. *Climate Change 1992: The Supplementary Report on the IPCC Scientific Assessment*. J. T. Houghton, G. J. Jenkins, and J. J. Ephraums (eds.). World Meteorological Organization and United Nations Environment Programme. Intergovernmental Panel on Climate Change. Cambridge University Press. Cambridge.

IPCC, 1996. *Climate Change 1995: The Science of Climate Change*. J. T. Houghton, L. B. Meira Filho, B. A. Callander, N. Harris, A. Kattenberg, and K. Maskell (eds.). Intergovernmental Panel on Climate Change. Cambridge University Press. Cambridge. 572 pp.

Jelgersma, S., M. Van der Zijp, and R. Brinkman. 1993. Sea-level rise and the coastal lowlands in the developing world. *Journal of Coastal Research* 9(4):958–972.

Karim, Z., S. G. Hussain, and M. Ahmed. 1996. Assessing impacts of climatic variations of food grain production in Bangladesh. *Water, Air, and Soil Pollution* 92:53–62.

Kellogg, W. W. and Z.-C. Zhao. 1988. Sensitivity of soil moisture to doubling of carbon dioxide in climate model experiments. Part I: North America. *Journal of Climate* 1:348–366.

Manabe, S. and R. T. Wetherald. 1986. Reduction in summer soil wetness induced by an increase in atmospheric carbon dioxide. *Science* 232:626–628.

Manabe, S. and R. T. Wetherald. 1987. Large-scale changes in soil wetness induced by an increase in atmospheric carbon dioxide. *Journal of the Atmospheric Sciences* 44(1):1211–1235.

Milliman, J. D. 1992. Sea-level response to climatic change and tectonics in the Mediterranean Sea. In Jeftic, L., J. D. Milliman, and G. Sestini. (eds.). *Climate Change and the Mediterranean: Environmental and Societal Impacts of Climatic Change and Sea-level Rise in the Mediterranean Region.* Edward Arnold. London. 673 pp.

Milliman, J. D., J. M. Broadus, and F. Gable. 1989. Environmental and economic implications of rising sea level and subsiding deltas: The Nile and Bengal examples. *Ambio* 18:340–345.

Monteith, J. L. 1965. Radiation and crops. *Experimental Agriculture Review* 1(4):241–251.

Parry, M. L., M. Blantran de Rozari, A. L. Chong, and S. Panich (eds.). 1992. *The Potential Socio-Economic Effects of Climate Change in South-East Asia.* United Nations Environment Programme. Nairobi. 126 pp.

Peterson, D. F. and A. A. Keller. 1990. Irrigation. In P. E. Waggoner (ed.). *Climate Change and U.S. Water Resources.* Wiley. New York. pp. 269–306.

Riebsame, W. E., K. M. Strzepek, J. L. Wescoat, Jr., G. L. Gaile, J. Jacobs, R. Leichenko, C. Magadza, R. Perritt, H. Phien, B. J. Urbiztondo, P. Restrepo, W. R. Rose, M. Saleh, C. Tucci, L. H. Ti, and D. Yates. 1995. Complex river basins. In K. Strzepek and J. Smith (eds.). *As Climate Changes: International Impacts.* Cambridge University Press. Cambridge. pp. 57–91.

Rind, D., R. Goldberg, J. Hansen, C. Rosenzweig, and R. Ruedy. 1990. Potential evapotranspiration and the likelihood of future drought. *Journal of Geophysical Research* 95(D7):9983–10004.

Rosenberg, N. J. 1996. Climate change and agricultural productivity. In S. J. Ghan, W. T. Pennell, K. L. Peterson, E. Rykiel, M. J. Scott, and L. W. Vail (eds.). *Regional Impacts of Global Climate Change: Assessing Change and Response at the Scales That Matter.* Battelle Press. Columbus, OH. pp. 135–168.

Rosenberg, N. J. 1993. Towards an integrated assessment of climate change: The MINK study. *Climatic Change* 24:1–173.

Rosenberg, N. J., B. A. Kimball, P. Martin, and C. F. Cooper. 1990. From climate and CO_2 enrichment to evapotranspiration. In P. E. Waggoner (ed.). *Climate Change and U.S. Water Resources.* Wiley. New York. pp. 151–175.

Rosenzweig, C. and D. Hillel. 1993. The dust bowl of the 1930s: Analog of greenhouse effect in the Great Plains? *Journal of Environmental Quality* 22(1):9–22.

Rosenzweig, C. and D. Hillel. 1994. *Egyptian Agriculture in the 21st Century.* IIASA. CP-94-12. Laxenburg, Austria. 29 pp.

Schaake, J. C. 1990. From climate to flow. In P. E. Waggoner (ed.). *Climate Change and U.S. Water Resources.* Wiley. New York. pp. 177–206.

Schwarz, M. and J. Gale. 1984. Growth response to salinity at high levels of carbon dioxide. *Journal of Experimental Botany* 35:193–196.

Wescoat, J. L., Jr. 1991. Managing the Indus River basin in light of climate change: Four conceptual approaches. *Global Environmental Change* Dec:381–396.

Wilhite, D. A. 1988. The Ogallala Aquifer and carbon dioxide: Are policy responses applicable? In M. H. Glantz (ed.). *Societal Responses to Regional Climate Change: Forecasting by Analogy.* Westview Press. Boulder, CO. pp. 353–373.

Zhao, Z.-C. and W. K. Kellogg. 1988. Sensitivity of soil moisture to doubling of carbon dioxide in climate model experiments. Part II: The Asian monsoon region. *Journal of Climate* 1:367–378.

Chapter 7   Analysis of Climate Change Impacts

Adams, R. M., C. Rosenzweig, R. M. Peart, J. T. Ritchie, B. A. McCarl, J. D. Glyer, R. B. Curry, J. W. Jones, K. J. Boote, and L. H. Allen, Jr. 1990. Global climate change and US agriculture. *Nature* 345:219–224.

Alcamo, J., G. J. J. Kreileman, M. Krol, and G. Zuidema. 1994. Modelling the global society-biosphere-climate system: Part 1: Model description and testing. *Water, Air and Soil Pollution* 76:1–35.

Antle, J. M. 1996. Methodological issues in assessing potential impacts of climate change on agriculture. *Agricultural and Forest Meteorology* 80:67–85.

Bach, W. 1979. The impact of increasing atmospheric CO_2 concentrations on the global climate: Potential consequences and corrective measures. *Environment International* 2:215–228.

Baier, W. 1973. Crop-weather analysis model: Review and model development. *Journal of Applied Meteorology* 12:937–947.

Barnola, J. M., D. Raynaud, Y. S. Korotkevich, and C. Lorius. 1987. Vostok ice core provides 160,000-year record of atmospheric CO_2. *Nature* 329:408–414.

Bazzaz, F. A. 1990. The response of natural ecosystems to the rising global CO_2 levels. *Annual Review of Ecological Systems* 21:167–196.

Biswas, A. K. 1980. Crop-climate models: A review of the state of the art. In J. Ausubel and A. K. Biswas (eds.). *Climatic Constraints on Human Activities*. II ASA Proceedings Ser. V. 10. Pergamon Press. Oxford. pp. 75–92.

Budyko, M. I. and Y. S. Sedunov. 1990. Anthropogenic climate changes. In H. J. Karpe and D. Otten (eds.). *Climate and Development: Proceedings of the World Congress on Climate and Development*. Springer-Verlag. New York. pp. 270–284.

Cline, W. R. 1992. *The Economics of Global Warming*. Institute for International Economics. Washington, DC. 399 pp.

Cramer, W. P. and A. M. Solomon. 1993. Climatic classification and future global redistribution of agricultural land. *Climate Research* 3:97–110.

Dickinson, R. E., R. M. Errico, F. Giorgi, and G. T. Bates. 1989. A regional climate model for the western U.S. *Climatic Change* 15:383–422.

Doorenbos, J. and A. H. Kassam. 1979. *Yield Response to Water*. FAO Irrigation and Drainage Paper 33. Food and Agriculture Organization of the United Nations. Rome.

Dregne, H. E. and N.-T. Chou. 1992. Global desertification, dimension, and costs. In H. E. Dregne (ed.). *Degradation and Restoration of Arid Lands*. Texas Tech University. Lubbock, TX.

Dudek, D. J. 1989. Climate change impacts upon agriculture and resources: A case study of California. In J. B. Smith and D. Tirpak (eds.). *The Potential Effects of Global Climate Change on the United States*. Appendix C-1. U.S. Environmental Protection Agency. Washington, DC. pp. 5-1–5-38.

Edmonds, J. A., H. M. Pitcher, N. J. Rosenberg, and T. M. L. Wigley. 1993. Design for the Global Change Assessment Model GCAM. Paper presented at the International Workshop on Integrated Assessment of Mitigation, Impacts and Adaptation to Climate Change, 13–15 October 1993, International Institute for Applied Systems Analysis. Laxenburg, Austria. 7 pp.

Fischer, G. and H. T. van Velthuizen. *Climate Change and Global Agricultural Potential Project: A Case Study of Kenya*. International Institute for Applied Systems Analysis. WP-96-71. Laxenburg, Austria. 96 pp.

Food and Agriculture Organization of the United Nations. 1978. *Report on the Agro-Ecological Zones Project. Vol. 1. Methodology and Results for Africa*. FAO. Rome. 158 pp.

Food and Agriculture Organization of the United Nations. 1993. *Agriculture: Towards 2010*. United Nations. Rome.

Giorgi, F. 1990. On the simulation of regional climate using a limited area model nested in a general circulation model. *Journal of Climate* 3:94–96.

Giorgi, F. and L. O. Mearns. 1991. Approaches to the simulation of regional climate change: A review. *Reviews of Geophysics* 29(2):191–216.

Giorgi, F., C. Shields Brodeur, and G. T. Bates. 1994. Regional climate change scenarios over the United States produced with a nested regional climate model: Spatial and seasonal characteristics. *Journal of Climate* 7:375–399.

Glantz, M. H. (ed.). 1988. *Societal Responses to Regional Climatic Change*. Westview Press. Boulder, CO. 428 pp.

Grotch, S. L. 1988. *Regional Intercomparisons of General Circulation Model Predictions and Historical Climate Data*. U.S. Department of Energy. DOE/NBB-0084. Washington, DC.

Hansen, J., A. Lacis, D. Rind, G. Russell, P. Stone, I. Fung, R. Ruedy, and J. Lerner. 1984. Climate sensitivity: Analysis of feedback mechanisms. In *Climate Processes and Climate Sensitivity*. J. E. Hansen and T. Takahashi (eds.). Maurice Ewing Series No. 5. American Geophysical Union. Washington, D.C. pp. 130–163.

Hayes, J. T., P. A. O'Rourke, W. H. Terjung, and P. E. Todhunter. 1982. A feasible crop yield model for worldwide international food production. *International Journal of Biometeorology* 26(3):239–257.

Hillel, D. 1977. *Computer Simulation of Soil-Water Dynamics: A Compendium of Recent Work*. International Development Research Centre. IDRC-082e. Ottawa. 214 pp.

Hulme, M. and S. Raper. 1993. An integrated framework to address climate change (ESCAPE) and further developments of the global and regional climate modules (MAGICC). Paper presented at the International Workshop on Integrated Assessment of Mitigation, Impacts and Adaptation to Climate Change, 13–15 October 1993, International Institute for Applied Systems Analysis. Laxenburg, Austria. 14 pp.

IBSNAT. 1990. *Proceedings of IBSNAT Symposium: Decision Support System for Agrotechnology Transfer*. 81st Annual Meeting of the American Society of Agronomy, Las Vegas. University of Hawaii. Honolulu.

IPCC. 1990. *Climate Change. The IPCC Scientific Assessment*. J. T. Houghton, G. J. Jenkins, and J. J. Ephraums (eds.). Intergovernmental Panel on Climate Change. Cambridge University Press. Cambridge. 365 pp.

IPCC. 1992. *Climate Change 1992. The Supplementary Report to the IPCC Scientific Assessment*. J. T. Houghton, B. A. Callander, and S. K. Varney (eds.). Intergovernmental Panel on Climate Change. Cambridge University Press, Cambridge. 200 pp.

IPCC. 1994. *IPCC Technical Guidelines for Assessing Climate Change Impacts and Adaptations*. T. R. Carter, M. L Parry, S. Nihioka, and H. Harasawa (eds.). Intergovernmental Panel on Climate Change. Department of Geography, University College London, UK and Center for Global Environmental Research, National Institute for Environmental Studies, Japan. 59 pp.

IPCC. 1996. *Climate Change 1995: The Science of Climate Change*. J. T. Houghton, L. B. Meira Filho, B. A. Callander, N. Harris, A. Kattenberg, and K. Maskell (eds.). Intergovernmental Panel on Climate Change. Cambridge University Press. Cambridge. 572 pp.

Jäger, J. and W. W. Kellogg. 1983. Anomalies in temperature and rainfall during warm Arctic seasons. *Climatic Change* 5:39–60.

Jones, J. W. and J. R. Ritchie. 1990. Crop growth models. In G. J. Hoffman et al. (eds.). *Management of Farm Irrigation Systems*. American Society of Agricultural Engineers Monograph. St. Joseph, MO.

Joyce, L. A. and R. N. Kickert. 1987. Applied plant growth models for grazing lands, forests, and crops. In K. Wisiol and J. D. Hesketh (eds.). *Plant Growth Modeling for Resource Management*. Vol. 1: *Current Models and Methods*. CRC Press. Boca Raton. pp. 17–55.

Kaiser, H. M., S. J. Riha, D. Wilks, D. G. Rossiter, and R. Sampath. 1993. A farm-level analysis of economic and agronomic impacts of gradual climate warming. *American Journal of Agricultural Economics* 75:387–398.

Kalkstein, L. S. (ed.). 1991. *Global Comparisons of Selected GCM Control Runs and Observed Climate Data*. U.S. Environmental Protection Agency. 21P-2002. Washington, DC. 251 pp.

Katz, R. W. 1977. Assessing the impact of climatic change on food production. *Climatic Change* 1:85–96.

Kellogg, W. W. and Z.-C. Zhao. 1988. Sensitivity of soil moisture to doubling of carbon dioxide in climate model experiments. Part I: North America. *Journal of Climate* 1:348–366.

Kida, H., T. Koide, H. Sasaki, and M. Chiba. 1991. A new approach to coupling a limited area model to a GCM for regional climate simulation. *Journal of the Meteorological Society of Japan* 69:723–728.

Kutzbach, J. E. and F. A. Street-Perrott. 1985. Milankovitch forcing of fluctuations in the level of tropical lakes from 18 to 0 kyr BP. *Nature* 317:130–134.

LeDuc, S. K. 1980. *Corn Models for Iowa, Illinois, and Indiana.* NOAA Center for Environmental Assessment Services. Columbia, MO.

Leemans, R. and A. M. Solomon. 1993. Modeling the potential change in yield and distribution of the earth's crops under a warmed climate. *Climate Research* 3:79–96.

Loomis, R. S., R. Rabbinge, and E. Ng. 1979. Explanatory models in crop physiology. *Annual Review of Plant Physiology* 30:339–67.

Lough, J. M., T. M. L. Wigley, and J. P. Palutikof. 1983. Climate and climate impact scenarios for Europe in a warmer world. *Journal of Climate and Applied Meteorology* 22:1673–1684.

MacCracken, M. C., M. I. Budyko, A. D. Hecht, and Y. A. Israel. 1990. *Prospects for Future Climate. A Special US/USSR Report on Climate and Climate Change.* Lewis Publishers. Chelsea, MI. 270 pp.

Manabe, S. and R. T. Wetherald. 1987. Large-scale changes in soil wetness induced by an increase in CO_2. *Journal of Atmospheric Science* 44:1211–1235.

Manne, A., R. Mendelsohn, and R. Richels. 1993. MERGE—A Model for Evaluating Regional and Global Effects of GHG Reduction Policies. Paper presented at the International Workshop on Integrated Assessment of Mitigation, Impacts and Adaptation to Climate Change, 13–15 October 1993, International Institute for Applied Systems Analysis. Laxenburg, Austria. 14 pp.

McGregor, J. L. and K. Walsh. 1993. Nested simulations of perpetual January climate over the Australian region. *Journal of Geophysical Research* 98:23283–23290.

Mearns, L. O., C. Rosenzweig, and R. Goldberg. 1992. Effect of changes in interannual climatic variability on CERES-Wheat yields: Sensitivity and $2 \times CO_2$ general circulation model studies. *Agricultural and Forest Meteorology* 62:159–189.

Mearns, L. O., C. Rosenzweig, and R. Goldberg. 1996. The effect of changes in daily and interannual climatic variability on CERES-Wheat: A sensitivity study. *Climatic Change* 32(3):257–292.

Mendelsohn, R., W. D. Nordhaus, and D. Shaw. 1994. The impact of global warming on agriculture: A Ricardian analysis. *American Economic Review* 84(4):753–771.

Newman, J. E. 1980. Climate change impacts on the growing season of the North American corn belt. *Biometeorology* 7:128–142.

Nordhaus, W. D. 1992. The DICE model: Background and structure of a dynamic integrated climate economy model of the economics of global warming. Cowles Foundation Discussion Paper No. 1009. New Haven, CT.

Parry, M. L., T. R. Carter, and N. T. Konijn (eds.). 1988. *The Impact of Climatic Variations on Agriculture.* Vol. 1: *Assessments in Cool Temperate and Cold Regions.* Kluwer Academic. Dordrecht. 876 pp.

Peart, R. M., J. W. Jones, R. B. Curry, K. Boote, and L. H. Allen, Jr. 1989. Impact of climate change on crop yields in the southeastern U.S. A. In J. B. Smith and D. Tirpak (eds.). *The Potential Effects of Global Climate Change on the United States.* Report to Congress). Appendix C-1. EPA-230-05-89-053. U.S. Environmental Protection Agency. Washington, DC. pp. 2-1–2-54.

Peck, S. C. and T. J. Teisberg. 1992. CETA. A model for carbon emissions trajectory assessment. *Energy Journal* 13(1):55–77.

Pepper, W. J., J. A. Leggett, R. J. Swart, J. Wassson, J. Edmonds, and I. Mintzer. 1992. *Emission Scenarios for the IPCC—An Update: Background Documentation on Assumptions, Methodology, and Results.* U.S. Environmental Protection Agency. Washington, DC.

Pittock, A. B. and M. J. Salinger. 1982. Towards regional scenarios for a CO_2-warmed earth. *Climatic Change* 4:23–40.

Ramirez, J., C. M. Sakamoto, and R. E. Jensen. 1975. Wheat. In *Impacts of Climate Change on the Biosphere*. CIAP Monograph 5. Part 2. Climatic Effects. pp. 4-37–4-90.

Richardson, C. W. 1981. Stochastic simulation of daily precipitation, temperature, and solar radiation. *Water Resources Research* 17:182–190.

Richardson, C. W. and D. A. Wright. 1984. *WGEN: A Model for Generating Daily Weather Variables*. U.S. Department of Agriculture Agricultural Research Service. ARS Publication 8. Washington, DC.

Rimmington, G. M. and D. S. Charles-Edwards. 1987. Mathematical descriptions of plant growth and development. In K. Wisiol and J. D. Hesketh (eds.). *Plant Growth Modeling for Resource Management* Vol. 1: *Current Models and Methods*. CRC Press. Boca Raton, FL. pp. 3–15.

Ritchie, J. T., B. D. Baer, and T. Y. Chou. 1989. Effect of global climate change on agriculture in the Great Lakes region. In J. Smith and D. Tirpak (eds.). *The Potential Effects of Global Climate Change on the United States*. Office of Policy, Planning, and Evaluation. U.S. Environmental Protection Agency. Washington, DC. Appendix C. pp. 1-1–1-21.

Robock, A. R. P. Turco, M. A. Harwell, T. P. Ackerman, R. Andressen, H.-S. Change, and M. V. K. Sivakumar. 1993. Use of general circulation model output in the creation of climate change scenarios for impact analysis. *Climatic Change* 23:293–335.

Rosenberg, N. J. 1982. The increasing CO_2 concentration in the atmosphere and its implication on agricultural productivity. II: Effects through CO_2-induced climatic change. *Climatic Change* 4:239–254.

Rosenberg, N. J. (ed.). 1993. *Towards an Integrated Impact Assessment of Climate Change: The MINK Study*. Reprinted from *Climatic Change* 2(1–2):1–173. Kluwer Academic. Dordrecht.

Rosenzweig, C. 1985. Potential CO_2-induced climate effects on North American wheat-producing regions. *Climatic Change* 7:367–389.

Rosenzweig, C. 1990: Crop response to climate change in the southern Great Plains: A simulation study. *Professional Geographer* 42(1):20–37.

Rosenzweig, C. and A. Iglesias (eds.). 1994. *Implications of Climate Change for International Agriculture: Crop Modeling Study*. U.S. Environmental Protection Agency. Washington, DC.

Rosenzweig, C. and M. L. Parry. 1994. Potential impact of climate change on world food supply. *Nature* 367:133–138.

Rosenzweig, C., M. L. Parry, G. Fischer, and K. Frohberg. 1993. *Climate Change and World Food Supply*. Environmental Change Unit. Report No. 3. Oxford University. Oxford. 28 pp.

Rosenzweig, C., J. Phillips, R. Goldberg, J. Carroll, and T. Hodges. 1997. Potential impacts of climate change on citrus and potato production in the U.S. *Agricultural Systems* 52:455–479.

Schlesinger, M. and Z. Zhao. 1988. Seasonal climate changes induced by doubled CO_2 as simulated by the OSU atmospheric GCM/mixed layer ocean model. Oregon State University. Climate Research Institute. Corvallis, OR.

Semenov, M. A. and J. R. Porter. 1995. Climatic variability and modelling of crop yields. *Agricultural and Forest Meteorology* 73:265–283.

Smith, J. B. and D. Tirpak (eds.). 1989. *The Potential Effects of Global Climate Change on the United States*. U.S. Environmental Protection Agency Report to Congress. EPA-230-05-89-050. Washington, DC. 409 pp.

Stockle, C. O., J. R. Williams, N. H. Rosenberg, and C. A. Jones. 1992a. Estimating the effect of carbon dioxide-induced climate change on growth and yield of crops. I: Modification to the EPIC model for climate change analysis. *Agricultural Systems* 38: 225–238.

Stockle, C. O., P. T. Dyke, J. R. Williams, C. A. Jones, and N. J. Rosenberg. 1992b. Estimating the effect of carbon dioxide-induced climate change on growth and yield of crops. II: Assessing the impacts on maize, wheat and soybean in the Midwestern U.S.A. *Agricultural Systems* 38: 239–256.

Strain, B. R. and J. D. Cure (eds.). 1985. *Direct Effects of Increasing Carbon Dioxide on Vegetation*. DOE/ER-0238. U.S. Department of Energy. Washington, DC.

Thompson, L. M. 1969a. Weather and technology in the production of corn in the U.S. cornbelt. *Agronomy Journal* 61:453–456.

Thompson, L. M. 1969b. Weather and technology in the production of wheat in the United States. *Journal of Soil and Water Conservation* 24:219–224.

Thompson, L. M. 1975. Weather variability, climate change, and grain production. *Science* 188:535–541.

Thompson, S. L. and D. Pollard. 1995a. A global climate model (GENESIS) with a land-surface-transfer scheme (LSX). Part I: Present climate simulation. *Journal of Climate* 8:732–761.

Thompson, S. L. and D. Pollard. 1995b. A global climate model (GENESIS) with a land-surface-transfer scheme (LSX). Part II: CO_2 sensitivity. *Journal of Climate* 8:1104–1121.

Tubiello, F. N., C. Rosenzweig, and T. Volk. 1995. Interactions of CO_2, temperature and management practices: Simulations with a modified version of CERES-Wheat. *Agricultural Systems* 49:135–152.

United Nations. 1996. *World Population Prospects: The 1996 Revision*. United Nations. New York.

United Nations. 1990. *Overall Socio-Economic Perspective of the World Economy to the Year 2000*. United Nations. New York.

United Nations Environment Programme. 1992. *World Atlas of Desertification*. Edward Arnold, Hodder, and Stoughton. Kent, UK.

U.S. Country Studies Program. 1994. *Guidance for Vulnerability and Adaptation Assessments*. Washington, DC.

Waggoner, P. E. 1983. Agriculture and a climate changed by more carbon dioxide. In *Changing Climate*. National Academy of Sciences Press. Washington, DC. pp. 383–418.

Wang, W.-C., M. P. Dudek, and X.-Z. Liang. 1992. Inadequacy of effective CO_2 as a proxy in the regional climate change due to other radiatively active gases. *Geophysical Resesarch Letters* 19:1375–1378.

Warrick, R. A. 1984. The possible impacts on wheat production of a recurrence of the 1930s drought in the U.S. Great Plains. *Climatic Change* 6:5–26.

Washington, W. M. and G. A. Meehl. 1984. Seasonal cycle experiment on the climate sensitivity due to a doubling of CO_2 with an atmospheric general circulation model coupled to a simple mixed-layer ocean model. *Journal of Geophysical Research* 89:9475–9503.

Wigley, T. M. L., P. D. Jones, K. R. Briffa, and G. Smith. 1990. Obtaining sub-grid-scale information from coarse-resolution general circulation model output. *Journal of Geophysical Research* 95:1943–1953.

Wigley, T. M. L. and S. C. B. Raper. 1992. Implications for climate and sea level of revised IPCC emissions scenarios. *Nature* 357:293–300.

Wilks, D. S. 1989. Statistical specification of local surface weather elements from large-scale information. *Theoretical and Applied Climatology* 40:119–134.

Wilks, D. S. 1992. Adapting stochastic weather generation algorithms for climate change studies. *Climatic Change* 22:67–84.

World Bank. 1992. *Population*. Washington, DC.

World Bank. 1993. *Income Projections*. Washington, DC.

World Coast Conference, 1993. *Preparing to Meet the Coastal Challenges of the 21st Century*. Conference Report. Coastal Zone Management Centre. The Hague, The Netherlands.

Crop Model References

Acock, B. et al. 1983. *The Soybean Crop Simulator GLYCIM: Model Documentation 1982*, report 2, U.S. Department of Energy, Carbon Dioxide Research Division, Office of Energy Research, Washington, DC.

Aggarwal, P. K. and F. W. T. Penning de Vries. 1989. Potential and water-limited yields in rice-based cropping systems in Southeast Asia. *Agricultural Systems* 30:49–69.

Arkin, G. F. et al. 1976. A dynamic grain sorghum growth model. *Transactions of the ASAE* 19(4):622–626, 630.

Baker, C. H. and R. D. Horrocks. 1976. CORNMOD, a dynamic simulator of corn production. *Agricultural Systems* 4:57–77.

Baker, D. N. et al. 1983. *GOSSYM: A Simulation of Cotton Crop Growth and Yield.* Technical Bulletin 1089. South Carolina Agricultural Experimental Station. Clemson, SC.

Boote, K. J. et al. 1985. Modeling growth and yield of groundnut. In *Proceedings of the International Symposium on Agrometeorology of Groundnut.* ICRISAT Sahelian Center. Niamey, Niger.

Brown, L. G. et al. 1985. *COTCROP: Computer Simulation of Growth and Yield.* Information Bulletin 69. Mississippi Agricultural and Forestry Experimental Station, Mississippi State University. Mississippi State, MS.

Buttler, I. W. 1989. Predicting water constraints to productivity of corn using plant-environmental simulation models. Ph.D. diss. Cornell University. Ithaca, NY.

Childs, S. W. et al. 1977. A simplified model of corn growth under moisture stress. *Transactions of the ASAE* 20(5):858–865.

Curry, R. B. et al. 1975. SOYMOD I: A dynamic simulator of soybean growth and development. *Transactions of the ASAE* 18(5):963–968.

Dennison, R. F. and R. S. Loomis. 1989. *An Integrative Physiological Model of Alfalfa Growth and Development* (complete citation unavailable).

Duncan, W. G. 1975. SIMAIZ: A model simulating growth and yield in corn. In D. N. Baker et al. (eds.). *An Application System Method to Crop Production.* Mississippi Agricultural and Forestry Experimental Station, Mississippi State University. Mississippi State, MS.

Fick, G. W. 1981. *ALSIM I (Level 2) User's Manual.* Agronomy mimeo 81-35. Department of Agronomy, Cornell University. Ithaca, NY.

Godwin, D. C. et al. 1990. *A User's Guide to CERES-Rice-V2.10.* International Fertilizer Development Center. Muscle Shoals, AL.

Godwin, D. C. and P. L. G. Vlek. 1985. Simulation of nitrogen dynamics in wheat cropping systems. In W. Day and R. K. Arkin (eds.). *Wheat Growth and Modeling.* Plenum Press. New York.

Griffin, T. S., B. S. Johnson, and J. T. Ritchie. 1993. *A Simulation Model for Potato Growth and Development: SUBSTOR-Potato Version 2.0.* Research Report Series 02. IBSNAT, Department of Agronomy and Soil Science, College of Tropical Agriculture and Human Resources, University of Hawaii. Honolulu.

Hoogenboom, G. et al. 1989. A computer model for the simulation of bean growth and development. In *Advances in Bean* (Phaseolus vulgaris *L.) Research and Production.* Publication No. 23. CIAT. Cali, Colombia.

Horie. T. 1988. Simulated rice yield under changing climatic conditions in Hokkado Island. In M. L. Parry et al. (eds.). *Assessment of Climate Effects on Agriculture.* Vol. 1. Kluwer Academic. Dordrecht.

Inman-Bamber, N.G. 1991. A growth model for sugar-cane based on a simple carbon balance and the CERES-maize water balance. *South African Journal of Plant Science* 8(2):93–99.

Jackson, B. S. et al. 1988. The cotton simulation model "COTTAM": Fruiting model calibration and testing. *Transactions of the ASAE* 31(3):846–854.

Jones, C. A. and J. R. Kiniry (eds.). 1986. *Ceres-Maize: A Simulation Model of Maize Growth and Development.* Texas A&M University Press. College Station, TX.

Jones, J. W. et al. 1989. SOYGRO *v5.42 Soybean Crop Growth Simulation Model: User's Guide.* Florida Agricultural Experimental Station Journal 8304. University of Florida, Gainesville.

Maas, S. J. and G. F. Arkin. 1980. *TAMW: A Wheat Growth and Development Simulation*

Model. Research Center Program and Model Development No. 80-3. Texas Agricultural Experimental Station, Blackland Research Center. Temple, TX.

Monteith, J. L. et al. 1989. RESCAP: A resource capture model for sorghum and pearl millet. In S. M. Virmani et al. (eds.). *Modeling the Growth and Development of Sorghum and Pearl Millet.* Research Bulletin 12. ICRISAT. Andhra Pradesh, India.

Morgan, T. H. et al. 1980. A dynamic model of corn yield response to water. *Water Resources Research* 16(10):59–64.

Newkirk, K. M. et al. 1989. *User Guide to VT-Maize Version 1.0 (R).* Virginia Water Resources Resource Center, Virginia Polytechnic Institute and State University. Blacksburg.

Ng, E., and R. S. Loomis. 1984. *Simulation of Growth and Yield of the Potato Crop.* PUDOC. Wageningen, The Netherlands.

Norman, J. M. and G. Campbell. 1983. Application of a plant-environment model to problems on irrigation. *Advances in Irrigation* 2:155–188.

Porter, J. R. 1993. AFRCWHEAT2: A model of the growth and development of wheat incorporating responses to water and nitrogen. *European Journal of Agronomy* 2:69–82.

Ritchie, J. T. 1985. A user-oriented model of the soil water balance in wheat. In W. Day and R. K. Arkin (eds.). *Wheat Growth and Modeling.* Plenum Press. New York.

Ritchie, J. T. and G. Alagarswamy. 1989. Simulation of sorghum and pearl millet phenology. In *Modeling the Growth and Development of Sorghum and Pearl Millet.* Research Bulletin 12. S. M. Virmani et al. (eds.). ICRISAT. Andhra Pradesh, India.

Ritchie, J. T. et al. 1989a. *Development of a Barley Yield Simulation Model.* USDA 86-CRST-2-2867. Michigan State University. East Lansing.

Ritchie, J. T. et al. 1989b. *A User's Guide to CERES-Maize-V2.10.* International Fertilizer Development Center. Muscle Shoals, AL.

Rosenthal, W. D. et al. 1989. *SORKAM: A Grain Sorghum Crop Growth Model.* MP-1669. Texas Agricultural Experiment Station, College Station, TX.

Stapper, M. 1984. *SIMTAG: A Simulation Model of Wheat Genotypes.* University of New England, Department of Agronomy and Soil Sciences. Armidale, New South Wales, Australia.

Stapper, M. and G. F. Arkin. 1980. *CORNF: A Dynamic Growth and Development Model for Maize* (Zea mays L.): *Program and Documentation.* No. 80-2. Texas Agricultural Experimental Station. College Station.

Stockle, C. O. and G. S. Campbell. 1989. Simulation of crop response to water and nitrogen: An example using wheat. *Transactions of the ASAE* 32(1):66–74.

van Keulen, H. and N. G. Seligman. 1987. *Simulation of Water Use, Nitrogen Nutrition, and Growth of Spring Wheat Crop.* Centre for Agricultural Publications and Documentation. Wageningen, The Netherlands.

Wilkerson, G. G. et al. 1983. Modeling soybean growth for crop management. *Transactions of the ASAE* 26(1):63–73.

Williams, J. R. et al. 1984. A modeling approach to determining the relationship between erosion and soil productivity. *Transactions of the ASAE* 27(1):129–144.

Young, J. H. et al. 1979. A peanut and development model. *Peanut Science* 6:27–36.

Chapter 8 Regions at Risk

Acock, B. and L. H. Allen, Jr. 1985. Crop responses to elevated carbon dioxide concentrations. In B. R. Strain and J. D. Cure (eds.). *Direct Effects of Increasing Carbon Dioxide on Vegetation.* DOE/ER-0238. U.S. Department of Energy. Washington, DC. pp. 53–97.

Adams, R. M., C. Rosenzweig, R. M. Peart, J. T. Ritchie, B. A. McCarl, J. D. Glyer, R. B. Curry, J. W. Jones, K. J. Boote, and L. H. Allen, Jr. 1990. Global climate change and U.S. agriculture. *Nature* 345(6272):219–222.

Buol, S. W., P. A. Sanchez, S. B. Weed, and J. M. Kimble. 1990. Predicted impact of climatic warming on soil properties and use. In B. A. Kimball, N. J. Rosenberg, and L. H. Allen, Jr. (eds.). *Impact of Carbon Dioxide, Trace Gases, and Climate Change on Global Agriculture.* ASA Special Publication No. 53. American Society of Agronomy. Madison, WI. pp. 71–82.

Dixon, R. K. (ed.). 1997 Climate Change Impacts and Response Options in Eastern and Central Europe. Special Issue. *Climatic Change* 36(1 and 2):1–232.

Food and Agriculture Organization. 1990. *FAO Yearbook. Production.* Vol. 44. Food and Agriculture Organization of the United Nations. Rome.

Hansen, J., G. Russell, D. Rind, P. Stone, A. Lacis, S. Lebedeff, R. Ruedy, and L. Travis. 1983. Efficient three-dimensional global models for climate studies: Models I and II. *Monthly Weather Review* 111(4):609–662.

Haws, L. D., H. Inoue, A. Tanaka, and S. Yoshida. 1983. Comparison of crop productivity in the tropics and temperate zone. In *Potential Productivity of Field Crops Under Different Environments.* International Rice Research Institute. Los Baños, Philippines. pp. 403–413.

Hillel, D. 1997. *Small-scale Irrigation for Arid Zones.* United Nations Food and Agriculture Organization. Rome.

Hillel, D. 1987. *The Efficient Use of Water in Irrigation: Principles and Practices for Improving Irrigation in Arid and Semiarid Regions.* World Bank Technical Paper No. 64. World Bank. Washington, DC. 107 pp.

IPCC. 1996. Agriculture. In R. T. Watson, M. C. Zinyowera, and R. H. Moss (eds.). *Climate Change 1995: Impacts, Adaptations and Mitigation of Climate Change: Scientific-Technical Analyses.* Contribution of Working Group II to the Second Assessment Report of the Intergovernmental Panel on Climate Change. Cambridge University Press, Cambridge. pp. 427–467.

Jodha, N. S. 1989. Potential strategies for adapting to greenhouse warming: Perspectives from the developing world. In N. J. Rosenberg, W. E. Easterling, III, P. R. Crosson, and J. Darmstadter (eds.). *Greenhouse Warming: Abatement and Adaptation.* Resources for the Future. Washington, DC. pp. 147–158.

Kellogg, W. W. and Z.-C. Zhao. 1988. Sensitivity of soil moisture to doubling of carbon dioxide in climate model experiments. I: North America. *Journal of Climate* 1:348–366.

Liverman, D. M. 1991. Global warming and Mexican agriculture: Some preliminary results. In J. Reilly and M. Anderson (eds.). *Economic Issues in Global Climate Change: Agriculture, Forestry, and Natural Resources.* Westview Press. Boulder, CO. pp. 332–352.

Liverman, D. M. and K. O'Brien. 1991. Global warming and climate change in Mexico. *Global Environmental Management* 1(4):351–364.

Manabe, S. and R. T. Wetherald. 1987. Large-scale changes in soil wetness induced by an increase in CO_2. *Journal of Atmospheric Science* 44:1211–1235.

Matthews, R. B., M. J. Kropff, D. Bachelet, and H. H. van Laar. (eds.). 1995. *Modeling the Impact of Climate Change on Rice Production in Asia.* CAB International in association with the International Rice Research Institute. Oxford, UK. 289 pp.

Mearns, L. O., C. Rosenzweig, and R. Goldberg. 1992. Effect of changes in interannual climatic variability on CERES-Wheat yields: Sensitivity and $2 \times CO_2$ general circulation model studies. *Agricultural and Forest Meteorology* 62:159–189.

Nicholson, S. E. 1989. African Drought: Characteristics, Causal Theories and Global Teleconnections. International Union of Geodesy and Geophysics and American Geophysical Union. AGU Monograph No. 52. Washington, DC.

Nishioka, S. H. Harasawa, H. Hashimoto, T. Ookita, K. Masuda, and T. Morita (eds.). 1993. *The Potential Effects of Climate Change in Japan.* Center for Global Environmental Research, National Institute for Environmental Studies, and Environment Agency of Japan. Tsukuba, Japan. 95 pp.

Nix, H. A. 1985. Agriculture. In R. W. Kates, J. H. Ausubel, and M. Berberian (eds.). *Climate Impact Assessment.* SCOPE 27. Wiley. New York.

Parry, M. L. 1978. *Climatic Change, Agriculture and Settlements.* Dawson. Folkestone, UK.

Parry, M. L., M. Blantran de Rozari, A. L. Chong, and S. Panich (eds.). 1992. *The Potential Socio-Economic Effects of Climate Change in South-East Asia.* United Nations Environment Programme. Nairobi. 126 pp.

Parry, M. L., T. R. Carter, and N. T. Konijn (eds.). 1988a. *The Impact of Climatic Variations in Agriculture.* Vol. 1: *Assessments in Cool Temperate and Cold Regions.* Kluwer Academic. Dordrecht. 876 pp.

Parry, M. L., T. R. Carter, and N. T. Konijn (eds.). 1988b. *The Impact of Climatic Variations on Agriculture.* Vol. 2: *Assessments in Semi-Arid Regions.* Kluwer Academic. Dordrecht. 764 pp.

Pearman, G. 1988. *Greenhouse: Planning for Climate Change.* CSIRO. Canberra. 752 pp.

Pierce, J. T. 1990. *The Food Resource.* Longman. New York. 334 pp.

Ribot, J. C., A. R. Magalhaes, and S. S. Panagides (eds.). 1996. *Climate Variability, Climate Change and Social Vulnerability in the Semi-arid Tropics.* Cambridge University Press. Cambridge. 175 pp.

Rind, D., R. Goldberg, J. Hansen, C. Rosenzweig, and R. Ruedy. 1990. Potential evapotranspiration and the likelihood of future drought. *Journal of Geophysical Research* 95(D7):9983–10,004.

Rose, E. 1989. Direct (physiological) effects of increasing CO_2 on crop plants and their interactions with indirect (climatic) effects. In J. B. Smith and D. Tirpak (eds.). *The Potential Effects of Global Climate Change on the United States.* Report to Congress. Appendix C-2. EPA 230-05–89-053. U.S. Environmental Protection Agency. Washington, DC. pp. 7-1–7-37.

Rosenzweig, C. 1985. Potential CO_2-induced climate effects on North American wheat-producing regions. *Climatic Change* 4:239–254.

Rosenzweig, C. and A. Iglesias. 1994. *Implications of Climate Change for International Agriculture: Crop Modeling Study.* U.S. Environmental Protection Agency. EPA 230-B-94-003. Washington, DC.

Rosenzweig, C. and D. Liverman. 1992. Predicted effects of climate change on agriculture: A comparison of temperate and tropical regions. In S. K. Majumdar, L. S. Kalkstein, B. Yarnal, E. W. Miller, and L. M. Rosenfeld. *Global Climate Change: Implications, Challenges and Mitigation Measures.* Pennsylvania Academy of Science. Easton, PA. pp. 346–362.

Rosenzweig, C. and M. L. Parry. 1994. Potential impacts of climate change on world food supply. *Nature* 367:133–138.

Rosenzweig, C., J. Phillips, R. Goldberg, J. Carroll, and T. Hodges. 1996. Potential impacts of climate change on citrus and potato production in the U.S. *Agricultural Systems* 52:455–479.

Schmandt, J. and J. Clarkson. 1992. Introduction: Global warming as a regional issue. In J. Schmandt and J. Clarkson (eds.). *The Regions and Global Warming: Impacts and Response Strategies.* Oxford University Press. New York. pp. 3–11.

Smith, J. B., S. Huq, S. Lenhart, L. J. Mata, I. Nemesova, and S. Toure (eds.). 1996. *Vulnerability and Adaptation to Climate Change: Interim Results from the U.S. Country Studies Program.* Kluwer Academic. Dordrecht.

Smith, J. B. and D. Tirpak (eds.). 1989. *The Potential Effects of Global Climate Change on the United States.* Report to Congress. EPA 230-05–89-053. U.S. Environmental Protection Agency. Washington, DC.

Stinner, B. R., R. A. J. Taylor, R. B. Hammond, F. F. Purrington, and D. A. McCartney. 1989. Potential effects of climate change on plant-pest interactions. In J. B. Smith and D. Tirpak (eds.). *The Potential Effects of Global Climate Change on the United States.* Report to Congress. Appendix C-2. EPA 230-05–89-053. U.S. Environmental Protection Agency. Washington, DC. pp. 8-1–8-35.

Swaminathan, M. S. and S. K. Sinha (eds.). 1986. *Global Aspects of Food Production.* Tycooly International. Oxford. pp. 417–449.

Thompson, L. M. 1975. Weather variability, climate change and food production. *Science* 188:534–541.

U.K. Department of the Environment. 1991. *The Potential Effects of Climate Change in the United Kingdom.* United Kingdom Climate Change Impacts Review Group. HMSO. London. 124 pp.

UNEP. 1996. *Handbook on Methods for Climate Change Impact Assessment and Adaptation Strategies.* Draft Version 1.3. United Nations Environment Programme and Institute for Environmental Studies, Vrije Universiteit Amsterdam. Nairobi, Kenya and Amsterdam, The Netherlands.

U.S. Country Studies Program. 1994. *Guidance for Vulnerability and Adaptation Assessments.* Washington, DC. Also published as: Benioff, R., S. Guill, and J. Lee. 1996. *Vulnerability and Adaptation Assessments: An International Handbook.* Kluwer Academic. Dordrecht.

Wilson, C. A. and J. F. B. Mitchell. 1987. A doubled CO_2 climate sensitivity experiment with a global climate model including a simple ocean. *Journal of Geophysical Research* 92(13):315–343.

WMO. 1979. *Proceedings of the World Climate Conference.* World Meteorological Organization. Geneva.

North America

Canada

Arthur, L. 1988. *The Implication of Climatic Change for Agriculture in the Prairie Provinces.* CCD 88–01. Atmospheric Environment Service. Environment Canada. Downsview.

Bootsma, A., W. Blackburn, R. Stewart, R. Muma, and J. Dumanski. 1984. *Possible Effects of Climatic Change on Estimated Crop Yields in Canada.* Technical Bulletin 1988–5E. Research Branch. Agriculture Canada. Ottawa.

Bootsma, A. and R. de Jong. 1988. Climate risk analyses of the prairie region. In J. Dumanski and V. Kirkwood (eds.). *Crop Production Risks in the Canadian Prairie Region in Relation to Climate and Land Resources.* Technical Bulletin 1988–5E. Research Branch. Agriculture Canada. Ottawa.

Brklacich, M. and B. Smit. 1992. Implications of changes in climatic averages and variability on food production opportunities in Ontario, Canada. *Climatic Change* 20(1):1–21.

Brklacich, M., R. Stewart, V. Kirkwood, and R. Muma. 1994. Effects of global climate change on wheat yields in the Canadian prairie. In C. Rosenzweig and A. Iglesias (eds.). *Implications of Climate Change for International Agriculture: Crop Modeling Study.* U.S. Environmental Protection Agency. EPA 230-B-94–003. Washington, DC.

Cohen, S., E. Wheaton, and J. Masterton. 1992. *Impacts of Climatic Change Scenarios in the Prairie Provinces: A Case Study from Canada.* SRC Publication No. E-2900–4-D-92. Saskatchewan Research Council. Saskatoon.

Singh, B. and R. Stewart. 1991. Potential impacts of a CO_2-induced climate change using the GISS scenario on agriculture in Quebec, Canada. *Ecosystems and Agriculture* 35:327–347.

Smit, B., M. Brklacich, R. B. Stewart, R. McBridge, M. Brown, and D. Bond. 1989. Sensitivity of crop yields and land resource potential to climatic change in Ontario, Canada. *Climatic Change* 14(2):153–174.

Stewart, R. 1990. Possible effects of climatic change on estimated crop yields in Canada: A review. In G. Wall and M. Sanderson (eds.). *Climatic Change: Implications for Water and Ecological Resources.* Department of Geography. Occasional Paper No. 11. University of Waterloo. Waterloo. pp. 275–284.

Williams, G., H. Jones, E. Wheaton, R. Stewart, and R. Fautley. 1988. Estimating the impacts of climatic change on agriculture in the Canadian prairies, the Saskatchewan case study. In M. Parry, T. Carter, and N. Konjin (eds.). *The Impact of Climatic Variations on*

Agriculture. Vol. 1: *Assessment in Cold Temperate and Cold Regions.* Kluwer Academic. Dordrecht. pp. 221–379.

United States

CAST. 1992. *Preparing U.S. Agriculture for Global Climate Change.* Council for Agricultural Science and Technology. Report 119. Ames, IA. 96 pp.

Crosson, P. 1989. Climate change and mid-latitudes agriculture: Perspectives on consequences and policy responses. *Climatic Change* 15:51–73.

Easterling, W. E. 1996. Adapting North American agriculture to climate change: A review. *Agricultural and Forest Meteorology* 80(1):1–53.

Kaiser, H., S. Riha, D. Wilkes, and R. Sampath. 1993. Adaptation to global climate change at the farm level. In H. M. Kaiser and T. E. Drennen (eds.). *Agricultural Dimensions of Global Climate Change.* St. Lucie Press. Delray Beach, FL. pp. 136–152.

Mendelsohn, R., W. D. Nordhaus, and D. Shaw. 1994. The impact of global warming on agriculture: A ricardian analysis. *American Economic Review* 84(4):753–771.

Rosenzweig, C., B. Curry, J. T. Ritchie, J. W. Jones, T.-Y. Chou, R. Goldberg, and A. Iglesias. 1994. The effects of potential climate change on simulated grain crops in the United States. In C. Rosenzweig and A. Iglesias. *Implications of Climate Change for International Agriculture: Crop Modeling Study.* U.S. Environmental Protection Agency. EPA 230-B-94–003. Washington, DC.

Mexico

Liverman, D. and K. O'Brien. 1991. Global warming and climate change in Mexico. *Global Environmental Change* 1(4):351–364.

Liverman, D., M. Dilley, K. O'Brien, and L. Menchaca. 1994. Possible impacts of climate change on maize yields in Mexico. In C. Rosenzweig and A. Iglesias. *Implications of Climate Change for International Agriculture: Crop Modeling Study.* U.S. Environmental Protection Agency. EPA 230-B-94–003. Washington, DC.

South America

Baethgen, W. E. 1994. Impact of climate change on barley in Uruguay: Yield changes and analysis of nitrogen management systems. In C. Rosenzweig and A. Iglesias. *Implications of Climate Change for International Agriculture: Crop Modeling Study.* U.S. Environmental Protection Agency. EPA 230-B-94–003. Washington, DC.

Baethgen, W. E. and G. O. Magrin. 1995. Assessing the impacts of climate change on winter crop production in Uruguay and Argentina using crop simulation models. In C. Rosenzweig, L. H. Allen, Jr., L. A. Harper, S. E. Hollinger, and J. W. Jones. (eds.). *Climate Change and Agriculture: Analysis of Potential International Impacts.* American Society of Agronomy. ASA Special Publication No. 59. Madison, WI.

de Siqueira, O. J. F., J. R. B. Farias, and L. M. A. Sans. 1994. Potential effects of global climate change for Brazilian agriculture: Applied simulation studies for wheat, maize, and soybeans. In C. Rosenzweig and A. Iglesias. *Implications of Climate Change for International Agriculture: Crop Modeling Study.* U.S. Environmental Protection Agency. EPA 230-B-94–003. Washington, DC.

Downing, T. E. 1992. *Climate Change and Vulnerable Places: Global Food Security and Country Studies in Zimbabwe, Kenya, Senegal, and Chile.* Research Report No. 1. Environmental Change Unit. University of Oxford. Oxford. 54 pp.

Sala, O. E. and J. M. Paruelo. 1994. Impacts of global climate change on maize production in Argentina. In C. Rosenzweig and A. Iglesias. *Implications of Climate Change for*

International Agriculture: Crop Modeling Study. U.S. Environmental Protection Agency. EPA 230-B-94-003. Washington, DC.

Europe

Bindi, M., G. Maracchi, and F. Miglietta. 1993. Effects of climate change on the ontomorphogenic development of winter wheat in Italy. In G. J. Kenny, P. A. Harrison, and M. L. Parry (eds.). *The Effects of Climate Change on Agricultural and Horticultural Potential in Europe.* Environmental Change Unit. University of Oxford. Oxford.

Carter, T. R., J. H. Porter, and M. L. Parry. 1991. Climatic warming and crop potential in Europe: Prospects and uncertainties. *Global Environmental Change* 1:291–312.

Delecolle, R., D. Ripoche, F. Ruget, and G. Gosse. 1994. Possible effects of increasing CO_2 concentration on wheat and maize crops in north and southeast France. In C. Rosenzweig and A. Iglesias. (eds.). *Implications of Climate Change for International Agriculture: Crop Modeling Study.* U.S. Environmental Protection Agency. EPA 230-B-94-003. Washington, DC.

Dellecolle, R., F. Ruget, G. Gosse, and D. Ripoche. 1995. Possible effects of climate change on wheat and maize crops in France. In C. Rosenzweig, L. H. Allen, Jr., L. A. Harper, S. E. Hollinger, and J. W. Jones (eds.). *Climate Change and Agriculture: Analysis of Potential International Impacts.* American Society of Agronomy. ASA Special Publication No. 59. Madison, WI.

Iglesias, A. and M. I. Minguez. 1995. Prospects for maize production in Spain under climate change. In C. Rosenzweig, L. H. Allen, Jr., L. A. Harper, S. E. Hollinger, and J. W. Jones (eds.). *Climate Change and Agriculture: Analysis of Potential International Impacts.* American Society of Agronomy. ASA Special Publication No. 59. Madison, WI.

Iglesias, A. and M. I. Minguez. 1997. Modelling crop-climate interactions in Spain: Vulnerability and adaptation of different agricultural systems to climate change. *Journal of Mitigation and Adaptation Strategies for Global Change* 1(3):273–288.

Kapetanaki, G. and C. Rosenzweig. 1996. Impact of climate change on maize yield in central and northern Greece: A simulation study with CERES-Maize. *Journal of Mitigation and Adaptation Strategies for Global Change* 1(3):251–271.

Kenny, G. J. and P. A. Harrison. 1992a. Thermal and moisture limits of grain maize in Europe: Model testing and sensitivity to climate change. *Climate Research* 2:113–129.

Kenny, G. J. and P. A. Harrison. 1992b. The effects of climate variability and change on grape suitability in Europe. *Journal of Wine Research* 3:163–183.

Kenny, G. J., P. A. Harrison, and M. L. Parry (eds.). 1993. *The Effect of Climate Change on Agricultural and Horticultural Potential in Europe.* Environmental Change Unit. University of Oxford. Oxford. 224 pp.

Le Houerou, H. N. 1992. Vegetation and land use in the Mediterranean basin by the year 2050: A prospective study. In J. Jeftic, J. D. Milliman, and G. Sestini. *Climatic Change and the Mediterranean.* United Nations Environment Programme. New York.

Morettini, A. 1972. Ambiente climatico e pedologico. *Olivicoltura.* pp. 219–246.

Olesen, J. E., F. Friis, and K. Grevsen. 1993. Simulated effects of climate change on vegetable crop production in Europe. In G. J. Kenny, P. A. Harrison, and M. L. Parry (eds.). *The Effect of Climate Change on Agricultural and Horticultural Potential in Europe.* Environmental Change Unit. University of Oxford. Oxford. 224 pp.

Parry, M. L., T. R. Carter, and N. T. Konijn (eds.). 1988. *The Impact of Climatic Variations in Agriculture.* Vol. 1: *Assessments in Cool Temperate and Cold Regions.* Kluwer Academic. Dordrecht. 876 pp.

Rosenzweig, C., F. Tubiello, and V. Gornitz. 1995. *Impact of Future Climate Change in Italy.* Ministero Dell'Ambiente. Servizio Inquinamento Atmosferico e Acustico e Le Industrie a Rischio. Rome 33 pp.

298 *Bibliography*

UK Department of the Environment. 1991. *The Potential Effects of Climate Change in the United Kingdom.* Climate Change Impacts Review Group. HMSO. London. 124 pp.

Former USSR and Eastern Europe

Menzhulin, G. V., L. A. Koval, and A. L. Badenko. 1995. Potential effects of global warming and carbon dioxide on wheat production in the Commonwealth of Independent States. In C. Rosenzweig, L. H. Allen, Jr., L. A. Harper, S. E. Hollinger, and J. W. Jones (eds.). *Climate Change and Agriculture: Analysis of Potential International Impacts.* American Society of Agronomy. ASA Special Publication No. 59.
Sirotenko, O. D., A. A. Velichko, V. A. Dolgiy-Trach, and V. A. Klimanov. 1991. Global warming and the agroclimatic resources of the Russian plain. *Soviet Geography* 32(5):337–384.

Africa

Downing, T. E. 1992. *Climate Change and Vulnerable Places: Global Food Security and Country Studies in Zimbabwe, Kenya, Senegal, and Chile.* Environmental Change Unit. Research Report No. 1. University of Oxford. Oxford, UK.
Fischer, G. and H. T. van Velthuizen. *Climate Change and Global Agricultural Potential Project: A Case Study of Kenya.* International Institute for Applied Systems Analysis. WP-96-71. Laxenburg, Austria. 96 pp.
Hillel, D. 1991. *Out of the Earth: Civilization and the Life of the Soil.* University of California Press. Berkeley. 321 pp.
Jallow, B. P. 1996. Vulnerability and Adaptation Assessment for The Gambia. In J. B. Smith, S. Huq, S. Lenhart, L. J. Mata, I. Nemesova, and S. Toure (eds.). *Vulnerability and Adaptation to Climate Change: Interim Results from the U.S. Country Studies Program.* Kluwer Academic. Dordrecht.
Kays, S. J. 1985. Physiology of yield in sweet potato. In J. C. Bouwkamp (ed.). *Sweet Potato Products: A Natural Resource for the Tropics.* CRC Press. Boca Raton, FL. 271 pp.
Magadza, C. H. D. 1994. Climate change: Some likely multiple impacts in southern Africa. *Food Policy* 19(2):165–191.
Muchena, P. 1994. Implications of climate change for maize yields in Zimbabwe. In C. Rosenzweig and A. Iglesias. *Implications of Climate Change for International Agriculture: Crop Modeling Study.* U.S. Environmental Protection Agency. EPA 230-B-94-003. Washington, DC.
Phillips, J. 1995. High temperature effects on sweet potato. American Society of Agronomy. Abstracts of Annual Meeting.
Schulze, R. E., G. A. Kiker, and R. P. Kunz. 1993. Global climate change and agricultural productivity in southern Africa. *Global Environmental Change* 4(1):329–349.
Sivakumar, M. V. K. 1993. Global climate change and crop production in the Sudano-Sahelian zone of west Africa. *International Crop Science* Vol. 1. Crop Science Society of America. Madison, WI.

Egypt

Bielorai, H. 1983. *Irrigation Research in the Institute of Soils and Water of the Volcani Center: Goals and Achievements.* Volcani Institute. Israel.
CAPMAS. 1993. *Statistical Year Book: Arab Republic of Egypt 1952–1992.* Central Agency for Public Mobilization and Statistics. Cairo.
Eid, H. M. 1994. Impact of climate change on simulated wheat and maize yields in Egypt.

In C. Rosenzweig and A. Iglesias (eds.). *Implications of Climate Change for International Agriculture: Crop Modeling Study*. U.S. Environmental Protection Agency. Washington, DC.

El-Shaer, M. H., C. Rosenzweig, A. Iglesias, H. M. Eid, and D. Hillel. 1996. Possible scenarios for Egyptian agriculture in the future. *Journal of Mitigation and Adaptation Strategies for Global Change* 1(3):233–250.

Hillel, D. 1994. *Rivers of Eden: The Struggle for Water and the Quest for Peace in the Middle East*. Oxford University Press. New York.

Nicholls, R. J. and S. P. Leatherman. 1994. Sea-Level Rise. In K. M. Strzepek and J. B. Smith (eds.). *As Climate Changes: International Impacts and Implications*. Cambridge University Press. Cambridge. pp. 92–123.

Rosenzweig, C. and D. Hillel. 1994. *Egyptian Agriculture in the 21st Century*. International Institute for Applied Systems Analysis. CP-94–12. Laxenburg, Austria. 29 pp.

Sestini, G. 1992. Implication of climatic changes for the Nile Delta. In L. Jeftic, J. D. Milliman, and G. Sestini (eds.). *Climatic Change and the Mediterranean*. Edward Arnold. London. pp. 535–601.

Strzepek, K. M., S. C. Onyeji, D. N. Yates, and M. Saleh. 1995. A socio-economic analysis of integrated climate change impacts on Egypt. In K. M. Strzepek and J. B. Smith (eds.). *As Climate Changes: International Impacts and Implications*. Cambridge University Press. Cambridge. pp. 180–200.

Asia

South and Southeast Asia

Escano, C. R. and L. V. Buendia. 1994. Climate impact assessment for agriculture in the Philippines: Simulation of rice yield under climate change scenarios. In C. Rosenzweig and A. Iglesias (eds.). *Implications of Climate Change for International Agriculture: Crop Modeling Study*. U.S. Environmental Protection Agency. EPA 230-B-94–003. Washington, DC.

Karim, Z., M. Ahmed, S. G. Hussain, and Kh. B. Rashid. 1994. Impact of climate change on the production of modern rice in Bangladesh. In C. Rosenzweig and A. Iglesias. *Implications of Climate Change for International Agriculture: Crop Modeling Study*. U.S. Environmental Protection Agency. EPA 230-B-94–003. Washington, DC.

Karim, Z., G. Hussain, and M. Ahmed. 1996. Assessing impacts of climatic variations on food-grain production in Bangladesh. *Water, Air, and Soil Pollution* 92:53–62.

Matthews, R. B., M. J. Kropff, D. Bachelet, and H. H. van Laar. 1995. *Modeling the Impact of Climate Change on Rice Production in Asia*. CAB International in association with the International Rice Research Institute. Oxford, UK. 289 pp.

Parry, M. L., M. Blantran de Rozari, A. L. Chong, and S. Panich (eds.). 1992. *The Potential Socio-Economic Effects of Climate Change in South-East Asia*. United Nations Environment Programme. Nairobi.

Qureshi, A. and A. Iglesias. 1994. Implications of global climate change for agriculture in Pakistan: Impacts on simulated wheat production. In C. Rosenzweig and A. Iglesias. *Implications of Climate Change for International Agriculture: Crop Modeling Study*. U.S. Environmental Protection Agency. EPA 230-B-94–003. Washington, DC.

Rao, D. G., J. C. Katyal, S. K. Sinha, and K. Srinivas. 1995. Impacts of climate change on sorghum productivity in India: Simulation study. In C. Rosenzweig, L. H. Allen, Jr., L. A. Harper, S. E. Hollinger, and J. W. Jones (eds.). *Climate Change and Agriculture: Analysis of Potential International Impacts*. American Society of Agronomy. ASA Special Publication No. 59. Madison, WI. pp. 325–337.

Rao, D. G. and S. K. Sinha. 1994. Impact of climate change on simulated wheat production in

India. In C. Rosenzweig and A. Iglesias. *Implications of Climate Change for International Agriculture: Crop Modeling Study.* U.S. Environmental Protection Agency. EPA 230-B-94–003. Washington, DC.

Ropelewski, C. F. and M. S. Halpert. 1987. Global and regional scale precipitation patterns associated with the El Niño/Southern Oscillation. *Monthly Weather Review* 115:1606–1626.

Tongyai, C. 1994. Impact of climate change on simulated rice production in Thailand. In C. Rosenzweig and A. Iglesias. *Implications of Climate Change for International Agriculture: Crop Modeling Study.* U.S. Environmental Protection Agency. EPA 230-B-94–003. Washington, DC.

East Asia

Horie, T. 1987. The effects on rice yields in Hokkaido. In M. L. Parry et al. (eds.). *The Impact of Climatic Variations on Agriculture.* Vol. 1: *Assessments in Cool Temperature and Cold Regions.* Kluwer Academic. Dordrecht. pp. 809–825.

Jin, Z., D. Ge, H. Chen, and J. Fang. 1995. Effects of climate change on rice production and strategies for adaptation in southern China. In C. Rosenzweig, L. H. Allen, Jr., L. A. Harper, S. E. Hollinger, and J. W. Jones (eds.). *Climate Change and Agriculture: Analysis of Potential International Impacts.* American Society of Agronomy. ASA Special Publication No. 59. Madison, WI. pp. 307–323.

Matthews, R. B., M. J. Kropff, and D. Bachelet, and H. H. van Laar. (eds.). 1995. *Modeling the Impact of Climate Change on Rice Production in Asia.* CAB International in association with the International Rice Research Institute. Oxford, UK. 289 pp.

Seino, H. 1995. Implications of climate change for crop production in Japan. In C. Rosenzweig, L. H. Allen, Jr., L. A. Harper, S. E. Hollinger, and J. W. Jones (eds.). *Climate Change and Agriculture: Analysis of Potential International Impacts.* American Society of Agronomy. ASA Special Publication No. 59. Madison, WI. pp. 293–306.

Uchijima, T. 1987. The effects of latitudinal shift of rice yield and cultivable area in Northern Japan. In M. L. Parry et al. (eds.). *The Impact of Climatic Variations on Agriculture.* Vol. 1: *Assessments in Cool Temperature and Cold Regions.* Kluwer Academic. Dordrecht. pp. 797–808.

Uchijima, T. and H. Seino. 1988. Probable effects of CO_2-induced climatic change on agroclimatic resources and net primary productivity in Japan. *Bulletin of National Institute Agro-Environmental Science* 4:67–88.

Oceania

Baer, B. D., W. S. Meyer, and D. Erskine. 1994. Possible effects of global climate change on wheat and rice production in Australia. In C. Rosenzweig and A. Iglesias (eds.). *Implications of Climate Change for International Agriculture: Crop Modeling Study.* U.S. Environmental Protection Agency. EPA 230-B-94–003. Washington, DC.

Blumenthal, C. S., F. Bekes, I. L. Batey, C. W. Wrigley, J. J. Moss, D. J. Mares, and E. W. R. Barlow. 1991. Interpretation of grain quality results from wheat variety trials with reference to hgih temperature stress. *Australian Journal of Agricultural Research* 42:325–334.

Campbell, B. D., G. M. McKeon, R. M. Gifford, H. Clark, D. M. Stafford Smith, P. C. D. Newton, and J. L. Lutze. 1995. Impacts of atmospheric composition and climate change on temperate and tropical pastoral agriculture. In G. Pearman and M. Manning (eds.). *Greenhouse 94.* CSIRO. Australia.

Hennessy, K. J. and K. Clayton-Greene. 1995. Greenhouse warming and vernalisation of high-chill fruit in southern Australia. *Climatic Change* 30(3):327–348.

Hobbs, J., J. R. Anderson, J. L Dillon, and H. Harris. 1988. The effects of climatic variations on agriculture in the Australian wheat belt. In M. L. Parry, T. R. Carter, and N. T. Konijn (eds.). *The Impact of Climatic Variations on Agriculture.* Vol. 2: *Assessments in Semi-Arid Regions.* Kluwer Academic. Dordrecht. pp. 665–753.

IPCC. 1996. Agriculture. In R. T. Watson, M. C. Zinyowera, and R. H. Moss (eds.). *Climate Change 1995: Impacts, Adaptations and Mitigation of Climate Change: Scientific-Technical Analyses.* Contribution of Working Group II to the Second Assessment Report of the Intergovernmental Panel on Climate Change. Cambridge University Press. Cambridge. pp. 427–467.

Kenny, G. J., R. A. Warrick, N. D. Mitchell, A. B. Mullan, and M. J. Salinger. 1995. CLIMPACTS: An integrated model for assessment of the effects of climate change on New Zealand ecosystems. *Journal of Biogeography* 22:883–895.

Salinger, M. J., M. W. Williams, J. M. Williams, and R. J. Martin. 1990. Agricultural resources. In *Climatic Change: Impacts on New Zealand. Implications for the Environment, Economy, and Society.* Ministry for the Environment, Wellington, New Zealand.

Singh, U., D. C. Godwin, and R. J. Morrison. 1990. Modelling the impact of climate change on agricultural production in the South Pacific. In P. J. Hughes and G. McGregor (eds.). *Global Warming-Related Effects of Agriculture and Human Health and Comfort in the South Pacific.* University of Papua New Guinea. South Pacific Regional Environment Programme and the United Nations Environment Programme. Port Moresby, Papua New Guinea. pp. 24–40.

Wang, Y. P. and R. M. Gifford. 1995. A model of wheat grain growth and its application to different temperature and carbon dioxide levels. *Australian Journal of Plant Physiology* 22:843–855.

Wang, Y. P., Jr. Handoko, and G. M. Rimmington. 1992. Sensitivity of wheat growth to increased air temperature for different scenarios of ambient CO_2 concentration and rainfall in Victoria, Australia—A simulation study. *Climatic Research* 2:131–149.

Wardlaw, I. F., I. A. Dawson, P. Munibi, and R. Fewster. 1989. The tolerance of wheat to high temperatures during reproductive growth. I. Survey procedures and general response patterns. *Australian Journal of Agricultural Research* 40:1–13.

Chapter 9 Global Assessments and Future Food Security

Appendini, K. and D. Liverman. 1994. Agricultural policy, climate change and food security in Mexico. *Food Policy* 19(2):149–164.

Bohle, H. G., T. E. Downing, and M. J. Watts. 1994. Climate change and social vulnerability: Toward a sociology and geography of food insecurity. *Global Environmental Change* 4(1):37–48.

Cane, M. A., G. Eshel, and R. W. Buckland. 1994. Forecasting Zimbabwean maize yield using eastern equatorial Pacific sea surface temperature. *Nature* 370:204–205.

Chen, R. S. and R. W. Kates. 1994a. Climate change and world food security. *Global Environmental Change* 4(1):3–6.

Chen, R. S. and R. W. Kates. 1994b. World food security: prospects and trends. *Food Policy* 19(2):192–208.

Christian, G. and J. Stack. 1992. *The Dimensions of Household Food Insecurity in Zimbabwe, 1980–1991.* Working Paper No. 5. Food Studies Group. Oxford University. Oxford.

Cramer, W. P. and A. M. Solomon. 1993. Climatic classification and future global redistribution of agricultural land. *Climate Research* 3:97–110.

Darwin, R., M. Tsigas, J. Lewandrowski, and A. Raneses. 1995. *World Agriculture and Climate Change: Economic Adaptation.* Report No. AER-709. United States Department of Agriculture Economic Research Service. Washington, DC.

Downing, T. E. 1992. *Climate Change and Vulnerable Places: Global Food Security and Country Studies in Zimbabwe, Kenya, Senegal, and Chile.* Research Report No. 1. Oxford University. Environmental Change Unit. Oxford.

Downing, T. E. (ed.). 1995. *Climate Change and World Food Security.* Springer-Verlag. Berlin. 662 pp.

FAO. 1984. *Fourth World Survey.* Food and Agriculture Organization of the United Nations. Rome.

FAO. 1987. *Fifth World Survey.* Food and Agriculture Organization of the United Nations. Rome.

FAO. 1988. *1987 Production Yearbook.* Food and Agriculture Organization of the United Nations. Rome.

FAO. 1996. *World Food Summit: Conference Statement and Technical Background Documents.* Food and Agriculture Organization of the United Nations. Rome, Italy.

Fischer, G., K. Frohberg, M. A. Keyzer, and K. S. Parikh. 1988. *Linked National Models: A Tool for International Food Policy Analysis.* Kluwer. Dordrecht.

Fischer, G., K. Frohberg, M. A. Keyzer, and K. S. Parikh, and W. Tims. 1990. *Hunger—Beyond the Reach of the Invisible Hand.* International Institute for Applied Systems Analysis, Food and Agriculture Project. Laxenburg, Austria.

Fischer, G., K. Frohberg, M. L. Parry, and C. Rosenzweig. 1994. Climate change and world food supply, demand and trade: Who benefits, who loses? *Global Environmental Change* 4(1):7–23.

Hansen, J., G. Russell, D. Rind, P. Stone, A. Lacis, S. Lebedeff, R. Ruedy, and L. Travis. 1983. Efficient three-dimensional global models for climate studies: Models I and II. *Monthly Weather Review* 111(4):609–662.

International Bank for Reconstruction and Development/World Bank. 1990. *World Population Projections.* Johns Hopkins University Press. Baltimore, MD.

International Benchmark Sites Network for Agrotechnology Transfer (IBSNAT). 1989. *Decision Support System for Agrotechnology Transfer Version 2.1 (DSSAT v2.1).* Department of Agronomy and Soil Science, Collection of Tropical Agriculture and Human Resources, University of Hawaii, Honolulu.

IPCC, 1990. *Climate Change: The IPCC Impacts Assessment.* W. J. McG. Tegart, G. W. Sheldon, and D. C. Griffiths (eds.). Australian Government Publishing Service. Canberra.

IPCC, 1996. Agriculture. In R. T. Watson, M. C. Zinyowera, and R. H. Moss (eds.). *Climate Change 1995: Impacts, Adaptations and Mitigation of Climate Change: Scientific-Technical Analyses.* Cambridge University Press. Cambridge. pp. 427–467.

Kane, S., J. Reilly, and J. Tobey. 992. An empirical study of the economic effects of climate change on world agriculture. *Climatic Change* 21(1):17–35.

Leemans, R. and A. M. Solomon. 1993. Modeling the potential change in yield and distribution of the earth's crops under a warmed climate. *Climate Research* 3:79–96.

Liverman, D. M. and K. O'Brien. 1991. Global warming and climate change in Mexico. *Global Environmental Change* 1(4):351–364.

Magadza, C. H. D. 1994. Climate change: Some likely multiple impacts in Southern Africa. *Food Policy* 19(2):165–191.

Manabe, S. and R. T. Wetherald. 1986. Reduction in summer soil wetness induced by an increase in atmospheric carbon dioxide. *Science* 232:626–628.

Muchena, P. 1994. Implications of climate change for maize yields in Zimbabwe. In C. Rosenzweig and A. Iglesias (eds.). *Implications of Climate Change for International Agriculture: Crop Modeling Study.* U.S. Environmental Protection Agency. EPA 230-B-94–003. Washington, DC.

Muchena, P. and A. Iglesias. 1995. Vulnerability of maize yields to climate change in different farming sectors in Zimbabwe. In C. Rosenzweig, L. H. Allen, Jr., L. A. Harper, S. E. Hollinger, and J. W. Jones (eds.). *Climate Change and Agriculture: Analysis of Potential International Impacts.* American Society of Agronomy. Special Publication No. 59. Madison, WI. pp. 229–239.

Peart, R. M., J. W. Jones, R. B. Curry, K. Boote, and L. H. Allen, Jr. 1989. Impact of climate change on crop yield in the Southeastern U.S. A. In J. B. Smith and D. A. Tirpak (eds.). *The Potential Effects of Global Climate Change on the United States.* EPA-230-05–89-050. U.S. Environmental Protection Agency. Washington, DC.

Reilly, J. and N. Hohmann. 1993. Climate change and agriculture: The role of international trade. *American Economic Association Papers and Proceedings* 83:306–312.

Reilly, J., N. Hohmann, and S. Kane. 1994. Climate change and agricultural trade: Who benefits, who loses? *Global Environmental Change* 4(1):24–36.

Rosenzweig, C., L. H. Allen, Jr., L. A. Harper, S. E. Hollinger, and J. W. Jones. (eds.). 1995a. *Climate Change and Agriculture: Analysis of Potential International Impacts.* ASA Special Publication No. 59. American Society of Agronomy. Madison, WI. 382 pp.

Rosenzweig, C. and A. Iglesias (eds.). 1994. *Implications of Climate Change for International Agriculture: Crop Modeling Study.* U.S. Environmental Protection Agency. EPA 230-B-94-003. Washington, DC.

Rosenzweig, C. and M. L. Parry. 1994. Potential impact of climate change on world food supply. *Nature* 367:133–138.

Rosenzweig, C., M. L. Parry, and G. Fischer. 1995b. Climate Change, World Food Supply and Vulnerable Regions. In K. Strzepek and J. Smith (eds.). *As Climate Changes: International Impacts and Implications.* Cambridge University Press. Cambridge. pp. 27–56.

Ruttan, V. W., D. E. Bell, and W. C. Clark. 1994. Climate change and food security: Agriculture, health and environmental research. *Global Environmental Change* 4(1):63–77.

Sen, A. 1981. *Poverty and Famines: An Essay on Entitlement and Deprivation.* Clarendon Press. Oxford. 257 pp.

Wilson, C. A. and J. F. B. Mitchell. 1987. A doubled CO_2 climate sensitivity experiment with a global climate model including a simple ocean. *Journal of Geophysical Research* 92(13):315–343.

World Bank. 1986. *World Bank, Poverty and Hunger: Issues and Options for Food Security in Developing Countries.* World Bank. Washington, DC.

Chapter 10   Adaptation, Economics, and Policy

Adams, R. M., R. A. Fleming, C.-C. Chang, B. A. McCarl, and C. Rosenzweig. 1995. A reassessment of the economic effects of global climate change on U.S. agriculture. *Climatic Change* 30(2):147–167.

Adams, R. M., C. Rosenzweig, R. Peart, J. T. Ritchie, B. McCarl, J. Glyer, R. Curry, J. Jones, K. Boote, and L. H. Allen, Jr. 1990. Global climate change and U.S. agriculture. *Nature* 345:219–224.

Aspen Global Change Institute. 1994. Report on the 1994 AGCI Summer Session II on Surprise and Global Environmental Change. Draft 2.1.

Bowes, M. D. and P. R. Crosson. 1993. Consequences of climate change for the MINK Economy: Impacts and Responses. *Climatic Change* 24:131–158.

CAST. 1992. *Preparing US Agriculture for Global Climate Change.* Council for Agricultural Science and Technology. Report 119. Ames, IA. 96 pp.

Dowlatabadi, H. and M. G. Morgan. 1993. Integrated assessment of climate change. *Science* 259:1813, 1932.

Easterling, W. E. 1996. Adapting North American agriculture to climate change: A review. *Agricultural and Forest Meteorology* 80:1–53.

Fischer, G., K. Frohberg, M. L. Parry, and C. Rosenzweig. 1994. Climate change and world food supply, demand and trade: Who benefits, who loses? *Global Environmental Change* 4(1):7–23.

Hillel, D. 1997. *Small-scale Irrigation for Arid Zones.* United Nations Food and Agriculture Organization. Rome.

IBSNAT. 1990. *Proceedings of IBSNAT Symposium: Decision Support System for Agrotechnology*

Transfer. 81st Annual Meeting of the American Society of Agronomy. Las Vegas. University of Hawaii. Honolulu.

IPCC. 1996. *Climate Change 1995: The Science of Climate Change.* J. T. Houghton, L. B. Meira Filho, B. A. Callander, N. Harris, A. Kattenberg, and K. Maskell (eds.). Intergovernmental Panel on Climate Change. Cambridge University Press. Cambridge. 572 pp.

Jackson, M. T. and B. V. Ford-Lloyd. 1990. Plant genetic resources—A perspective. In M. T. Jackson, B. V. Ford-Lloyd, and M. L. Parry (eds.). *Climatic Change and Plant Genetic Resources.* Belhaven Press. London. pp. 1–17.

Kaiser, H. M., S. Riha, D. Wilkes, and R. Sampath. 1993. Adaptation to global climate change at the farm level. In H. M. Kaiser and T. E. Drennen (eds.). *Agricultural Dimensions of Global Climate Change.* St. Lucie Press. Delray Beach, FL. pp. 136–152.

Lewandrowski, J. K. and R. J. Brazee. 1993. Farm programs and climate change. *Climatic Change* 23:1–20.

Mendelsohn, R., W. D. Nordhaus, and D. Shaw. 1994. The impact of global warming on agriculture: A ricardian analysis. *American Economic Review* 84(4):753–771.

Mendelsohn, R., W. Nordhaus, and D. Shaw. 1996. Climate impacts on aggregate farm value: Accounting for adaptation. *Agricultural and Forest Meteorology* 80:55–66.

National Academy of Sciences Committee on Science, Engineering, and Public Policy. 1992. *Policy Implications of Greenhouse Warming.* National Academy Press. Washington, DC.

Office of Technology Assessment. 1993. *Preparing for an Uncertain Climate.* Vol. 1. OTA-O-567. U.S. Congress. U.S. Government Print Office. Washington, DC.

Parry, M. L., T. R. Carter, and M. Hulme. 1996. What is a dangerous climate change? *Global Environmental Change* 6(1):1–6.

Reilly, J., N. Hohmann, and S. Kane. 1994. Climate change and agricultural trade: Who benefits, who loses? *Global Environmental Change* 4(1):24–36.

Riebsame, W. E., K. M. Strzepek, J. L. Wescoat, Jr., G. L. Gaile, J. Jacobs, R. Leichenko, C. Magadza, R. Perritt, H. Phien, B. J. Urbiztondo, P. Restrepo, W. R. Rose, M. Saleh, C. Tucci, L. H. Ti, and D. Yates. 1995. Complex river basins. In K. M. Strzepek and J. B. Smith (eds.). *As Climate Changes: International Impacts and Implications.* Cambridge University Press. Cambridge. pp. 57–91.

Rosenzweig, C. and M. L. Parry. 1994. Potential impact of climate change on world food supply. *Nature* 367:133–138.

Rosenzweig, C. and D. Hillel. 1995. Potential impacts of climate change on agriculture and food supply. *Consequences* 1(2):23–32.

Rosenzweig, C., M. L. Parry, and G. Fischer. 1995. Climate Change, World Food Supply and Vulnerable Regions. In K. Strzepek and J. Smith (eds.). *As Climate Changes: International Impacts and Implications.* Cambridge University Press. Cambridge. pp. 27–56.

Rosenzweig, C., M. L. Parry, G. Fischer, and K. Frohberg. 1993. *Climate Change and World Food Supply.* Environmental Change Unit. Research Report No. 3. Oxford University. Oxford. 28 pp.

Ruttan, V. W. 1991. Review of climate change and world agriculture. *Environment* 33:25–29.

Squire, G. R. 1990. Effects of changes in climate and physiology around the dry limits of agriculture in the tropics. In M. T. Jackson, B. V. Ford-Lloyd, and M. L. Parry (eds.). *Climatic Change and Plant Genetic Resources.* Belhaven Press. London. pp. 116–147.

U.S. Country Studies Program. 1994. *Guidance for Vulnerability and Adaptation Assessments.* U.S. Countries Studies Management Team. Washington, DC.

FIGURE CREDITS

Chapter 1

Figure 1.1   With permission of D. Hillel and C. Rosenzweig.

Figure 1.2   Reprinted from S. D. Gedzelman, *The Science and Wonders of the Atmosphere*, copyright 1980, with kind permission from John Wiley & Sons.

Figure 1.3   Reprinted from J. T. Houghton, G. J. Jenkins, and J. J. Ephraums (eds.). *Climate Change: The IPCC Scientific Assessment*, copyright 1990, with kind permission from the Intergovernmental Panel on Climate Change.

Figure 1.4   Reprinted from J. Hansen, I. Fung, A. Lacis, D. Rind, S. Lebedeff, R. Ruedy, and G. Russell, "Global climate changes as forecast by Goddard Institute for Space Studies three-dimensional model," *Journal of Geophysical Research*, 93: 9341–9364, copyright 1988, with kind permission from the American Geophysical Union.

Figure 1.5   Drawn from data of Keeling et al. on the World-Wide Web from CDIAC, NDP-001r, 1995 Oak Ridge National Laboratory.

Figure 1.6   Reprinted from R. A. Houghton and D. L. Skole, "Carbon," in B. L. Turner II, W. C. Clark, R. W. Kates, J. F. Richards, J. T. Matthews, and W. B. Meyer (eds.), *The Earth as Transformed by Human Action*, copyright 1990, with kind permission from Cambridge University Press.

Figure 1.7   Reprinted from R. A. Houghton and D. L. Skole, "Carbon," in B. L. Turner II, W. C. Clark, R. W. Kates, J. F. Richards, J. T. Matthews, and W. B. Meyer (eds.), *The Earth as Transformed by Human Action*, copyright 1990, with kind permission from Cambridge University Press.

Figure 1.8   Reprinted from R. T. Watson, "Atmospheric Ozone," in U.S. Environmental Protection Agency, *Effects of Changes in Stratospheric Ozone and Global Climate.* Vol. 1.: *Overview.* August 1986.

Figure 1.9   Reprinted from J. T. Houghton, G. J. Jenkins, and J. J. Ephraums (eds.), *Climate Change: The IPCC Scientific Assessment*, copyright 1990, with kind permission from the Intergovernmental Panel on Climate Change.

Figure 1.10   Reprinted from World Meteorological Organization, *The Physical Basis of Climate and Climate Modelling*, GARP Publications, Series No. 16, copyright 1975, with kind permission from the World Meteorological Organization.

Figure 1.11   Reprinted from R. E. Dickinson, "Impact of human activities on climate—a framework," in W. C. Clark and R. E. Munn (eds.), *Sustainable Development of the Biosphere*, copyright 1986, with kind permission from Cambridge University Press.

Figure 1.12   Reprinted from J. Hansen, G. Russell, D. Rind, P. Stone, A. Lacis, S. Lebedeff, R. Ruedy, and L. Travis, "Efficient three-dimensional global models for climate studies: Models I and II," *Monthly Weather Review*, 111(4): 609–662, copyright 1983, with kind permission from the American Meteorological Society.

Figure 1.13   Reprinted from P. J. Robinson, "Comparisons of Rand climatology and GCM outputs for Australia and tropical Asia," in L. S. Kalkstein (ed.), *Global Comparisons of Selected GCM Control Runs and Observed Climate Data*, United States Environmental Protection Agency, April 1991.

Figure 1.14   With permission of C. Rosenzweig and D. Hillel.

Figure 1.15   Reprinted from J. T. Houghton, L. G. Meira Filho, B. A. Callander, N. Harris, A. Kattenberg, and K. Maskell (eds.), *Climate Change 1995: The Science of Climate Change*, copyright 1996, with kind permission from the Intergovernmental Panel on Climate Change.

Figure 1.16   Reprinted with kind permission from J. Hansen.

Figure 1.17   Reprinted from J. T. Houghton, L. G. Meira Filho, B. A. Callander, N. Harris, A. Kattenberg, and K. Maskell (eds.), *Climate Change 1995: The Science of Climate Change*, copyright 1996, with kind permission from the Intergovernmental Panel on Climate Change.

Figure 1.18   Reprinted from J. T. Houghton, L. G. Meira Filho, B. A. Callander, N. Harris, A. Kattenberg, and K. Maskell (eds.), *Climate Change 1995: The Science of Climate Change*, copyright 1996, with kind permission from the Intergovernmental Panel on Climate Change.

Figure 1.19   Reprinted from J. Hansen, A. Lacis, R. Ruedy, M. Sato, and H. Wilson, "How sensitive is the world's climate?," *National Geographic Research and Exploration* 9(2:): 142–158, copyright 1993, with kind permission from *National Geographic Research & Exploration*. Recent data printed with kind permission from J. Hansen.

Chapter 2

Figure 2.1   Reprinted from R. T. Watson, M. C. Zinyowera, R. H. Moss, and D. J. Dokken (eds.), *Climate Change 1995: Impacts, Adaptations and Mitigation of Climate Change: Scientific-Technical Analyses*, copyright 1996, with kind permission from the Intergovernmental Panel on Climate Change.

Figure 2.2   Reprinted from R. A. Houghton, J. E. Hobbie, J. M. Melillo, B. Moore, B. J. Peterson, G. R. Shaver, and G. M. Woodwell, "Changes in the carbon content of terrestrial biota and soils between 1860 and 1980: A net release of CO_2 to the atmosphere," *Ecological Monographs* 53(3), copyright 1983, with kind permission from the Ecological Society of America.

Figure 2.3   Drawn from data of Food and Agriculture Organization of the United Nations, Rome.

Figure 2.4   Reprinted from R. A. Houghton, J. E. Hobbie, J. M. Melillo, B. Moore, B. J. Peterson, G. R. Shaver, and G. M. Woodwell, "Changes in the carbon content of terrestrial biota and soils between 1860 and 1980: A net release of CO_2 to the atmosphere," *Ecological Monographs* 53(3), copyright 1983, with kind permission from the Ecological Society of America.

Figure 2.5   Drawn from data of the Food and Agriculture Organization of the United Nations, Rome.

Figure 2.6   Reprinted from R. Sedjo, *Environment* 31(1): 15–20, with permission of the Helen Dwight Reid Educational Foundation. Published by Heldref Publications, 1319 Eighteenth St., NW, Washington, D.C. 20036–1802, copyright 1989.

Figure 2.7   Reprinted from World Resources Institute [in collaboration with the United Nations Environment Programme and the United Nations Development Programme], copyright 1990, World Resources, 1990–1991, Oxford University Press, with kind permission from the World Resources Institute.

Figure 2.8   Reprinted from L. M. Burke and D. A. Lashof, "Greenhouse gas emissions related to agriculture and land-use practices," in B. A. Kimball, N. J. Rosenberg, and L. H. Allen, Jr. (eds.), *Impact of Carbon Dioxide, Trace Gases, and Climate Change on Global Agriculture*, copyright 1990, American Society of Agronomy Special Publication no. 53, with kind permission from Crop Science Society of America.

Figure 2.9   Reprinted from W. H. Patrick Jr. and R. D. DeLaune, "Chemical and biological redox systems affecting nutrient availability in the coastal wetlands," *Geoscience and Man* 18: 131–137, copyright 1977, with kind permission from Geoscience Publications.

Figure 2.10   Reprinted from Y. Takai and E. Wada, "Methane formation in water logged paddy soils and its controlling factors," in H. W. Scharpenseel, M. Schomaker, and A. Ayoub (eds.), *Soils on a Warmer Earth*, copyright 1990, with kind permission from Elsevier Science - NL, Sara Burgerhartstraat 25, 1055 KV Amsterdam, The Netherlands.

Figure 2.11   Reprinted from M. A. K. Khalil, R. A. Rasmussen, M. X. Wang, and L. Ren, "Methane emissions from rice fields in China," *Environmental Science and Technology* 25: 979–981, copyright 1991, with kind permission from the American Chemical Society.

Figure 2.12   Reprinted from M. A. K. Khalil, R. A. Rasmussen, M. X. Wang, and L. Ren, "Methane emissions from rice fields in China," *Environmental Science and Technology* 25: 979–981, copyright 1991, with kind permission from the American Chemical Society.

Figure 2.13   Reprinted from L. M. Burke and D. A. Lashof, "Greenhouse gas emissions related to agriculture and land-use practices," in B. A. Kimball, N. J. Rosenberg, and L. H. Allen, Jr. (eds.), *Impact of Carbon Dioxide, Trace Gases, and Climate Change on Global Agriculture*, copyright 1990, American Society of Agronomy Special Publication no. 53, with kind permission from Crop Science Society of America.

Figure 2.14   Reprinted from N. C. Brady, *The Nature and Properties of Soils*, 10th ed., copyright 1990, with kind permission from Prentice-Hall Inc.

Figure 2.15   Reprinted from R. Wollast, "Interactions between major biogeochemical cycles in marine ecosystems," in G. E. Likens (ed.), *Some Perspectives of the Major Biogeochemical Cycles*, copyright 1981, from John Wiley & Sons.

Figure 2.16   Reprinted from R. B. Jackson IV, "On estimating agriculture's net contribution to atmospheric carbon," *Water, Air, and Soil Pollution* 64: 121–137, copyright 1992, with kind permission from Kluwer Academic Publishers.

Figure 2.17   Reprinted from *Greenhouse Gas Inventory Reporting Instructions*, Vol. 1, copyright 1994, with kind permission from the Intergovernmental Panel on Climate Change.

Chapter 3

Figure 3.1   With permission from the *Annual Review of Ecology and Systematics*, Vol. 21, copyright 1990, by Annual Reviews Inc.

Figure 3.2   Reprinted from A. Fakhri Bazzaz and E. D. Fajer, "Plant life in a CO_2-rich world," *Scientific American* (January 1992): 68–74, with kind permission from Scientific American.

Figure 3.3   Reprinted from S. Akita and D. N. Moss, "Photosynthetic responses to CO_2 and light by maize and wheat leaves adjusted for constant stomatal apertures," *Crop Science* 13: 234–237, copyright 1973, with kind permission from Crop Science Society of America.

Figure 3.4   Reprinted from J. I. L. Morison, "Sensitivity of stomata and water use efficiency to high CO_2," *Plant, Cell, and Environment* 8: 467–474, copyright 1985, with kind permission from Blackwell Science Ltd.

Figure 3.5   Reprinted from J. M. Norman and T. J. Arkebauer, "Predicting canopy photosynthesis and light use efficiency from leaf characteristics," in K. J. Boote and R. S. Loomis (eds.), *Modeling Crop Photosynthesis—from Biochemistry to Canopy*, CSSA Special Publication no. 19, copyright 1991, with kind permission from Crop Science Society of America.

Figure 3.6   Reprinted with kind permission from H. Rogers.

Figure 3.7   Reprinted from A. Pisek, W. Larcher, A. Vegis, and K. Napp-Zin, "The normal temperature range," in H. Precht, J. Christopherson, H. Hensel, and W. Larcher, *Temperature and Life*, copyright 1973, with kind permission from Springer- Verlag.

Figure 3.8   Reprinted from I. F. Wardlaw, "The physiological effects of temperature on plant growth," *Proceedings of the Agronomy Society of New Zealand* 9: 39–48, copyright 1979.

Figure 3.9   Reprinted from A. Pisek, W. Larcher, A. Vegis, and K. Napp-Zin, "The normal temperature range," in H. Precht, J. Christopherson, H. Hensel, and W. Larcher, *Temperature and Life*, copyright 1973, with kind permission from Springer-Verlag.

Figure 3.10   Reprinted from B. Acock and M. C. Acock, "Modeling approaches for predicting crop ecosystem responses to climate change," in D. R. Buxton, R. Shibles, R. A. Forsberg, B. L. Blad, K. H. Asay, G. M. Paulsen, and R. F. Wilson (eds.), *International Crop Science I*, copyright 1993, with kind permission from Crop Science Society of America.

Figure 3.11   Reprinted from J. B. Kiniry, W. D. Rosenthal, B. S. Jackson, and G. Hoogenboom, "Predicting leaf development of crop plants," in T. Hodges (ed.), *Predicting Crop Phenology*, copyright 1991, with kind permission from Crop Science Society of America.

Figure 3.12   With permission of C. Rosenzweig and F. Tubiello.

Figure 3.13   Redrawn from D. M. Gates, *Biophysical Ecology*, copyright 1980, with kind permission from Springer-Verlag.

Figure 3.14   Reprinted from T. C. Hsaio, E. Ferreres, and D. W. Henderson, "Water stress and dynamics of growth and yield of crop plants," in O. Lange, L. Kappen, and E. D. Schultze (eds.), *Water and Plant Life: Problems and Modern Approaches*, with kind permission from Springer-Verlag.

Figure 3.15   Reprinted from A. Bauer, *Effect of Water Supply and Seasonal Distribution on Spring Wheat Yields*, Bulletin 490, North Dakota Agricultural Experiment Station, Fargo, 1972.

Chapter 4

Figure 4.1   Reprinted from D. T. Patterson, "Implications of global climate change for impact of weeds, insects, and plant diseases," in D. R. Buxton, R. Shibles, R. A. Forsberg, B. L. Blad, K. H. Asay, G. M. Paulsen, and R. F. Wilson (eds.), *International Crop Science I*, copyright 1993, with kind permission from Crop Science Society of America.

Figure 4.2   Reprinted from D. T. Patterson, "Implications of global climate change for impact of weeds, insects, and plant diseases," in D. R. Buxton, R. Shibles, R. A. Forsberg, B. L. Blad, K. H. Asay, G. M. Paulsen, and R. F. Wilson (eds.), *International Crop Science I*, copyright 1993, with kind permission from Crop Science Society of America.

Figure 4.3   Reprinted from W. J. Mattson and R. J. Haack, "The role of drought in outbreaks of plant-eating insects," *BioScience* 37(2): 110–118, copyright 1987, with kind permission from American Institute of Biological Sciences.

Figure 4.4   Reprinted from B. Stinner, in J. B. Smith and D. A. Tirpak (eds.), *The Potential Effects of Global Climate Change on the United States*, United States Environmental Protection Agency, copyright 1989.

Chapter 5

Figure 5.1   Reprinted from the International Geosphere-Biosphere Program, *Report no. 5*, copyright 1989, with kind permission from the International Geosphere-Biosphere Program.

Figure 5.2   Reprinted from S. W. Buol, P. A. Sanchez, S. B. Weed, and J. M. Kimble, "Predicted impact of climatic warming on soil properties and use," in B. A. Kimball, N. J. Rosenberg, and L. H. Allen, Jr. (eds.), *Impact of Carbon Dioxide, Trace Gases, and Climate Change on Global Agriculture*, copyright 1990, with kind permission from Crop Science Society of America.

Figure 5.3   Reprinted from S. W. Buol, P. A. Sanchez, S. B. Weed, and J. M. Kimble, "Predicted impact of climatic warming on soil properties and use," in B. A. Kimball, N. J. Rosenberg, and L. H. Allen, Jr. (eds.), *Impact of Carbon Dioxide, Trace Gases, and Climate Change on Global Agriculture*, copyright 1990, with kind permission from Crop Science Society of America.

Figure 5.4   Reprinted from the International Geosphere-Biosphere Program, *Report no. 5*, copyright 1989, with kind permission from the International Geosphere-Biosphere Program.

Figure 5.5   Reprinted from J. W. Raich and W. H. Schlesinger, "The global carbon dioxide flux in soil respiration and its relationship to vegetation and climate,"

Tellus 44B: 81–99, copyright 1992, with kind permission from the Swedish Geophysical Society.

Figure 5.6   Reprinted from W. H. Schlesinger, J. P. Winkler, and J. P. Megonigal, "Soils and the global carbon cycle," in T. M. L. Wigley and D. S. Schimel (eds.), *The Carbon Cycle*, copyright 1997, with kind permission from Cambridge University Press.

Figure 5.7   Reprinted from S. W. Buol, P. A. Sanchez, S. B. Weed, and J. M. Kimble, "Predicted impact of climatic warming on soil properties and use," in B. A. Kimball, N. J. Rosenberg, and L. H. Allen, Jr. (eds.), *Impact of Carbon Dioxide, Trace Gases, and Climate Change on Global Agriculture*, copyright 1990, with kind permission from Crop Science Society of America.

Figure 5.8   Reprinted with permission from R. Lal, "Global soil erosion by water and carbon dynamics," in R. Lal, J. Kimble, E. Levine, and B. A. Steward (eds.), *Soils and Global Change*, Advances in Soil Science, copyright 1995, Lewis Publishers, an imprint of CRC Press, Boca Raton, Florida.

Chapter 6

Figure 6.1   Reprinted from S. Manabe and R. T. Wetherald, "Large-scale changes in soil wetness induced by an increase in atmospheric carbon dioxide," *Journal of the Atmospheric Sciences* 44(1): 1211–1235, copyright 1987, with kind permission from the American Meteorological Society.

Figure 6.2   With permission of C. Rosenzweig and D. Hillel.

Figure 6.3   Drawn from data from World Resources Institute, 1994–95.

Figure 6.4   Reprinted from D. F. Peterson and A. A. Keller, "Irrigation," in P. E. Waggoner (ed.), *Climate Change and U.S. Water Resources*, copyright 1990, with kind permission from John Wiley & Sons.

Figure 6.5   Reprinted from P. H. Gleick, "Vulnerability of water systems," in P. E. Waggoner (ed.), *Climate Change and U.S. Water Resources*, copyright 1990, with kind permission from John Wiley & Sons.

Figure 6.6   Reprinted from W. L. Powers, "The Ogallala's Bounty Evaporates," *Science of Food and Agriculture* 5(3), September 1987, Council for Agricultural Science and Technology, Ames, Iowa.

Figure 6.7   Reprinted from W. E. Riebsame, K. M. Strzepek, J. L. Wescoat Jr., G. L. Gaile, J. Jacobs, R. Leichenko, C. Magadza, R. Perritt, H. Phien, B. J. Urbitztondo, P. Restrepo, W. R. Rose, M. Saleh, C. Tucci, L. H. Ti, and D. Yates, "Complex river basins," in K. M. Strzepek and J. B. Smith (eds.), *As Climate Changes: International Impacts and Implications*, copyright 1995, with kind permission from Cambridge University Press.

Figure 6.8   Reprinted from W. E. Riebsame, K. M. Strzepek, J. L. Wescoat Jr., G. L. Gaile, J. Jacobs, R. Leichenko, C. Magadza, R. Perritt, H. Phien, B. J. Urbitztondo, P. Restrepo, W. R. Rose, M. Saleh, C. Tucci, L. H. Ti, and D. Yates, "Complex river basins," in K. M. Strzepek and J. B. Smith (eds.), *As Climate Changes: International Impacts and Implications*, copyright 1995, with kind permission from Cambridge University Press.

Figure 6.9   Reprinted with kind permission from V. Gornitz.

Chapter 7

Figure 7.1   Reprinted with permission from C. Rosenzweig.

Figure 7.2   Reprinted from J. M. Lough, T. M. L. Wigley, and J. P. Palutikof, "Climate and climate impact scenarios for Europe in a warmer world," *Journal of Climate and Applied Meteorology* 22: 1673–1684, copyright 1983, with kind permission from the American Meteorological Society.

Figure 7.3   Reprinted with permission from C. Rosenzweig.

Figure 7.4   Reprinted with permission from C. Rosenzweig.

Figure 7.5   Reprinted with permission from C. Rosenzweig.

Figure 7.6   Reprinted from D. J. Dudek, "Climate change impacts upon agriculture and resources: A case study of California," in J. B. Smith and D. A. Tirpak (eds.), *The Potential Effects of Global Climate Change on the United States*, United States Environmental Protection Agency, June 1989.

Figure 7.7   Reprinted from R. Benioff, S. Guill, and J. Lee (eds.), *Vulnerability and Adaptation Assessments: An International Handbook*, copyright 1996, U.S. Country Studies Program and Kluwer Academic Publishers.

Figure 7.8   Reprinted from R. M. Adams, J. D. Glyer, and B. A. McCarl, "The economic effects of climate change on U.S. agriculture: A preliminary assessment," and C. Rosenzweig, "Agriculture," in J. B. Smith and D. A. Tirpak (eds.), *The Potential Effects of Global Climate Change on the United States*, United States Environmental Protection Agency, June 1989.

Chapter 8

Figure 8.1   Reprinted with permission from D. Hillel.

Figure 8.2   Reprinted with permission from D. Hillel.

Figure 8.3   Reprinted from H. Eid, "Impact of climate change on simulated wheat and maize yields in Egypt," in C. Rosenzweig and A. Iglesias (eds.), *Implications of Climate Change for International Agriculture: Crop Modeling Study*, United States Environmental Protection Agency, June 1994.

Figure 8.4   Reprinted from K. M. Strzepek, S. C. Onyeji, D. N. Yates, and M. Saleh, "A socio-economic analysis of integrated climate change impacts on Egypt," in K. M. Strzepek and J. B. Smith (eds.), *As Climate Changes: International Impacts and Implications*, copyright 1995, with kind permission from Cambridge University Press.

Figure 8.5   Reprinted from K. M. Strzepek, S. C. Onyeji, D. N. Yates, and M. Saleh, "A socio-economic analysis of integrated climate change impacts on Egypt," in K. M. Strzepek and J. B. Smith (eds.), *As Climate Changes: International Impacts and Implications*, copyright 1995, with kind permission from Cambridge University Press.

Chapter 9

Figure 9.1   Reprinted from G. Fischer, K. Frohberg, M. A. Keyzer, and K. S. Parikh, *Linked National Models: A Tool for International Food Policy Analysis*, copyright 1988, with kind permission from Kluwer Academic Publishing Co.

Figure 9.2   Reprinted with permission from C. Rosenzweig.

Figure 9.3   Reprinted with permission from C. Rosenzweig.

Figure 9.4	Reprinted with permission from C. Rosenzweig.
Figure 9.5	Reprinted with permission from C. Rosenzweig.
Figure 9.6	Reprinted with permission from C. Rosenzweig.
Figure 9.7	Reprinted with permission from C. Rosenzweig.
Figure 9.8	Reprinted from R. S. Chen and R. W. Kates, "World food security: prospects and trends," *Food Policy* 19(2): 192–208, copyright 1994, with kind permission from Elsevier Science Ltd, the Boulevard, Langford Lane, Kidlington 0X5 1GB, UK.
Figure 9.9	Reprinted from P. Muchena, "Implications of climate change for maize yields in Zimbabwe," in C. Rosenzweig and A. Iglesias (eds.), *Implications of Climate Change for International Agriculture: Crop Modeling Study*, United States Environmental Protection Agency, June 1994.
Figure 9.10	Reprinted from P. Muchena, "Implications of climate change for maize yields in Zimbabwe", in C. Rosenzweig and A. Iglesias (eds.), *Implications of Climate Change for International Agriculture: Crop Modeling Study*, United States Environmental Protection Agency, June 1994.
Figure 9.11	Reprinted with permission from C. Rosenzweig.
Figure 9.12	Reprinted with permission from C. Rosenzweig.

Chapter 10

Figure 10.1	Reprinted from G. R. Squire, "Effects of changes in climate and physiology around the dry limits of agriculture in the tropics," in M. T. Jackson, B. V. Ford-Lloyd, and M. L. Parry (eds.), *Climatic Change and Plant Genetic Resources*, copyright 1990, Belhaven Press, a division of Pinter Publishers.
Figure 10.2	Reprinted with permission from G. Fischer, K. Frohberg, M. L. Parry, and C. Rosenzweig, "Climate change and world food supply, demand and trade: Who benefits, who loses?" *Global Environmental Change* 4(1): 7–23, © 1994 with kind permission from Elsevier Science Ltd., the Boulevard, Langford Lane, Kidlington 0X5, IGB, UK.
Figure 10.3	Reprinted with permission from C. Rosenzweig.

INDEX